Mixing Secrets for the Small Studio

Discover how to achieve release-quality mixes even in the smallest studios by applying power-user techniques from the world's most successful producers.

Mixing Secrets for the Small Studio is the best-selling primer for small-studio enthusiasts who want chart-ready sonics in a hurry. Drawing on the back-room strategies of more than 160 famous names, this entertaining and down-to-earth guide leads you step-by-step through the entire mixing process. On the way, you'll unravel the mysteries of every type of mix processing, from simple EQ and compression through to advanced spectral dynamics and "fairy dust" effects. User-friendly explanations introduce technical concepts on a strictly need-to-know basis, while chapter summaries and assignments are perfect for school and college use.

- Learn the subtle editing, arrangement, and monitoring tactics that give industry insiders their competitive edge, and master the psychological tricks that protect you from all the biggest rookie mistakes.
- Find out where you *don't* need to spend money, as well as how to make a limited budget really count.
- Pick up tricks and tips from leading-edge engineers working on today's multi-platinum hits, including Derek "MixedByAli" Ali, Michael Brauer, Dylan "3D" Dresdow, Tom Elmhirst, Serban Ghenea, Jacquire King, the Lord-Alge brothers, Tony Maserati, Manny Marroquin, Noah "50" Shebib, Mark "Spike" Stent, DJ Swivel, Phil Tan, Andy Wallace, Young Guru, and many, many more . . .

Now extensively expanded and updated, including new sections on mix-buss processing, mastering, and the latest advances in plug-in technology.

Mike Senior is a professional engineer who has worked with Wet Wet Wet, The Charlatans, Reef, Therapy, and Nigel Kennedy. He specializes in adapting the techniques of top producers for those working in small studios and on a tight budget. He has transformed dozens of amateur productions for *Sound On Sound* magazine's popular "Mix Rescue" column, proving time and again that you can achieve commercial-grade results with affordable gear—once you know how!

Sound on Sound Presents series . . .

Series Editorial Advisor: Dave Lockwood

Other title in the series:

The Studio SOS Book: Solutions and Techniques for the Project Recording Studio
By Paul White, Hugh Robjohns, and Dave Lockwood

The SOS Guide to Live Sound: Optimising Your Band's Live-Performance Audio
By Paul White

Recording Secrets for the Small Studio
By Mike Senior

The Music Producer's Survival Guide: Chaos, Creativity, and Career in Independent and Electronic Music, 2nd Edition
By Brian M. Jackson

Mixing Secrets for the Small Studio, 2nd Edition
By Mike Senior

Mixing Secrets for the Small Studio

Second Edition

Mike Senior

Routledge
Taylor & Francis Group

NEW YORK AND LONDON

Second edition published 2019
by Routledge
711 Third Avenue, New York, NY 10017

and by Routledge
2 Park Square, Milton Park, Abingdon, Oxon, OX14 4RN

Routledge is an imprint of the Taylor & Francis Group, an informa business

First edition published by Focal Press 2011

Library of Congress Cataloging-in-Publication Data
Names: Senior, Mike author.
Title: Mixing secrets for the small studio / Mike Senior.
Description: Second edition. | New York, NY : Routledge, 2018. | Series: Sound on Sound presents | Includes bibliographical references and index.
Identifiers: LCCN 2018011345| ISBN 9781138556362 (hardback : alk. paper) | ISBN 9781138556379 (pbk. : alk. paper) | ISBN 9781315150017 (ebook)
Subjects: LCSH: Popular music—Production and direction. | Sound studios.
Classification: LCC ML3790 .S43 2018 | DDC 781.49—dc23
LC record available at https://lccn.loc.gov/2018011345

ISBN: 978-1-138-55636-2 (hbk)
ISBN: 978-1-138-55637-9 (pbk)
ISBN: 978-1-315-15001-7 (ebk)

Typeset in Giovanni
by Swales & Willis Ltd, Exeter, Devon, UK

Visit the companion website for this book: *http://www.cambridge-mt.com*

To my parents

Contents

ACKNOWLEDGMENTS ix

INTRODUCTION xi

PART 1 • Hearing and Listening 1

CHAPTER 1 Using Nearfield Monitors 3
CHAPTER 2 Supplementary Monitoring 35
CHAPTER 3 Low-End Damage Limitation 53
CHAPTER 4 From Subjective Impressions to Objective Results 63

PART 2 • Mix Preparation 87

CHAPTER 5 Essential Groundwork 89
CHAPTER 6 Timing and Tuning Adjustments 99
CHAPTER 7 Comping and Arrangement 119

PART 3 • Balance 131

CHAPTER 8 Building the Raw Balance 133
CHAPTER 9 Compressing for a Reason 165
CHAPTER 10 Beyond Compression 187
CHAPTER 11 Equalizing for a Reason 197
CHAPTER 12 Beyond EQ 223
CHAPTER 13 Frequency-Selective Dynamics 237
CHAPTER 14 The Power of Side Chains 255
CHAPTER 15 Toward Fluent Balancing 261

PART 4 • Sweetening to Taste 269

CHAPTER 16 Mixing with Reverb 271
CHAPTER 17 Mixing with Delays 297
CHAPTER 18 Stereo Enhancements 305
CHAPTER 19 Master-Buss Processing, Automation, and
Endgame 321
CHAPTER 20 Conclusion 367

APPENDIX 1   **Music Studios and the Recording Process:**
An Overview..369
APPENDIX 2   **Who's Who: Selected Discography** 379
APPENDIX 3   **Quote References**..407
APPENDIX 4   **Picture Credits**...419

INDEX 421

Acknowledgments

This second edition is largely a response to direct feedback from readers, so I'd like to thank all those who queried, heckled, and generally peppered me with thorny mixing questions ever since the first edition appeared back in 2011. The hundreds of updates and clarifications that you've inspired have enhanced the clarity and usefulness of this new edition immeasurably.

In addition, I'd like specifically to thank all the interviewers who have done an immense service to us all by shedding so much light on top-level studio practice: Michael Barbiero, Matt Bell, Bill Bruce, Richard Buskin, Dan Daley, Tom Doyle, Maureen Droney, Tom Flint, Keith Hatschek, Sam Inglis, Dave Lockwood, Howard Massey, Bobby Owsinski, Andrea Robinson, Simon Sherbourne, and Paul Tingen. Paul Tingen deserves special praise for his dogged pursuit of the hottest current hitmakers for *Sound on Sound*'s "Inside Track" series. I'm grateful as well to Russ Elevado, Warren Huart, Roey Izhaki, Jason Moss, Roger Nichols, Eric "Mixerman" Sarafin, and Mike Stavrou for their own insightful teachings on the subject of mixdown, as well as to Dave Pensado and all the guests who so graciously share their expertise on the splendid Pensado's Place web TV series. Many thanks are also owed to Philip Newell, Keith Holland, and Julius Newell for permission to reproduce the results of their superb NS10M research paper; to Phil Ward for alerting me to the perils of speaker porting; and to Simon-Claudius Wystrach, Roberto Détrée, and Munich's Mastermix Studios for allowing me to photograph their speakers.

In developing the first edition of this text for publication, I was assisted a great deal by Matt Houghton and Geoff Smith, as well as the entire editorial department at *Sound on Sound* magazine. In preparing the new edition, I'm very grateful for the draft-text feedback I received from Simon Gordeev, Matt Leigh, Daniel Plappert, and especially Simon-Claudius Wystrach. Thanks also to Raghav Venkatesan and Preethi Shankaran for helping to update the Appendices, and to Lara Zoble and the whole team at Taylor & Francis for their patience and expertise in bringing this project to fruition.

What I could not have predicted back in 2011, was that the book's associated web site (www.cambridge-mt.com/ms-intro.htm) would take on such a life of its own as an independent educational resource for learners and tutors worldwide. Indeed, it's current health and continued expansion are only possible because of ongoing technical assistance from Mike Zufall and Indigo Technologies, and through the generous financial support of the site's public-spirited patrons. (If you'd also like to help fund the site, please surf to www.cambridge-mt.com/support.htm)

Above all, I'd like to thank my wonderful wife and daughters for their unwavering love and support—and for steadfastly refusing to take me too seriously!

WHAT YOU'LL LEARN FROM THIS BOOK

This book will teach you how to achieve release-quality mixes on a budget within a typical small-studio environment by applying power-user techniques from the world's most successful producers. Using these same methods, I've carried out dozens of mix makeovers for *Sound on Sound* magazine's popular "Mix Rescue" series, working on mass-market gear in various home, project, and college studios. If you head over to www.cambridge-mt.com/MikeSenior. htm, you can find before/after audio comparisons for every one of these remixes, and this book is a one-stop guide to pulling off these kinds of night-and-day transformations for yourself.

WHAT YOU WON'T LEARN

This book will not teach you how to operate any specific brand of studio gear—that's what equipment manuals are for! The information here is deliberately "platform neutral," so that you can make just as much use of it whether you're on Cubase, Digital Performer, Live, Logic, Pro Tools, Reaper, Reason, Sonar, or any other software digital audio workstation (DAW) platform. And although I've made the assumption that the majority of cost-conscious mix engineers will now be working in software, most of my advice also carries over equally well to hardware setups, give or take a patch cord or two. Indeed, my own background is in computerless environments, so I know from experience that great results are attainable there too.

WHAT YOU NEED TO KNOW ALREADY

Although I've done my best to make this book accessible to studio newbies, there is nonetheless some basic background knowledge that you'll need to understand to get the best out of what I'll be writing about. In particular, I'm assuming that you:

- already understand something about the fundamental physics, measurement, and perception of sound;
- have some idea of the main stages involved in the multitrack production process;
- can identify the main functional components of hardware and software recording studios.

Many modern musicians will already have absorbed this stuff without realizing it, just by coming into contact with other like-minded people and following the activities of their favorite commercial artists. However, if you feel you might benefit from a quick refresher on any of that, or you'd like to clarify my usage of some of the essential technical terms involved, then check out **Appendix 1**, where I've provided a super-condensed overview of this material.

HOW TO USE THIS BOOK

Because this book has been specifically designed as a step-by-step primer, you'll get best results if you work through it from beginning to end. Later sections rely on material covered in earlier chapters, so some aspects of the discussion may not make proper sense if you just dip in and out. At the end of each chapter there is a **Cut to the Chase** section, which allows you to review a summary of each chapter's main "secrets" before proceeding further. Underneath it is an **Assignment** section, which suggests a number of practical activities to consolidate your understanding of each chapter, and these could also serve as coursework tasks within a more formal education framework. The **Web Resources** box leads to a separate website containing an extensive selection of related links and multimedia files, all of which may be freely used for educational purposes.

This book is based on my own wide-ranging research into the studio practices of more than 160 world-famous engineers, drawing on more than 5 million words of first-hand interviews. The text therefore includes hundreds of quotes from these high-fliers. If you don't recognize someone's name, then look it up in **Appendix 2** to get an idea of the most high-profile records they've worked on—you'll almost certainly have heard a few of those! If you'd like to check out any quote in its original context (which I'd heartily recommend), then follow the little superscript number alongside it to **Appendix 3**, where there's full reference information for each one. Finally, if you have any further questions or feedback, feel free to email me at ms@cambridge-mt.com.

PART 1
Hearing and Listening

Probably the most reliable way to waste your time in a small studio is by trying to mix before you can actually hear what you're doing. Without dependable information about what's happening to your audio, you're basically flying blind, and that can get messy. In the first instance, you'll face a frustratingly uphill struggle to get a mix that sounds good in your own studio, and then you'll invariably find that some of your hard-won mixes simply collapse on other playback systems, so that you're left unsure whether any of the techniques you've learned along the way are actually worth a brass farthing. You'll be back to square one, but with less hair.

Relevant advice from professional engineers is perhaps unsurprisingly thin on the ground here. After all, most pros have regular access to expensive high-end speaker systems in purpose-designed rooms with specialist acoustic treatment. However, even the hottest names in the industry don't always get to work in the glitziest of surroundings, and if you look carefully at their working methods, they have actually developed various tactics that enable them to maintain consistent high-quality results even under difficult circumstances. These same tricks can be applied effectively in small studios too. So much so, in fact, that as long as you take care with gear choice and studio setup, it's perfectly possible to produce commercially competitive mixes in a domestic environment with comparatively affordable equipment. Indeed, all of my remixes for *Sound on Sound* magazine's monthly "Mix Rescue" column have been carried out under exactly such restrictions.

But even God's own personal control room won't help you mix your way out of a wet paper bag unless you know how to *listen* to what you're hearing. In other words, once you're presented with a bunch of information about your mix, you need to know how to make objective decisions about that data, irrespective of your own subjective preferences, because that's the only way of repeatedly meeting the demands of different clients or different sectors of the music market. Do the cymbals need EQ at 12kHz? Does the snare need compression? How loud should the vocal be, and are the lyrics coming through clearly enough? These are the kinds of important mix questions that neither your listening system nor your mixing gear can answer—it's you, the engineer, who has to listen to the raw audio facts, develop a clear opinion about what needs to be changed, and then coax the desired improvements out of whatever equipment you happen to have at your disposal.

Most people who approach me because they're unhappy with their mixes think that it's their processing techniques that are letting them down, but in my experience the real root of their problems is usually either that they're not able to hear what they need to, or else that they haven't worked out how to listen to what they're hearing. So instead of kicking off this book by leaping headlong into a treatise on EQ, compression, or some other related topic, I'd like to begin by focusing on hearing and listening. Until you get a proper grip on those issues, any discussion of mixing techniques is about as useful as a chocolate heatsink.

CHAPTER 1
Using Nearfield Monitors

1.1 CHOOSING YOUR WEAPONS

Choosing the equipment that allows you to hear (or "monitor") your mix signal is not a task to be taken lightly, because it's the window through which you'll be viewing everything you do. For those on a strict budget, however, the unappetizing reality is that monitoring is one of those areas of audio technology where the amount of cash you're prepared to splash really makes a difference. This is particularly true with regard to your studio's primary monitoring system, which needs to combine warts-and-all mix detail with a fairly even frequency response across the biggest possible slice of the 20Hz to 20kHz audible frequency spectrum—a set of characteristics that doesn't come cheap.

That said, when choosing the stereo loudspeakers that will fulfill these duties in all but the most constrained studios, there's a lot you can do to maximize your value for money. First off, furniture-rattling volume levels aren't tremendously important for mixing purposes, despite what you might guess from seeing pics of the dishwasher-sized beasts mounted into the walls of famous control rooms—most mix engineers use those speakers mainly for parting the visiting A&R guy's hair! "There just aren't many situations where the main monitors sound all that good," says Chuck Ainlay. "The mains in most studios are intended primarily for hyping the clients and playing real loud."[1] "I don't use the big monitors in studios for anything," says Nigel Godrich, "because they don't really relate to anything."[2] You'll get a more revealing studio tool at a given price point if you go for something where the designers have spent their budget on audio quality rather than sheer power. As it happens, the most high-profile mix engineers actually rely almost exclusively on smaller speakers set up within a couple of meters of their mix position (commonly referred to as nearfield monitors). If you sensibly follow their example in your own studio, you shouldn't need gargantuan speaker cones and rocket-powered amplifiers, even if you fancy making your ears water.

> ## SURROUND MONITORING
>
> Before acquiring a multispeaker surround setup for a small studio, I'd advise thinking it through pretty carefully. Until you can reliably get a great stereo mix, I for one see little point in spending a lot of extra money complicating that learning process. In my experience, a limited budget is much better spent achieving commercial-quality stereo than second-rate surround, so I make no apologies for leaving the topic of surround mixing well alone and concentrating instead on issues that are more directly relevant to most small-studio denizens.

Another simple rule of thumb is to be wary of hi-fi speakers, because the purpose of most hi-fi equipment is to make everything sound delicious, regardless of whether it actually is. This kind of unearned flattery is the last thing you need when you're trying to isolate and troubleshoot sneaky sonic problems. I'm not trying to say that all such designs are inevitably problematic in the studio, but most modern hi-fi models I've heard are just too tonally hyped to be of much use, and maintenance issues are often a concern with more suitable pre-1990s systems. Speakers with built-in amplification (usually referred to as "active" or "powered') are also a sensible bet for the home studio: they're more convenient and compact; they take the guesswork out of matching the amplifier to your model of speaker; they're normally heavier, which increases the inertia of the cabinet in response to woofer excursions; and many such designs achieve performance improvements by virtue of having separate matched amplifiers for each of the speaker's individual driver units.

Beyond those issues, a lot of monitor choice is about personal preference, and there's nothing wrong with that. Some people prefer bright aggressive-sounding monitors, others restrained and understated ones, and neither choice is wrong as such. The main thing to remember is that no monitors are truly "neutral," and every professional engineer you ask will have his or her own personal taste in this department. Part of the job of learning to mix is getting accustomed to the way your own particular speakers sound, so don't get too uptight about minute differences in tone between speakers. Go for something that appeals to you, and then concentrate on tuning your ears to how your chosen model responds in your own control room. "You've got to be careful about getting new monitors," advises Dave Way. "You've got to break them in and get to know them before you start to rely on them."[3] Part of doing this involves referring to a set of reference recordings with which you're familiar (discussed more in Chapter 4).

> No monitors are truly "neutral," and every professional engineer you ask will have his or her own personal taste in this department. Part of the job of learning to mix is getting accustomed to the way your own particular speakers sound.

1.1.1 Ported Speakers and Frequency Response

I have one further piece of advice to offer when choosing monitors, but I've deliberately held it in reserve, because I want to give it special attention. It's this: the less money you have to spend, the more you should beware ported monitors. Such speakers are sometimes also referred to as "bass reflex" or "reflex loaded" designs, and they incorporate some kind of hole or vent in the speaker cabinet that encourages the whole box to resonate in sympathy with the speaker's drivers. The main purpose of this resonance is to increase the low-frequency output, an aspect of a small speaker's performance that is naturally restricted based on its limited woofer size. By using a port to compensate for the woofer's natural low-end roll-off, manufacturers can have a wider flat region on their published frequency–response graph, as well as giving the speaker a louder, beefier sound that'll help impress Joe Public's wallet in the shops. Figure 1.1 illustrates the basic effect of porting on a typical small-studio monitor's low-end frequency response. The solid line on the graph shows the kind of response you'd expect of a fairly typical small ported speaker, with the output remaining within a ±3dB window down to maybe 55Hz. If you defeated the speaker's port by blocking it, however, you'd find that the response changed to something like that shown by the dotted line: the trace now drifts out of the ±3dB window almost an octave higher, just above 100Hz.

So what's so bad about using a port to widen a speaker's frequency response? The problem is that porting also has several less well-advertised side effects

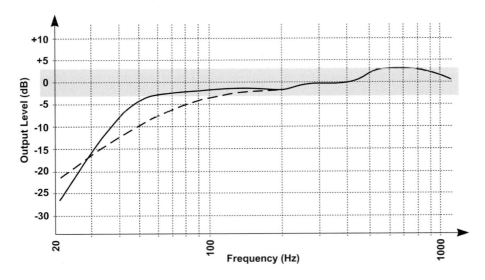

FIGURE 1.1
The solid line on this graph shows the kind of sub-1kHz frequency response plot you might expect for a small and budget-friendly ported studio monitor. The dotted line shows how the response changes when the port is artificially blocked, thereby defeating the cabinet's low-frequency resonance. The shaded region indicates the ±3dB "flat" region of the speaker's quoted frequency–response specification.

FIGURE 1.2
Some affordable two-way ported nearfield monitors (*top to bottom*): the ADAM A7X has dual circular ports either side of the woofer, the KRK Rokit 8 G3 has a port slot under the woofer, the Behringer B2031A has a narrow port slot on each side of the tweeter, and the Yamaha HS8 features a single circular port on the rear panel.

that can easily conspire to hamstring you at mixdown. Given the widespread use of porting in budget nearfield monitors, it's important to understand what these side effects of porting are. On the one hand, this knowledge makes it easier to evaluate monitors objectively when making a purchase; on the other hand, it better equips you to work around potential porting gremlins when the choice of monitors is beyond your control—for example, in a college facility or a friend's home studio, or if you've already spent your speaker budget before reading this book! So bear with me while I look at this issue in more detail.

The first problem with porting can already be seen in Figure 1.1: although the port stops the response dropping off until 50Hz, the output takes a real nosedive beyond that. This means that although the speaker's overall low-frequency output is boosted by the port, the relationship between the sub-50Hz levels and the rest of the signal is seriously skewed at the same time, which makes it trickier to make judgments about instruments with important low-frequency components. So assuming, for the sake of example, that you're playing back the sound of a bass instrument that is completely consistent in its low-frequency levels, the perceived volume of its fundamental frequency will still dance around alarmingly as the notes change pitch, depending on how far the fundamental slips down the steep frequency–response roll-off.

Bear in mind that the lowest fundamental from a typical bass guitar is around 41Hz, whereas pianos, organs, and synths are just some of the sources that will happily generate fundamentals in the 20 to 40Hz bottom octave. In contrast to the fundamental, however, the first harmonic of these bass notes lies an octave above, typically in the much flatter frequency–response region above 50Hz, so it'll be tough going to decide whether there's the right amount of each of these frequencies respectively. And, of course, if we step back into the wild again, where untamed rampaging bass parts are often anything but consistent, how are you expected to judge when your mix processing has actually reined them in properly?

Kick drums are equally complicated to deal with. Let's say that you're comparing the kick level in your own mix to something on a favorite commercial record, but your kick drum has loads of energy at 30Hz, whereas the comparison track's kick is rich in the 50Hz region. Because the speaker is effectively recessing the 30Hz region by 12dB compared to the 50Hz region, you're likely to fade your own kick drum up too high, only to discover a rumbling mess lurking underneath your mix in other monitoring environments. Although the loss of low end on an unported monitor is also a problem, it's much easier to compensate for this mentally while mixing, because the relative levels of neighboring low-frequency bands are more representative.

1.1.2 Killer Side Effects of Porting

These porting anomalies, however, are only the tip of the iceberg, because frequency–response graphs only show how speakers respond to constant full-frequency noise, a test signal that is nothing like the varied and fast-moving waveforms of music. Much more troublesome is the way that porting hinders the monitor's ability to track moment-to-moment changes in the mix signal. Specifically, the port causes any spectral energy at its resonant frequency to ring on for a short time, and while it's this resonant buildup that generates the port's flattering low-frequency level boost for a constant noise test signal, the same quality also adds short resonant tails to fleeting percussive attack noises (often referred to as transients), such that they can seem louder and less punchy than they actually are. Sounds that stop abruptly suffer a similar problem, with the port ringing on after they've finished. In this case, the resonance not only disguises the true decay attributes of the sound itself, but it can also make it difficult to judge the character and level of short-duration studio effects (such as modulated delays and reverb), which are often very useful at mixdown.

Another possible problem with ported speakers is that the ringing of the port can dominate over the real fundamental frequencies of low bass notes, making them difficult to distinguish from each other. Speaker reviewers sometimes refer to this phenomenon as "one-note bass," and it adds unwelcome uncertainty to tuning judgments at the low end. A commercial recording that I find particularly good for revealing this occurrence is Skunk Anansie's "Infidelity" (from the band's album *Stoosh*), where the meandering bass line quickly becomes murky and ill-defined in the presence of low-end monitoring resonances. (The track is also good for testing the frequency response of a monitoring system, as only the most extended response can do justice to that particular kick drum's almost seismic low-frequency rumble.)

Were the port-ringing consistent across the audio spectrum, you could mentally compensate for it perhaps, but of course it's not: it's more or less severe depending on how much of a given transient's energy resides around the porting frequency. Furthermore, I've so far taken for granted that the port has only one resonant frequency. In reality, however, it's difficult to stop the thing resonating at a whole range of higher frequencies too, which leads to unpredictable time-smearing artifacts right across the frequency spectrum. So it's not just bass instruments that you may be unable to judge reliably, but everything else too! Although it's perfectly possible for speaker designers to use careful internal cabinet design and damping to tame all but the desired low-frequency port resonance, that does cost them money, so this is where more affordable designs can really come a cropper.

Of course, a simple frequency–response graph leaves you blissfully ignorant of any of this stuff, because it only has axes for frequency and level. If you want to lay bare resonance side effects, then you need to add a third dimension to your frequency–response graph: time. Fortunately, there is a type of graph that does exactly that, called a spectral decay or "waterfall" plot. It reveals what happens to a speaker's output when a constant full-range test signal is suddenly switched off—as the graph develops in time (in other words moving from the background into the foreground, speaking three-dimensionally), you can see how much different frequencies ring on.

The left-hand column of Figure 1.3 shows waterfall plots for three well-designed small nearfield monitors. The top graph is for an unported model, whereas the two lower plots are for ported designs. You can see the low end of the ported models ringing on, as you'd expect, but otherwise the midrange and high end stop quickly without any obvious resonant trails. Compare this with the waterfall plots in the right-hand column of Figure 1.3, measured from three budget

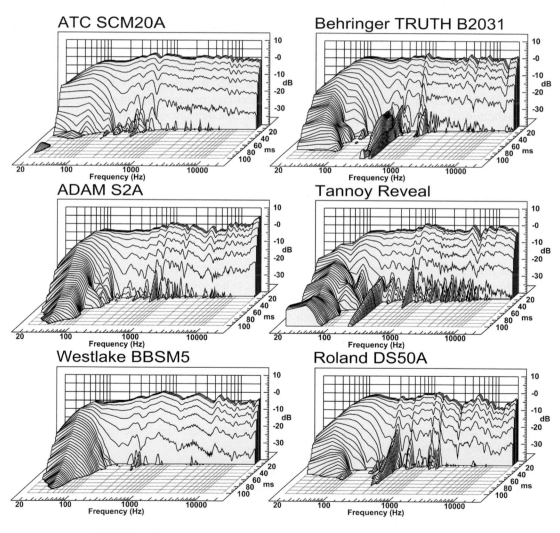

FIGURE 1.3
Waterfall plots for six different sets of studio monitors.

ported nearfields. What they clearly illustrate is that each speaker has prominent resonances well up into the midrange—although it has to be said that other factors can contribute unwanted midrange resonances to speakers as well, so it's not just porting side effects you're seeing here. The less well-controlled a monitor's resonances, the less easily you can mix with it. Unfortunately, few loudspeaker manufacturers provide waterfall plots as standard for their speakers (I wonder why?), but that hasn't stopped a few researchers and reviewers publishing their own measurements unofficially, and I've included links to some of those information sources within this chapter's web resources.

But even that's not the end of the story: ports can also produce turbulence noise, which obscures other parts of your mix; compression artifacts, which mess with the apparent level of bass instruments as you adjust the monitoring volume; and distortion, which misleadingly gives extra midrange body to bass instruments, making them seem more audible in the mix than they should be. If you want to hear what I'm talking about, download the LFSineTones audio file from this chapter's web resources, and listen to its low-frequency sine-wave tones through a budget ported monitor. Particularly on the lowest frequencies you'll usually hear a good dose of fluttering port noise and low-level distortion harmonics overlaid on what should be pure tones. Need any further convincing? Then consider the fact that two of the most influential mixing speakers in the history of audio production are unported designs: the Yamaha NS10 and the Auratone 5C Super Sound Cube. (You can see the waterfall plots for these speakers in Figure 1.5, and although neither has a particularly flat frequency response, both are extraordinarily well-behaved as far as resonances are concerned.)

All of which brings me back to my main point: the less money you're going to spend on monitors, the more you should approach ported models armed with holy water and cloves of garlic! In my experience, you'll have to part with well over £1500 ($2000) for a pair of ported nearfields that can reliably deliver what you need to mix competitively, whereas I don't think you need to spend that much on unported designs to get similar mixing muscle, just so long as you're willing to work with lower overall volume levels. (For my most up-to-date recommendations of specific nearfield systems, as well as practical tips on what to listen for when auditioning monitors, check out this chapter's web resources page.) Before I'm labelled as some kind of dogmatic portophobe, though, let me add that once you're comfortably above that kind of price range I think ported monitors are often no less capable of delivering great mixes than unported models, and the choice between the two designs becomes more a question of personal preference than anything else.

> The less money you're going to spend on monitors, the more you should approach ported models armed with holy water and cloves of garlic!

PASSIVE RADIATORS AND TRANSMISSION LINES

Not all monitor speakers can be categorized clearly as ported or unported, as several project-studio monitor manufacturers now use passive radiators (effectively dummy speaker cones that vibrate in sympathy with the woofer) to achieve ported-style bass enhancement from a sealed cabinet. Another midway design is PMC's transmission-line system, whereby the external port hole feeds a damped internal ducting network designed to reduce the problematic side effects of porting. However, the waterfall plots in Figure 1.4 suggest to me that these strategies are only of limited use in overcoming the resonance issues of porting, a suspicion borne out in my own personal experience of such designs.

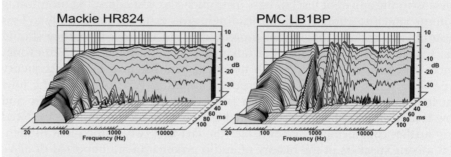

FIGURE 1.4
Waterfall plots for the Mackie HR824 and PMC LB1BP monitors.

1.1.3 Speaker Stands and Other Mounting Hardware

You can fork out for the fanciest monitors you like, but unless you set them up sensibly in your room, you might as well have spent most of that money on doughnuts for all the good it'll do your sound. I've visited a large number of small studios, and one thing the majority have in common is that their owners have underestimated the importance of monitor installation, with the result that the monitoring sounds only a fraction as good as it should, given the cost of the speakers. So let's look at ways you can maximize the quality of the sound, whatever speakers you're using.

For a start, the speaker cabinets should be as firmly fixed as possible, because if they move at all in sympathy with the woofer excursions it'll mess with how the low end of the mix is represented. How exactly you decide to mount the boxes will depend on the physical limitations you have to work with in your particular setup, but my recommendation is to use dedicated speaker stands, as these typically give a much better sound than desks and shelves and can be moved around the room more easily than heavy-duty wall brackets. Stands don't need to be exorbitantly expensive either, as long as they are solid enough to keep the speaker still. In fact, you can easily build decent ones yourself if you're handy with your woodwork and use suitably chunky raw materials.

The first thing the mounting hardware has to do is present as much inertia as possible, so the speaker cabinet moves as little as possible in reaction to woofer excursions. If you think the stands/brackets themselves aren't man enough for this job, it's worth trying to add weight to them in the first instance by putting a piece of paving slab underneath each speaker. A bit of rubber matting can also help, by improving the mounting platform's grip on the speaker cabinet. The other main thing your speaker mountings need to do is minimize the transmission of the speaker's physical vibrations into other resonant objects. If your speaker sets off a physical resonance within its stand, for example, it can skew your perception of the mix just as much as any resonances inherent in the speaker design

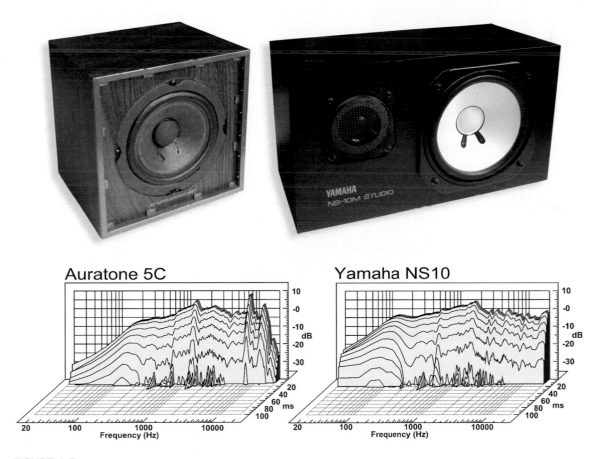

FIGURE 1.5
Two of the most revered mixing speakers are unported designs: the Auratone 5C Super Sound Cube (*left*) and the Yamaha NS10 (*right*). Below them you can see their waterfall plots.

FIGURE 1.6
Resting your nearfield monitors on some heavy bricks or tiles can help increase the speaker cabinet's inertia.

itself. This is one reason why putting speakers on domestic shelves or desks can cause insurmountable monitoring difficulties—it's often surprising how readily these vibrate in sympathy. The LFSineTones audio file is good for revealing resonances, so play it back at a reasonably high volume to see whether you hear any furniture humming along. Try also resting a finger on your mounting hardware (stand, bracket, or whatever) and check whether you feel any obvious vibrations.

One advantage of dedicated speaker stands is that they often have a hollow frame. It can be filled with sand, which is good for damping resonances and also increases the stand's inertia. However, other solutions to the resonance problem include high-density foam platforms wedged between the speaker and the mounting surface (such as Auralex's Mo Pads) or little inverted mounting spikes in a similar location (for example Sound Network's China Cones). Primacoustic's Recoil Stabilizer is another popular option, as it incorporates not only a foam base but also a rubberized steel top platform for extra grip and inertia. However, although these kinds of studio widgets can yield noticeable improvements if your monitors are sitting on a normal shelf or desk; my feeling about them is that they're unlikely to justify the extra outlay if you've already sensibly invested £100 ($150) or so in sand-filled speaker stands or similarly sturdy mounting hardware.

1.2 POSITIONING THE SPEAKERS

Whatever you actually sit the speakers on, their exact positioning is also critical to getting good audio reproduction. You should try wherever possible to aim the speakers directly at the listening position. A speaker's frequency response is measured on axis (i.e., from directly in front of it), so if you listen off axis, you won't be hearing what the designer intended you to—high frequencies are more directional than low frequencies, so high-end details in particular tend to suffer. Moving around your listening room should amply demonstrate these effects with any full-bandwidth music mix, but if you want to hear the phenomenon at its starkest, then try listening to a constant full-range test signal (such as the PinkNoise file downloadable from this chapter's web resources) through just one of your speakers. These aren't just minuscule sonic niceties we're talking about. High frequencies are also easily shadowed by physical objects, so make sure you can actually see the drivers you're listening to.

PHASE AND COMB FILTERING

I've touched on the ideas of phase and comb filtering in the main text, but because they have so many ramifications when mixing, it's worth looking at the subject more closely. The best way to start thinking about phase is first to consider a sine-wave signal, the simple audio waveform from which all complex musical sounds can theoretically be built. A sine wave generates only a single audio frequency, according to how many times its waveform shape repeats in a second. For example, a 1kHz sine wave repeats its waveform 1000 times per second, with each waveform repetition lasting 1ms. Imagine that you have two mixer channels, each fed from the same sine-wave source at the same frequency. The peaks and troughs of the two waveforms will be exactly in line, and mixing them together will simply produce the same sine wave, only louder. In this situation we talk about the two sine waves being "in phase" with each other.

If you gradually delay the audio going through the second channel, however, the peaks and troughs of the two sine waves shift out of alignment. Because of the unique properties of sine waves, the combination of the two channels will now still produce a sine wave of the same frequency, but its level will be lower than if the two channels were in phase, and we say that "partial phase cancellation" has occurred. When the second channel is delayed such that its peaks coincide exactly with the first channel's troughs (and vice versa), the two waveforms will combine to produce silence. At this point we say that the waveforms are completely "out of phase" with each other and that "total phase cancellation" has occurred.

When total phase cancellation occurs, you sometimes hear engineers say that the signals are "180 degrees out of phase." This is a phrase that's not always used correctly, and it can therefore be a bit confusing. To describe the phase relationship between two identical waveforms, mathematicians often quantify the offset between them in degrees, where 360 degrees equals the duration of each waveform repetition. Therefore, a zero-degree phase relationship between two sine waves makes them perfectly in phase, whereas a 180-degree phase relationship puts them perfectly out of phase, resulting in total phase cancellation. All the other possible phase relationships put the waveforms partially out of phase with each other, resulting in partial phase cancellation. What's confusing about the "180 degrees out of phase"

(continued)

term is that it is sometimes used to refer to a situation where the second channel's waveform has been flipped upside down, so that the peaks become troughs and vice versa—a process more unambiguously referred to as polarity reversal. This scenario also results in silence at the combined output, hence the common confusion in terminology, but it's very important to realize that the total phase cancellation here is brought about by inverting one of the waveforms, not by delaying it.

Now let's scale things back up to deal with real-world sounds, made up as they are of heaps of different sine waves at different frequencies, each one fading in and out as pitches and timbres change. If we feed, say, a drum loop to our two mixer channels, instead of a single sine wave, any delay in the second channel will have a dramatic effect on the tonality of the combined signal, rather than just altering its level. This is because for a given delay, the phase relationships between sine waves on the first channel and those on the second channel depend on the frequency of each individual sine wave. So, for example, a 0.5ms delay in the second channel will put any 1kHz sine-wave components (the waveforms of which repeat every 1ms) completely out of phase with those on the first channel, resulting in total phase cancellation. On the other hand, any 2kHz sine-wave components (the waveforms of which repeat every 0.5ms) will remain perfectly in phase. As the frequency of the sine-wave components increases from 1kHz to 2kHz, the total phase cancellation becomes only partial, and the level increases toward the perfect phase alignment at 2kHz.

Of course, above 2kHz the sine-wave components begin partially phase canceling again, and if you're quick with your mental arithmetic you'll have spotted that total phase cancellation will also occur at 3kHz, 5kHz, 7kHz, and so on up the frequency spectrum, whereas at 4kHz, 6kHz, 8Hz, and so on the sine-wave components will be exactly in phase. This produces a characteristic series of regularly spaced peaks and troughs in the combined frequency response of our drum loop—an effect called comb filtering. A delay of just 0.000025s (a 40th of a millisecond) between the two channels will cause total phase cancellation at 20kHz, but you'll also hear partial phase cancellation at frequencies below this. As the delay increases, the comb filter response marches further down the frequency spectrum, trailing its pattern of peaks and troughs behind it, which themselves get closer and closer together. However, when the delay times reach beyond about 25ms or so (depending on the sound in question), our ears start to discern the higher frequencies of the delayed signal as distinct echoes, rather than as a timbral change, and as the delay time increases, phase cancellation is restricted to progressively lower frequencies.

Although it should now be clear that the tonal effects of comb filtering can be disastrous if two identical signals are combined with a delay between them, most real-world comb filtering at mixdown is actually much less severe, either because the out-of-phase signals aren't completely identical, or because they're at very different levels, or both.

Aiming the speakers isn't just about the horizontal plane either, because vertical alignment is usually even more important, for a couple of reasons. The first is that on most nearfield monitors, the cabinet is profiled around the tweeter to create what's called a waveguide, which is designed to horizontally disperse the jet of high frequencies more widely and thereby increase the size of the optimum listening area (or "sweet spot"). Although waveguides can be quite effective at this, they don't usually do the same job for the vertical high-frequency dispersion and can even make it narrower. But the second reason is that most nearfield monitors have more than one driver in them, with each driver in a different vertical position. A dedicated bit of circuitry or DSP (called a crossover) within the speaker splits the incoming signal's frequency range between the different drivers at factory-specified boundaries (called crossover frequencies). Although ideally the crossover should therefore prevent any over-lap between the frequency output of the different drivers, the truth is that there is inevitably a small spectral region around each crossover frequency where two drivers are both contributing significant levels at the same time. If the dis-tance from each driver to the listening position isn't the same, then the signals from the different drivers will arrive at the listening position at different times (or "out of phase" in geek-speak), and this gives rise to a potentially serious frequency-cancellation effect called comb filtering.

Although manufacturers typically do their best to keep crossover regions pretty narrow to minimize the effect of comb filtering, most affordable nearfield monitors have only two drivers, which means that any comb filtering between the woofer and the tweeter happens in the worst possible place from a mixing standpoint: right in the mid-frequency region, where our hearing is most sensitive. If you want to get a handle on the extent of the damage here, try this experiment. Play my PinkNoise file through a sin-gle nearfield speaker with vertically spaced drivers, and listen to it first of all directly on axis, from about two feet away. Now drift alternately about six inches to each side while keeping your vertical position constant. You'll hear a small change in tone on most speakers because of the high-frequency direc-tionality I mentioned earlier. Once you're used to that change, drift up and down by about six inches instead, and the tonal change will likely be much more noticeable. Although the effects of comb filtering between your speaker drivers won't usually be as obviously apparent in their own right when you're listening to a real-world mix, that doesn't mean they aren't there, and the ripples they put into the fre-quency response treacherously undermine your ability to judge both the tone and level balance of critical sounds in the midrange—things such as lead vocals, snare drums, and guitars.

> Most affordable nearfield monitors have only two drivers, which means any comb filtering between the woofer and the tweeter happens in the worst possible place from a mixing standpoint: right in the mid-frequency region, where our hearing is most sensitive.

1.2.1 Stereo Monitoring

A small studio's nearfield monitors will usually provide the most reliable source of information about a mix's stereo image, but for them to do this effectively they have to be set up so that the distance between the speakers

SPEAKERS ON THEIR SIDES?

There is a persistent myth among small-studio owners that putting speakers on their sides is the more "pro" method. True, a quick web surf will furnish you with countless pictures of *Battlestar Galactica*-style control rooms in which nearfield speakers are visible perched sideways on top of the console's meterbridge, but that setup has little to do with monitoring fidelity and everything to do with getting the nearfields out of the way of the big main monitors and maximizing the view through the control-room window during recording dates.

If you separate your speaker's drivers horizontally by placing the cabinets on their sides, then you have to be much more careful to keep your horizontal listening position absolutely consistent (both side-to-side and front-to-back) if you're going to avoid being stung by crossover comb-filtering effects. With vertical speakers, on the other hand, you'll only get these crossover problems if you move your head vertically, which gives you greater freedom of movement while you're working.

Add to this that the waveguides in most nearfield monitors are designed to broaden the high-frequency sweet spot when the speaker is vertical. If you flip that waveguide on its side it narrows the sweet spot instead, as well as bouncing more high-frequency energy off nearby desk and ceiling surfaces—something that's not sensible from an acoustics point of view.

equals the distance from each of the speakers to the listening position. This is because the human brain has evolved a powerful instinct that causes us to perceive the source of an incoming sound as being located on the same side as the ear that sound reaches first. (Presumably natural selection favored those cavemen who could work out where the saber-toothed tiger growls were coming from.) This instinct means that you don't have to move your head very far out of the sweet spot before the majority of the stereo image folds down into the closer speaker.

Pulling the speakers too far apart is a frequent mistake too, as it destabilizes the center of the stereo image, making balance judgments tricky for the important sounds that are typically placed there. One common reason people space their speakers overly wide is to accommodate dual computer displays side-by-side, but there are a couple of workarounds for that: either mount one display above the other instead; or lift the speakers up above the level of the display screens, pointing downwards towards the listening position. If anything, it's better to err on the side of placing the speakers too close together, because the narrowed stereo picture this produces is a lot easier to work with at mixdown than unstable central imaging. You'll also get the best stereo image if you try to position the whole monitoring setup such that the room is fairly symmetrical around the listener's line of sight, in order to retain image balance despite the impact of any sonic reflections from room boundaries and furniture.

One final pitfall to avoid is having your speakers out of polarity with each other. Normally any sound at the center of the stereo image should be coming out of both speakers at an equal level and with all the speaker drivers moving back and forth in sync. However, it's surprisingly common for less experienced studio users to mistakenly wire up their rig such that one driver is pushing toward the listener while the other is pulling away, such that the left-channel waveform is effectively inverted compared to the right-channel waveform, a situation usually referred to as having your speakers "out of phase," or to be more accurate "out of polarity." (To clarify: two audio waveforms are out of phase when there's a time delay between them; they're out of polarity when one of the waveforms is inverted, such that peaks become troughs and vice versa.) If you've got this problem, you'll get a very odd stereo listening experience, which feels a bit like having your brains sucked out your ear, and it also makes both stereo positioning and level balance difficult to judge.

To check for this effect, download the StereoTest audio file from this chapter's web resources and listen to it through your system. It contains a repeating pattern of four noise-burst test signals: the first only in the left channel, the second only in the right channel, the third in both channels, and the fourth in both channels, but with the right channel out of polarity with the left channel. The former pair of noise bursts will confirm that your speakers are indeed operating in stereo and are connected the right way round; the latter pair of noise bursts should make it fairly obvious if your speakers are out of polarity with each other; the third burst should be much more clearly central in the stereo image than the fourth if all is well. If you do discover that your speakers are out of polarity, then the finger of blame will almost certainly point to some facet of your audio wiring after the outputs of your studio's playback system, the most common culprit being that the positive and negative terminals of passive speakers haven't been connected correctly with the respective terminals on their amplifier.

COAXIAL MONITORS

Most affordable nearfields have physically separated drivers, so different frequency ranges emanate from different locations. A few manufacturers, though, use special 'coaxial' or 'dual-concentric' units that cleverly combine the different drivers so that the whole spectrum effectively emerges from a single point. Typical advantages of this include minimizing comb-filtering between drivers, improving the speaker's off-axis tonality, and sharpening stereo imaging. However, I don't think physically separated drivers fundamentally prevent you delivering dependable pro-grade mixes, so I wouldn't suggest that budget-conscious engineers pay any kind of premium for coaxial designs unless for reasons of personal taste.

1.3 DEALING WITH ACOUSTIC REFLECTIONS

Room acoustics is one area of small-studio design that's woefully neglected in the main. It's not that budding engineers don't pay lip service to the worthiness of acoustic design in principle, it's just that acoustic treatment materials are probably the most unexciting way to spend money on your studio. They make no sound. They have no twinkly lights. They can give you splinters. But make no mistake: room treatment really separates the sheep from the goats when it comes to mix results. Spike Stent minces no words: "You can have the best equipment in the world in your control room, but if the room sounds like shit, you're onto a hiding to nothing."[4] The room acoustics are at least as important to the sound of your monitoring as your speaker hardware is. Permit me to shout that again for you: THE ROOM IS AT LEAST AS IMPORTANT AS THE SPEAKERS! So that means engineers looking for a fast track to commercial-sounding mixes should plough as much money into dealing with their room acoustics as they do into buying their monitors.

> Acoustic treatment materials are probably the most unexciting way to spend money on your studio. But make no mistake: room treatment really separates the sheep from the goats when it comes to mix results.

Now I'm the first person to acknowledge that few people can afford to engage specialist acousticians to build their studio from the ground up. Indeed, my experience is that small setups almost universally find themselves squished unceremoniously into a shed, basement, spare bedroom, attic, or office unit, where the user faces a battle even to adjust the interior decor, let alone the construction and placement of the walls. However, even if you're relegated to the cupboard under the stairs, it's still vital that you make what acoustics improvements you can. The good news here is that you can work wonders with off-the-shelf products and a bit of do-it-yourself (DIY), even on a limited budget and in rental accommodation where your construction options are limited. When you consider that a typical untreated domestic acoustic environment will, in my experience, render roughly two thirds of the money you spent on your speakers wasted, there's simply no excuse for inaction if you're serious about your craft.

The first main set of acoustics problems that you'll need to lock horns with when working on nearfield monitors is that sound doesn't just squirt directly out of the front of each speaker and then stop when it hits your ears. It also sprays around in all other directions to a greater or lesser extent, bouncing off the walls, ceiling, and floor, as well as any other reasonably solid object in the room. This can create a situation where a sonic reflection arrives at the mixing position shortly after the sound that traveled there directly—in other words, you end up listening to two versions of roughly the same sound slightly out of phase. As I mentioned in Section 1.2, this kind of combination of out-of-phase signals can cause nasty comb-filtering ripples in the perceived frequency response—and not just in your speakers' relatively restricted crossover regions, but across the entire frequency spectrum.

The most troublesome reflectors you'll encounter will be the furniture you're working on, whether that's a large mixing console or the bedside table propping up your laptop, as well as any room boundaries within about three meters of your listening position. (To be honest, sonic reflections coming from farther than three meters away aren't a primary concern in domestic environments, because their greater time delays and diminished levels are unlikely to cause major comb-filtering problems.) In professional setups the room surfaces around the monitoring position are often cleverly angled to fire all the main sound reflections from the speakers away from the listening position, but that kind of approach is rarely viable in the smaller rectangular rooms of more modest studios. A more practical alternative in most cases is to place absorptive acoustic treatment at all the main reflection points in order to reduce the level of the reflected sound, and hence the severity of the comb filtering.

USING A SUBWOOFER

There are certain styles of music for which the very lowest frequencies are extremely important, so if that's the market you're working in then you'll need a speaker system that can actually reproduce them. In the words of Trina Shoemaker, "Forty Hertz exists now. . . . We have real low end in today's recordings, so you have to work with it."[5] "That bottom octave is absolutely vital in any kind of club music," adds James Wiltshire of The Freemasons. "The only way to really monitor it is on a full-range system."[6] One common way of extending the low-frequency reach of budget nearfield systems is to supplement a smaller pair of stereo "satellite" speakers with a single additional subwoofer to create a so-called 2.1 system. Although this would seem to compromise the stereo presentation of the mix, we actually hear stereo primarily from higher frequencies, so that's not a problem in practice. In fact, you can usually position subwoofers well off-center without unbalancing the stereo picture. An advantage of the 2.1 approach is that you don't get phase-cancellation issues between drivers at the low end, and you also get a certain amount of flexibility to position the subwoofer independently of the nearfield satellites to reduce room-mode problems (a serious acoustics issue we'll discuss in detail in Section 1.4). However, although many manufacturers of 2.1 systems suggest that you can place the subwoofer pretty much wherever you like, I'd personally advise restricting yourself to placements where the subwoofer and satellites are all an equal distance from your ears; otherwise you'll compromise the relative timing of low-frequency transients. Also, as with any speaker, beware of porting side effects on cheaper products.

1.3.1 Acoustic Foam in Moderation

A cost-effective absorber here is open-cell acoustic foam of the type manufactured by companies such as Auralex, and a meter-square area of such treatment covering each of the reflection points can make a big difference to the clarity of the speaker tone. In a typical small setup, that means you'd put patches of treatment on each wall, on the ceiling, on the wall behind the monitors, and on the wall behind the listener if that's within range. Some further foam on your work surface can be a bonus too, but clearly your options may be limited here! In general, the thicker your absorptive material, the lower down the frequency spectrum it's going to be able to absorb, so don't just use the cheapest foam you can get, because it'll probably only be 5cm thick or less; a 10cm thickness will do a much more respectable job. If you have to cut treatment costs, try using a pattern of thinner and thicker foam tiles, rather than going for thinner foam across the board.

You can easily glue acoustic foam directly to walls or ceilings with an appropriate spray adhesive, but if you do that you'll have the devil's own job getting it off again if you ever want to transplant your studio setup at a later date. It's therefore a much better idea to glue the foam to some kind of lightweight ply-board instead, so you can then hang the foam like a picture in whichever room you happen to want to work in—a great advantage if you have to get results in untreated college facilities or on other people's home rigs. You could also try angling the foamed boards to bounce some of the reflected frequencies away from the sweet spot, just as in purpose-designed rooms. (*Sound On Sound* magazine's editor, Paul White, suggests another elegant and thrifty wall-mounting method: if you glue a couple of old CD-ROMs to the back of the acoustic foam, then you can use their central spindle holes to hang it directly from panel pins.)

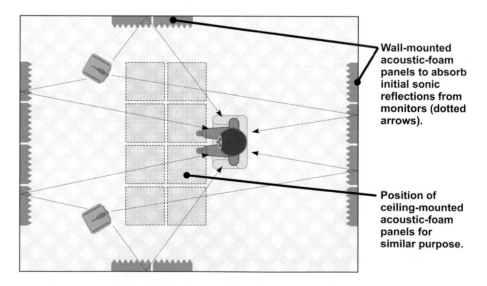

Wall-mounted acoustic-foam panels to absorb initial sonic reflections from monitors (dotted arrows).

Position of ceiling-mounted acoustic-foam panels for similar purpose.

FIGURE 1.7
If sound from your speakers reflects off control-room surfaces back toward the monitoring sweet spot, it can cause all sorts of frequency–response problems for the listener. This diagram shows the wall reflection points for a typical small rectangular control room and how you can use acoustic foam panels to reduce the problem. Don't forget that the ceiling reflects sound too!

Above all, however, resist the temptation to cover the entire room in foam—I can't tell you how often college studios in particular succumb to this seductive error. It is quite simply a recipe for disaster, because it hoovers up the top half of the room's reverberation, making for an extremely unnatural working environment. Although it makes good sense to damp down strong early reflections that can potentially comb filter your frequency response, you also want your monitoring environment to bear at least a passing resemblance to real-world living rooms and offices. If you plaster your whole studio in foam, you'll basically be mixing for people in padded cells—perhaps not the most lucrative demographic to set your sights on! What's more, the economics of covering such a large surface forces most foam fanatics to compromise on treatment thickness, which just sucks the life out of the upper frequencies, giving a superficial impression of acoustic control while the rest of the spectrum merrily runs riot. "I've heard a lot of home recordings," says Keith Olsen, "and there's a lot of what they think is deadening, because it takes out all the top end. That might save the neighbors from calling the cops, but the bottom end and midrange is still real ambient; it bounces around the room, and you get phase destruction."[7]

> If you plaster your whole studio in foam, you'll basically be mixing for people in padded cells—perhaps not the most lucrative demographic to set your sights on!

Because acoustic absorption like this is best used in moderation, it shouldn't break the bank, and if you're sensibly dividing your budget equally between the speakers and the acoustics, then even entry-level speakers should justify this kind of outlay. Still, if for whatever reason you can't afford proper acoustic treatment, you'll find that even soft furnishings such as thick curtains, blankets, and duvets can be of some benefit in damping those reflections if rigged appropriately. One more tip in this instance, though, would be to try to leave a few inches of air gap between the curtains/blankets and your room boundaries, as that has a broadly similar effect to increasing the thickness of the treatment. (You can pull this stunt with acoustic foam too by sticking small foam spacer blocks behind the main foam panels, as long as you're fairly open to the idea of covering yourself in glue and swearing a lot during the process.)

1.3.2 Boundary Effects

There's one further reflection issue to consider: a constructive low-frequency interference commonly referred to as the boundary effect. As you move a speaker closer to a room boundary, it reduces the delay between the direct and reflected sounds arriving at the listening position, making them less and less out of phase. This means that the comb filtering reduces and you start to get just a simple level reinforcement as the reflected sound adds power to the direct sound. However, for two reasons this reinforcement occurs primarily at low frequencies: first, their longer wavelengths shift less out of phase as a result of a given delay, and second, low frequencies are better transmitted by the speaker off-axis anyway. So if you place your speakers right up next to a wall, you'll get up to 3dB of bass boost, and this can rise to 6dB if you tuck them into a room corner where the effects of two boundaries gang up.

One solution is to EQ the speaker's output to compensate for the low-end tip-up. Indeed, a lot of active monitors aimed at compact setups have a little low-cut switch round the back for just this purpose. However, although this is one of the only situations where EQ can usefully bail out your acoustics, I'd still advise against placing your speakers right up against a wall if you can help it, because even with a sensible thickness of acoustic foam on that surface, there is still likely to be enough reflected sound arriving at the listening position to give significant comb-filtering problems in the midrange. Furthermore, if you're using monitors with ports at the rear of the cabinet, the proximity of the boundary is likely to increase turbulence as air zips in and out of the port opening, leading to extra noise and low-frequency distortion.

WHAT ABOUT DIFFUSION?

Another way to reduce the comb-filtering impact of early reflections is to use acoustic diffusers on reflective surfaces to scatter their reflections in lots of different directions. The downside of using diffusers in the small studio, though, is that they're usually more expensive than acoustic foam for a given area of coverage. However, that doesn't mean that diffusion has no part to play in project studios, because it turns out that things such as CD racks and bookshelves can work quite well, as long as they're stocked fairly irregularly. (If ever there were a good excuse for having a messy studio, then that's it!) Eric Schilling is a big fan of using a bookshelf like this: "It has mass, and each book has a different depth and size. The concept is brilliant in its simplicity."[8] So do try to position shelves like these usefully if you can—the wall behind the monitoring position is a particularly good bet, because that otherwise takes a fairly large area of acoustic foam to treat effectively, which risks overdeadening the room's high-frequency reverberation.

1.4 TACKLING ROOM RESONANCES

Although acoustic reflection problems can make mincemeat of monitoring accuracy, the remedies I've suggested are cost-effective, fairly simple to implement, and effective enough that comb filtering shouldn't stand in the way of you achieving commercial-level mixes. It's hardly surprising, then, that the more switched-on small-studio owners have often already implemented something along these lines. However, there is another equally important aspect of room acoustics, which is more difficult to tackle and so is often simply ignored by budget-conscious operators: room resonance.

1.4.1 Understanding the Problem

To understand how room resonances work, it helps to bring to mind how a guitar string resonates. At its lowest resonant frequency (called the first mode), the string is stationary at both ends and moves most at its middle point—or to use the technical terms, there are nodes at the ends of the string and an anti-node in the middle. However, the string also has a second resonant mode at twice this frequency, giving a vibration with three nodes, such that the string appears to be vibrating in two equal-length sections. Tripling the first mode's frequency gives you a third mode with four nodes, quadrupling it gives you a fourth mode with five nodes, and so on up the spectrum.

The reason it's useful to keep this image in mind is that the body of air between any parallel room boundaries has a similar series of resonant modes (sometimes also called "standing waves") at frequencies dictated by the distance between the surfaces, although the positions of the nodes and antinodes are actually swapped for air-pressure resonances, as illustrated in Figure 1.8. A quick-and-dirty way to work out the resonant frequency of the first room

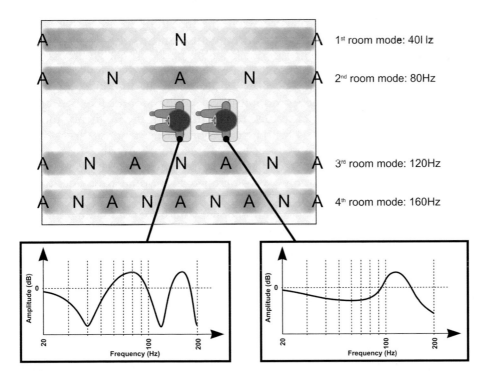

FIGURE 1.8
This diagram demonstrates what room resonances can do to your monitoring system's apparent frequency response. This picture shows the first four front-back room modes for a room measuring around 4.3m long. These occur at 40Hz, 80Hz, 120Hz, and 160Hz. The pressure nodes and antinodes are marked with "N" and "A," respectively, and although they are shown one above the other here for the sake of clarity, it's important to realize that they are actually overlaid on one another, all occurring simultaneously across the whole width of the room. The two frequency–response plots demonstrate the effects of these modes on the monitoring system's apparent frequency response at two different listening positions only about 75cm apart.

mode between a given pair of parallel boundaries is by dividing 172 by the distance in meters between them. Subsequent room modes will then be at multiples of that frequency, just as in our guitar-string example. So if the ceiling of your studio is 2.42m above the floor, then you'd expect the first room mode in that dimension to be at around 71Hz, the second at 142Hz, the third at 213Hz, and so forth.

Each room mode will generate its own regularly spaced series of nodes and antinodes between the room boundaries, and if this means that there's a node in your monitoring sweet spot, you'll hear a drastic frequency–response dip at that room mode's resonant frequency, whereas if there's an antinode at the listening position, you'll hear a significant level boost at that frequency instead. Because each pair of parallel room surfaces will contribute its own independent series of room modes, and most rectangular domestic rooms offer three pairs of parallel surfaces, small studios typically find themselves liberally peppered with nodes and antinodes at different frequencies.

So what does this mean in practice? Well, the first thing to say is that even a single room mode can easily push its resonant frequency 20dB out of kilter, so only a flying pig is likely to find a listening position that gives a faithful spectral balance when several room modes are active at the same time. Plus, if you move around the room at all while listening, the monitoring system's apparent frequency response will writhe around like a twerking python as I've tried to illustrate with the frequency plots in Figure 1.8. To be fair, room modes tend to affect primarily the lower half of the audio spectrum, by virtue of the fact that higher-frequency resonances are much more easily damped by normal room decor, but the remaining sub-1kHz disaster area is more than enough to scupper your hopes of making objective decisions about a mix.

> A single room mode can easily push its resonant frequency 20dB out of kilter, so only a flying pig is likely to find a listening position that gives a faithful spectral balance when several room modes are active at the same time.

Every room is different, though, so try this experiment to get a realistic idea of what the room modes are doing to your own monitoring. Play back the LFSineTones audio file through your system again and listen carefully from the sweet spot, comparing the relative levels of the pure sine-wave tones as they march in semitone steps up the bottom three octaves of the audio spectrum. If your studio is anything like most small, untreated control rooms, you'll probably find that some tones almost disappear, whereas others practically go into orbit! Table 1.1 shows roughly which frequencies occur at which times in the file, so grab a pencil and make a note of which are the most wayward tones while you're listening at the sweet spot. Now move your listening position a couple of feet away and try that little experiment again—it'll be a whole different story, and you'll probably find that some of the tones that were anemic before are now over-prominent, and vice versa.

Now it would be quite reasonable to say that sine-wave tones aren't much like a real musical signal, so it's also worthwhile to focus on what your room is doing to the evenness of bass lines on commercial tracks that you know to have been very well-produced in this department. If you want a suggestion here, then try the song "All Four Seasons," produced and engineered by Hugh Padgham for Sting's album *Mercury Falling*. The bass part on this track is wide ranging, but it is also extremely consistent, so all the notes should sound fairly even on any mix-ready monitoring system. If they don't, then you've seriously got to ask yourself how you're planning to judge bass balances in your own mixing work.

1.4.2 Some Practical Remedies

A common professional solution to low-end resonance problems is building oddly shaped rooms that minimize the number of parallel boundary surfaces, and if you've got builders in the family then by all means follow that route. However, for the rest of us the choice of which room we choose to mix in can also make a big difference. For a start, it's sensible to avoid small rooms where possible, because their resonances trespass further up the frequency spectrum than those of larger spaces. Also try to find a room where the dimensions aren't matched too closely, otherwise multiple similar room modes will gang up on the same frequency ranges, and that's just asking for trouble. Some home studios come spectacularly unstuck here, because they've been banished to a small 2.5m-wide cubic spare room where the powerful combination of room modes in all three dimensions conspires to make a complete dog's dinner not just of the low end, but also of the midrange.

Whatever room you're in, you can reduce the impact of the room modes if you avoid setting up your listening position (or indeed your monitors) exactly halfway between any of the room boundaries, where you're likely to get the worst pileup of nodes and antinodes. However, as I've already mentioned, offcenter setups bring with them the danger of a lopsided stereo balance because of unmatched room reflections, so it's advisable not to push the sweet spot too far left or right of center. That said, a stereo imbalance causes far fewer mixing problems in my experience than room-mode issues, so I'd personally prioritize room-mode treatment if push comes to shove.

Table 1.1 LFSineTones Audio File Map

Track Time	Frequency	Pitch
0:00	24Hz	F
0:01	25Hz	F#
0:02	26Hz	G
0:03	27Hz	G#
0:04	28Hz	A
0:05	29Hz	A#
0:06	31Hz	B
0:07	33Hz	C
0:09	35Hz	C#
0:10	37Hz	D
0:11	39Hz	D#
0:12	41Hz	E
0:13	44Hz	F
0:14	47Hz	F#
0:15	49Hz	G
0:16	52Hz	G#
0:17	55Hz	A
0:18	59Hz	A#
0:19	62Hz	B
0:20	65Hz	C
0:22	69Hz	C#
0:23	73Hz	D
0:24	77Hz	D#
0:25	82Hz	E
0:26	87Hz	F
0:27	92Hz	F#
0:28	98Hz	G
0:29	105Hz	G#
0:30	111Hz	A
0:31	117Hz	A#
0:32	123Hz	B
0:33	131Hz	C
0:35	139Hz	C#
0:36	147Hz	D
0:37	156Hz	D#
0:38	165Hz	E
0:39	175Hz	F
0:40	185Hz	F#
0:41	196Hz	G
0:42	208Hz	G#
0:43	220Hz	A
0:44	233Hz	A#
0:45	247Hz	B
0:46	262Hz	C

Rooms with lightweight walls can work to your advantage by allowing more low frequencies to escape, rather than resonating within—assuming that your neighbors don't take a dim view of this sound leakage! By the same token, concrete-walled basement rooms should be approached with caution, because low frequencies can have real trouble escaping that kind of environment and you'll have your work cut out trying to tame the resulting room modes. However, most small studios simply have to take whatever room happens to be free, so in any case you need to know how to make the best of things using acoustic treatment.

1.4.3 Mineral-Fiber Bass Traps

The best all-purpose tactic is to damp down the room modes as much as you can using low-frequency absorbers, often called bass traps. The downside here, though, is that bass traps need to be dense and bulky to do their job properly. As Eric Schilling notes, foam simply isn't up to the task: "Most people think that treating a room simply means going to a music store and buying foam. But if it's essentially a square room, it doesn't matter if you have some foam in the corner and a few pieces on the wall—you still won't be able to hear bass to save your life!"[9] The most commonly used alternative is large slabs of high-density mineral fiber, which offer much better low-frequency absorption. Placing the panels close to a given room boundary provides broadband absorption of all the associated dimension's room modes, and (much as with foam) you get an increase in the low-frequency absorption if you leave an air gap behind the slab—a foot or so if possible. Because mineral-fiber bass traps are typically more expensive than foam panels, one common trick is to place them across wall-to-wall and wall-to-ceiling corners, where they can treat modes in two dimensions simultaneously, and in small rooms this has the added advantage of intruding less on your workspace.

Normally not all dimensions of a room are equally troublesome, so it's a good idea to try to work out which are the biggest culprits. To do this, first listen carefully to my LFSineTones audio file while referring to Table 1.1, taking care to identify the most problematic frequencies—anywhere you can hear a big peak or trough in the apparent levels of the tones. Then divide 172 by each of these frequencies in turn to give a list of measurements in meters, and look with suspicion on any room dimension that works out as a simple multiple of any of those measurements. Once you know which dimensions are causing the biggest problems, you can concentrate your acoustic treatment resources more effectively.

Several companies now offer ready-made bass traps incorporating high-density mineral fiber, but the small studios I've been into usually need around a dozen 10cm-thick slabs to make their monitoring workable, and with off-the-shelf products that can seem rather an unpalatable battering of one's credit card. That's understandable, though, argues Mick Glossop: "Controlling

CAN EQUALIZATION CORRECT THE EFFECTS OF ROOM MODES?

Because room modes cause low-end frequency–response changes, it's tempting to think that EQ might be able to offer a solution to this kind of resonance problem—not least because an increasing number of manufacturers now offer EQ-based "room correction" software ostensibly for that purpose. The idea with these algorithms is that they measure the monitoring system's frequency response using a special test signal and a calibrated microphone, and then calculate an EQ curve to attempt to compensate for the nonlinearities they detect. However, for anyone serious about mixing on a budget, I think such systems are a red herring, for two different reasons. First, staying exactly in the sweet spot all the time isn't actually conducive to getting work done in a real studio, and if you wander out of the sweet spot, the frequency response will quickly go out of alignment again, as demonstrated in Figure 1.8. Even if you can somehow clamp yourself into the sweet spot, anyone working in the room with you will hear something completely different. More crucially, though, room resonances don't just affect frequency response; they also cause specific frequencies to ring on in time, with all the detrimental consequences we've already explored in relation to monitor porting back in Section 1.1.2. Equalizers are themselves resonant too, which only compounds the issue. So can equalization usefully combat room modes? For my money, no.

That's not to say that EQ can't fulfill a useful role in adapting the subjective tonality of the system to your personal tastes, and a lot of small nearfield monitors now have simple tone controls conveniently built into them for this purpose, or else let you adjust the relative output levels of the different drivers. My main recommendation is simply to restrict yourself to broad-brush EQ changes (a bit less brightness, say, or a touch more midrange), rather than getting too finicky and risking unwanted processing artifacts that may mess with your mixing decisions down the line. Remember that many celebrated mixing speakers are far from "flat" in the frequency domain, and whatever EQ you put in your monitoring chain, the reality is that you're still going to have to spend time learning how your speakers sound before you can get the best out of them.

very low frequencies costs more money, because it's big movements of air, and you need mass and big trapping systems to do that."[10]

However, the good news is that a bit of DIY can save you serious money, because it's possible to build reasonable bass trapping on your own at a fraction of the cost. What you need to find are 10cm-thick mineral-fiber panels with a density of around 50 to 100kg/m^3. They're used for insulation as much as for acoustics, so ask at an insulation company if you can't locate a dedicated supplier of acoustics materials in your area. A word of caution, though:

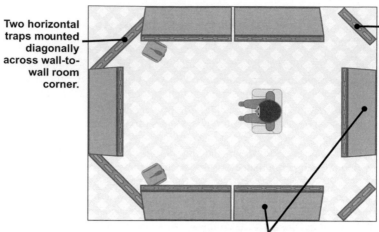

Two horizontal traps mounted diagonally across wall-to-wall room corner.

Vertical trap mounted diagonally across wall-to-wall room corner.

Horizontal traps mounted diagonally across wall-to-ceiling room corners.

FIGURE 1.9
This diagram shows how you might sensibly distribute a dozen 10cm × 60cm × 120cm mineral-fiber acoustic panels within a modestly sized control room to tame typical room-mode problems.

mineral fiber is a skin and breathing irritant, so be sure to wear appropriate protective mask, goggles, and gloves when handling it. By the same token, you'll want to cover each panel in some kind of acoustically neutral fabric to stop the mineral fibers from shedding all over the place—the kind of material commonly used to cover office dividers is quite a good bet, especially because you can usually get it in a wide range of colors. (For maximum Feng Shui, you could use the same material to cover your acoustic foam too.) If you're less than handy with a needle and thread, some manufacturers do little ready-made fabric sacks for the purpose that are fairly cost-effective too.

> A bit of DIY can save you serious money, because it's possible to build reasonable bass trapping on your own at a fraction of the cost of off-the-shelf products.

Because high-density mineral-fiber panels are about as stiff as a typical foam mattress, you can just use picture hooks and string to hold them in place in your studio if you're happy for them to sag a little over time under their own weight. If you want a tidier look, though, then there's nothing stopping you from building a simple wooden frame around them (either inside or outside the fabric), and covering it in chicken wire to hold the slab in place more assertively. Alternatively, if you're as much of a danger to yourself with a saw in your hand as I am, then there are also companies who sell easy-assembly DIY frame kits that fit around standard 10cm × 60cm × 120cm slabs to create a slick final product.

1.4.4 Limp-Mass Bass Traps

Now mineral-fiber slabs can do a lot of good, but even quite thick ones become less effective as you head down below 100Hz, so in a solidly built room with powerful sub-100Hz resonances, there's a limit to how much they can help. For example, there's one 6m × 4m × 2.5m basement room I've used that resonated powerfully at 30Hz and 60Hz along its long front-back dimension, despite the presence of two dozen 10cm × 60cm × 120cm mineral-fiber bass traps. In such cases, there's one other type of treatment that can be worth a shot: a limp-mass trap. This involves loosely hanging a large, heavy, and impermeable sheet a little distance from one room boundary, where it can damp low-frequency air movements. Some kind of rubberized barrier matting with a density of around 5kg/m² is a common choice for this task, and it helps if it has a reinforced backing so that you can if necessary hang it by its edge at ceiling level without it tearing or deforming under its own considerable weight. Failing that, old carpets can deliver something like the same result if mounted in a similar way and may be a cheaper option if you can get hold of them as scrap.

Lower frequencies need larger areas of treatment, so if you feel the need for limp-mass trapping for a troublesome room mode in one particular dimen sion, then you should think in terms of trying to treat pretty much the whole of one of the relevant room boundaries, even though this will inevitably

TOO MUCH BASS TRAPPING?

Although I've heaped scorn on the idea of covering every available control-room surface in acoustic foam, there's little benefit in similar restraint when it comes to bass trapping in small studio rooms—the more the merrier. But hang on a minute! Didn't I say earlier that too much acoustic foam could make a control room less like a real-world listening environment? Shouldn't that also apply to bass trapping? Well, yes it does. The difference, however, is that leaving your control room's modes to their own devices (as in most real-world playback environments) will actually make your low-end monitoring less representative, because any other room will have a different set of modes at a different set of frequencies. Better to tame your control-room modes as well as you can so that you've a clearer idea of the actual low-frequency balance, irrespective of any resonances of the end user's listening system.

One practical problem you may encounter when installing a lot of mineral-fiber bass trapping, though, is that it will also absorb high frequencies as well and can result in too dead a sound in that region of the spectrum—much the same problem as you get when too much acoustic foam has been installed. In these cases you may actually need to reintroduce some high-frequency reflection by fixing small areas of hard surface to the faces of some of the bass traps. Small picture frames, old CD-ROMs, or just random bits of hardboard are all possibilities here, although you should avoid covering more than about a third of the trap's surface area in this way or else you'll start to interfere unduly with its low-end absorption.

reduce the area you can use for your studio gear. Because the size of air gap behind the trapping adjusts its absorptive properties, it's a good idea to mount the matting on some kind of movable wooden frame if possible, so that you can use trial and error to strike the best balance between resonance reduction and loss of workspace. This kind of trapping is a lot less predictable than simple mineral-fiber absorbers, because it is itself to some extent resonant, so be prepared to spend a bit of time refining the setup to get the best out of it. Some variations worth considering are adjusting the fixing points for the matting, as well as hanging drapes, thin boards, or mineral-fiber panels in parallel. It's not an exact science, but if you're faced with heinous low-end resonance problems and a meager budget, then it can nonetheless be a viable bacon-saver. In the specific basement room I mentioned earlier, putting in free-hanging sheets of barrier matting across most of the width of the room and about a meter away from the rear wall was able to bring the worst low-end problems under control, and the loss of that workspace was a small price to pay for usable monitoring.

1.5 WHEN IS MY MONITORING GOOD ENOUGH?

There's little doubt that the standard of the nearfield system is one of the most significant quality bottlenecks in the small-studio production process. "The biggest mistakes that happen in project studios," says Al Schmitt, "are the result of the monitoring systems."[11] For this reason, improving your monitoring is perhaps the smartest way to spend money, and nothing else in your studio will give a better return on your investment. Indeed, once you can hear what you're doing, you'll find you're able to produce competitive end results with ridiculously affordable mixing gear—something I've demonstrated repeatedly by using the Cockos *Reaper* DAW and freeware plug-ins to transform underperforming mixes for *Sound On Sound* magazine's "Mix Rescue" column. Monitoring hardware will also outlast most of your other studio gear. How likely is it, for instance, that you'll be happy using your current computer, audio interface, control surface, software, plug-ins, or virtual instruments ten years from now? Decent loudspeakers and acoustic treatment will give decades of useful service—give or take the odd service and a handful of mothballs!

> ### FLUTTER ECHO
>
> Although the main resonance problems in studio rooms are at low frequencies, you can also get higher-frequency resonances too, often referred to as flutter echoes. The simplest test for these is clapping your hands in the sweet spot and listening for any hint of unnatural metallic-sounding buzz as the sound dies away in the room. If you're suspicious, then another patch of acoustic foam on one or both of the offending parallel surfaces should put an end to it with minimal fuss because high frequencies are so easily absorbed.

MULTI-PURPOSE MONITORS

Something that a lot of small-studio users don't appreciate is that loudspeakers often need to fulfill several different functions in the studio, above and beyond being a mixing tool. As such, it's important to realize that although the cost-effective mixing systems I recommend in this book will enable you to deliver a commercial-sounding mix, they won't necessarily work too well for other tasks. For example, you don't need massive playback volume to mix effectively, so the cheapest mix-ready monitoring systems won't allow you to crank up the playback level to enthuse the musicians on a tracking session, say. If you want true mixing capability *and* thunderous volumes, then expect to roughly double the loudspeaker prices I suggest here. In a similar vein, mixing speakers don't have to be a particularly pleasant listen to get the job done (as the popularity of the Yamaha NS10 amply attests!), so the most budget-friendly mix-capable setups may not much inspire on-site musical collaborators or convince visiting clients that your work sounds fantastic, regardless of the actual sound quality of your final mixdown files. If you want a gorgeous subjective timbre from your mixing speaker, that'll cost you extra too.

Nevertheless, studio owners only ever have so much money to go around, so they are understandably keen to spend no more than they really need in order to reliably achieve commercial-level mix results. In this spirit, let me suggest the kind of system that, in my experience, can deliver that level of performance while remaining on the right side of the point of diminishing returns. As I've already mentioned, ported nearfields only tend to become really reliable for mixing above the £1500 ($2000) mark, whereas there are cheaper unported systems that hit this usefulness level, as long as you don't need to stun small prey with your playback volume. A pair of solid speaker stands filled with sand will do such a set of speakers justice without the need for any extra mounting gizmos, and four or five square meters of 10cm thick acoustic foam will usually be ample to bring early reflections and flutter echoes under control. Unless you have solid concrete room boundaries or your studio is stuck in a basement, a dozen 10cm × 60cm × 120cm mineral-fiber bass traps should be enough to give you usably even bass response as well. Basically, with that lot in the bag your nearfield monitoring shouldn't be any excuse for turning in duff mix work.

Now I realize that I'm effectively suggesting you spend around £2500 ($3200) on monitors and acoustic treatment, which is a small fortune for most small-studio owners. Indeed, I got plenty of flak from readers of the first edition of this book who felt I was implying that they shouldn't bother trying to mix without immediately splashing that kind of cash on their monitoring system. So let me clarify. I'm *not* saying it's impossible to create a commercially competitive mix in a lesser monitoring environment—after all, there have been plenty of great-sounding commercial releases mixed under adverse conditions. However, I do firmly believe that you'll only deliver competitive mixes *reliably every time* if you can adequately hear what you're doing, which means investing seriously in your monitoring environment. Hand on heart, I honestly feel I'd be doing a disservice to anyone wishing to achieve consistent professional-grade results if I pretended otherwise.

It's perfectly possible to begin studying mixing with lesser-quality nearfield monitoring, though, because even the cheapest setup will tell you *something* useful about what's going on in your mix, so you can at least work out how to manipulate and control those elements of the sound. (The next three chapters explain how to optimize your results under those conditions, as well as when working on headphones.) But by the same token, you'll never learn to confidently handle all the different attributes of a mix unless you can hear them properly, so the fastest way to fully develop your mixing skillset is to funnel as much of your budget as possible into improving your monitoring situation. Furthermore, I've noticed that the vagaries of low-budget monitoring systems often encourage mixing students to adopt counterproductive habits that hamper their progress, so investing as heavily as you can in monitoring early on will reduce the amount of rueful "unlearning" you'll have to do further down the road.

Bear in mind, though, that even if you follow this chapter's recommendations to the letter, your monitoring system will still be far from perfect, and there are plenty of high-end speaker designers, acousticians, and mastering engineers who'd suck their teeth at it in disdain. But perfection is not the goal here, because the trick to achieving reliable commercial-grade results on a budget is to wring the maximum amount of useful information from cost-effective nearfield monitors, while employing cheaper and more specialized monitoring equipment to fill the remaining gaps in your understanding of the mix. Indeed, if the nearfield rig I've suggested is still way beyond your current means and you have no alternative but to work with hopelessly compromised nearfield monitoring, the truth of the matter is that you can still achieve surprisingly good mix results by relying more heavily on such supplementary monitoring tools. So what are these additional bits of gear? All will be explained in the next chapter, but before we get to that, let's quickly recap my main recommendations so far.

CUT TO THE CHASE

- A nearfield monitoring system is a good choice for small-studio mixing. Spend as much as you can afford on the speakers, because quality costs, and if your budget is tight then be wary of ported designs. When choosing a system for mixing purposes, favor studio monitors over hi-fi speakers, active models over passive, and accuracy over volume. No monitoring system is truly neutral-sounding, so you'll only get the best out of your nearfields once you've become accustomed to how they sound in your specific circumstances.

- Whatever speakers you use, mount them securely on solid, nonresonant surfaces, preferably away from room boundaries. If the speakers have more than one driver, then the cabinets should be oriented so that the drivers are equal distances from the listener and angled toward the listening position. In nearly all cases it's better for multidriver speakers to be vertically rather than horizontally aligned. For stereo listening, there should be the same distance between the listener and each of the speakers as between the speakers themselves, but if anything err on the side of a narrower speaker spacing. Make sure you check your speakers for polarity.
- Give some thought to the room you use for your studio, and ensure that you spend at least as much money on acoustic treatment as you do on monitors. Strategic use of acoustic foam can effectively target early-reflection and flutter-echo problems, but be careful not to overdo it. High-density mineral-fiber slabs can provide a fairly foolproof remedy for low-frequency room resonances, but if these don't prove effective enough in your room, then try supplementing those with additional limp-mass trapping. Don't waste your time trying to correct acoustic resonance problems with equalization, because the benefits of this are minimal in practice.

Assignment

- Invest as much money as you can in your nearfield speaker system, spending roughly the same amount on acoustic treatment as on the speakers themselves.
- Make the best of whatever system you can afford (or have access to) by making sure that the speakers are solidly mounted and sensibly positioned and that the room is appropriately treated.

Web Resources

On this book's companion website you'll find a selection of resources to support this chapter, including:

- the LFSineTones, PinkNoise, and StereoTest audio test files;
- my most up-to-date personal recommendations for small-studio nearfield monitoring systems, as well as links to suppliers of some of the acoustic treatment products I've discussed;
- further reading on related subjects including: monitor porting and placement; the Yamaha NS10 and Auratone 5C; phase and polarity; and how to build your own high-quality acoustic diffusers.

www.cambridge-mt.com/ms-ch1.htm

CHAPTER 2

Supplementary Monitoring

As I touched on at the end of Chapter 1, aiming for perfect nearfield monitoring in the small studio is a fool's errand. With the best will in the world, you'll always have problems, not least because it's neither practical nor comfortable to confine yourself to a tiny listening sweet spot for hours on end in order to maintain the best stereo imaging, minimize comb-filtering problems between the speaker drivers, and stay clear of room-mode danger zones. However, even if you've got some medieval torture instrument clamping your head into the optimal location, you'll still face a problem inherent in all stereo systems—namely, that your speaker cabinets are positioned at the left and right extremes of the stereo field, whereas many of the most important parts of your mix will be center stage, where there is no actual hardware. In other words, your lead vocals, kick, snare, and bass will all typically appear to be hovering in thin air in front of you, an illusion called a "phantom image," which is achieved by feeding equal levels of each of these tracks to both speakers at once.

The difficulty with phantom images as far as real-world mixing is concerned is that they invariably feel less stable than the "real" images at the extremes. Part of the reason for this is that normal manufacturing tolerances prevent any pair of remotely affordable speakers from being exactly matched as far as frequency response is concerned, which smears the frequency spectrum of centrally panned sounds haphazardly across the stereo image to a small but significant extent. The effects of typical small-room resonance modes and asymmetrical early reflections only make things worse, and then there's the fact that small movements of the listener's head will affect the sound of phantom images much more than that of real images because of comb filtering between the two speakers.

So what's the extent of the damage in your room? Fire up my StereoTest audio file again (the one we already used in Section 1.2.1) and listen to how the "real" left/right images of the first two noise bursts are narrower and more precise than the phantom central image of the third burst. (You can ignore the fourth burst in the pattern for this test.) You'll hear this effect to some extent even on high-spec monitoring systems, but in budget setups it's like night and day.

A real-world musical example that demonstrates this same thing quite well is the track "American Life" from Madonna's album of the same name. Here the verse lead vocal is very narrow and panned dead center, whereas the synth that appears 45 seconds in and the acoustic guitar that appears 15 seconds later are panned to opposite stereo extremes. Every small-studio system I've ever clapped ears on renders the synth and guitar images much more stable than the vocal's.

> The big drawback with carefully configured studio nearfields is that they don't actually sound anything like the majority of real-world playback systems.

The final big drawback with carefully configured studio nearfields is that they don't actually sound anything like the majority of real-world playback systems, so you can't expect them to tell you how your mix will fare once it's out in the big wide world. Yes, you should get a fairly good idea what multiplex popcorn munchers and chin-stroking hi-fi tweeds will hear, but you'll be pretty much in the dark regarding the low-fidelity, low-bandwidth playback of most mainstream listening devices.

So the most important thing to remember with your nearfields is to rely on them for what they're best at:

- for investigating what's going on across the whole frequency spectrum, especially at the frequency extremes;
- for judging the impact of mix processing on the sheer quality of your sounds;
- for evaluating and adjusting the stereo image;
- for understanding how your mix will sound on more high-fidelity listening systems;
- for impressing the pants off the artist, client, or anyone else in the room with you!

For a number of other monitoring requirements, however, there are more modest systems that can actually outperform even a decent small-studio nearfield rig, simply by virtue of being better adapted to specific duties. This chapter introduces some of these additional systems, specifically the ones you can't afford to do without if you're seriously planning on competing with the big guns. Let's start by looking at the most powerful of these, as epitomized by one of the most famous mixing speakers in studio history, the Auratone 5C Super Sound Cube.

2.1 THE GHOST OF THE AURATONE 5C SUPER SOUND CUBE

Despite the Auratone 5C's unassuming appearance and subjectively unappealing sound, it's been pivotal to the mixes of numerous hit records, including Adele's *21*, Dire Straits's *Brothers In Arms*, and Michael Jackson's *Thriller*, the biggest-selling album of all time. "I love Auratones!" enthuses Jackson's long-time producer and engineer Bruce Swedien. "You know what Quincy [Jones] calls them? The Truth Speakers! There's no hype with an Auratone. . . . Probably 80 percent of the mix is done on Auratones, and then I'll have a final listen or two on the big speakers."[1] Tom Elmhirst echoed this when talking about

FIGURE 2.1
The best-selling album ever, Michael Jackson's *Thriller*, was primarily mixed on Auratones.

his work for Adele: "When I mix I'll be jumping around for the first couple of hours playing the track loudly via my [nearfields], and once I know the bottom end is rocking, I'll mix with low volume on the Auratones for the rest of the day. If I can make a mix work on the Auratones, I know I'm flying."[2] Needless to say, you can still rely on most large studios having at least one set of Auratones on hand for those freelance engineers who appreciate what they have to offer—no matter how star-spangled the main and nearfield systems. These are all persuasive reasons to bring the magic of the Auratone to bear in your own mixing work.

But that's where there's a big old fly in the ointment: the classic Auratones aren't manufactured any more, and much of the remaining secondhand stock has now succumbed to the frailties of old age. So what can you do? Well, as long as you understand the nature of this speaker's extraordinary mixing abilities, and the design features responsible for them, then you can use this knowledge to track down a modern replacement that offers similar benefits.

2.1.1 Midrange Focus

First things first: the Auratone is portless, and we've already discussed at length the implications of that in Chapter 1. The next crucial attribute is the single small driver, and for several reasons. First, its restricted ability to reproduce both high and low frequencies focuses your attention on the midrange of your mix, which is the frequency region our ears are most sensitive to and which tends to carry the most important musical information. "The real perspective lives in that range," says Jack Joseph Puig. "It doesn't live in the highs, it doesn't live in the lows. That's what really speaks to the heart."[3] Bob Katz concurs: "The midrange is the key. If you lose the midrange, you lose it all."[4]

The midrange is also crucial because it's the frequency region of your mix that is most likely to survive the journey to your target listener's ears. "The midrange is . . . most common to every system in the world," says Puig. "Some system might have five woofers, one might not have any, one might have ten tweeters, but all of them have midrange, and so *that* has got to be right."[5] For one thing, you need a big listening system to produce much in the way of low frequencies, so small real-world playback devices (mobile devices, clock radios, and a lot of TVs, music-player docking stations, and portable wireless speakers too) won't give you much output in the sub-200Hz region. A huge number of mass-market playback systems also share the Auratone's one-cone design, which inherently detracts from high-frequency efficiency. Just think of all the small boomboxes, TVs, and multimedia speakers of this type that litter most offices and households. Now add in the arguably more widespread multispeaker announcement and piped-music systems that assail you while you're blowing froth off your skinny latte, stepping aboard a 747 to Malaga, or pressing the button for the 17th floor—more than one driver per speaker in large audio distribution networks like these would seriously bump up the installation cost, so it's pretty rare.

But it's not just an end-user's hardware that compromises the frequency extremes; it's the way it's used, too. If you recall that high frequencies are very directional, easily shadowed, and swiftly absorbed by normal soft furnishings, it's little surprise that extended high frequencies rarely trouble the public, simply because few people actually listen to speakers on axis. In fact, it's more than likely that their hi-fi speakers are concealed behind the sofa or their portable radio is tucked away on that top shelf at the other side of the office. At the other end of the spectrum, the high levels of low-frequency background noise in cars, planes, public spaces, and some workplaces will obscure most of that portion of a mix. (Yes, I know that some people pimp their rides with weapons-grade subwoofers, but those guys are probably mostly deaf into the bargain.) It should be clear by now why the Auratone's design is so adept at demonstrating which bits of your mix will reach the greatest number of listeners. In a nutshell, if your mix doesn't stand up against the competition on an Auratone, then you probably haven't done your job properly!

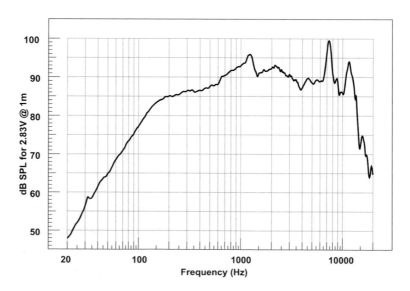

FIGURE 2.2
The frequency response of the Auratone 5C Super Sound Cube, which clearly shows its midrange bias.

2.1.2 Resistance to Acoustics and Comb-Filtering Problems

There are other practical advantages to the Auratone's single-driver construction that are especially relevant to small studios. The first is that the restricted low-frequency output barely tickles the room modes in most cases, with the result that low-end resonances don't interfere nearly as much with your mix decisions, especially when it comes to balancing bass instruments. The speaker isn't actually very loud overall either, and that reduces the impact of early reflections from the room as well.

Another real benefit of the single-cone design is that no crossover electronics are required to split the signal between different speaker drivers. Any processing between your mixer's output and your speakers has the potential to introduce side effects that can mislead you while mixing, and in the case of crossovers the worst gremlins tend to manifest themselves around the crossover frequency. Most affordable nearfield monitors have their crossovers in the 2 to 3kHz midrange region, so the razor-sharp midrange focus of a crossoverless Auratone provides a valuable alternate perspective.

By the same token, Auratones won't punish you with interdriver comb filtering if you slouch a bit in your studio chair. What's more, if you restrict yourself to using just a single Auratone, rather than a pair, you'll also get no comb filtering from side-to-side movements, so you're free to stroll around the control room with comparative impunity—you'll get some high-end loss off-axis, but that actually does very little to undermine the Auratone's essential usefulness.

FIGURE 2.3
A large number of real-world speakers have only one driver, including shopping-center piped-music systems, small radios/TVs, and mobile devices.

2.1.3 Balance Judgments and Mono Compatibility

Of course, if you want to listen to a stereo mix from just the one Auratone, then it forces you to listen back in mono. Rather than being a disadvantage, though, this only adds to the appeal. For a start, single-speaker mono suffers none of the phantom-image instability that plagues even decent nearfield stereo setups, simply because it doesn't create any phantom images at all—the physical location of the speaker is unequivocally the source for everything you hear. Central sounds therefore feel just as solid as those at the stereo extremes, so it becomes possible to judge exact level balances in your mix with pinpoint accuracy, regardless of stereo positioning. "The reason why I mix in mono is to gauge the level of everything," says Derek 'MixedByAli' Ali. "Not the actual mix—I'm gauging the level everything sits in the mix."[6]

However, you need to be aware that summing the two channels of a stereo mix to mono shifts the overall balance of the sounds. Partly this is an inevitable side effect of the way that stereo speakers create their illusory panorama—two sounds that appear equally loud in stereo will actually appear to be at different volumes in mono if they're positioned at different distances from the center of the stereo image. A good rule of thumb here is that if you listen to a stereo mix in mono, then the levels of central sounds will increase by roughly 3dB relative to sounds located at the edges of the picture. To hear this effect in action on your own system, download the MonoBalanceShift audio file from this chapter's web resources. It's made up of a pattern of four short brass-band chords that are differentiated only by their position in the stereo image: chord 1 is central; chord 2 is at the left-hand extreme; chord 3

WHAT, NO NS10S?!

I mentioned in Chapter 1 that Yamaha's NS10s are far and away the most celebrated mixing speakers in the professional music world, and two of the main reasons for this also apply to the Auratone: the sealed-box cabinet (with its smooth low-frequency roll-off, excellent transient response, and low distortion) and the midrange-heavy frequency balance (which reflects the tonal deficiencies of many small real-world playback systems). However, if you want the maximum bang for your buck in the small studio, I wouldn't personally recommend investing in a pair of these classic speakers for yourself. First of all, NS10s won't really give you much more useful information than the two monitoring systems I've so far discussed: a reasonable full-range nearfield system should easily match the capabilities of NS10s when it comes to judging stereo imaging and tonal quality, whereas an Auratone will deliver much the same closed-box accuracy and midrange emphasis. The second reason for considering the NS10 a nonessential purchase is that it doesn't obviate the need for either a decent nearfield system or an Auratone. It can't deliver the frequency extremes in the way reasonable nearfield systems now can (which is presumably why Chris Lord-Alge[7] uses his NS10s with an additional subwoofer), and its mix-balancing abilities can't match those of a single Auratone because of the same phasing complications that afflict all multidriver stereo speaker systems. Now I'm not saying that the NS10s don't deserve their kingpin status; it's just that they aren't as well suited to the small studio as they are to the large commercial setups where they beautifully bridge the gap between main wall-fitted monster monitors and any Auratone/grotbox in use. That said, if you already have NS10s in your setup, then there's no need to get rid of them—just add a suitable subwoofer (again, something unported if possible) and you've got a very creditable full-range 2.1 nearfield system.

One final word of caution: although Yamaha stopped making the NS10 many years ago, they now produce a range of budget-friendly nearfield monitors that mimic their celebrated forebear's distinctive "white-cone" cosmetics. That's about as far as the similarity goes, though, as far as I'm concerned, because those new speakers are all ported and, to my ears at least, don't provide significantly better time-domain performance than other leading ported designs in their price bracket.

is back in the center; chord 4 is at the right-hand extreme; and then the four-chord pattern starts again from chord 1. The levels of the chords are such that their loudness should appear to remain fairly constant as they trot backward and forward across the stereo field—assuming that you've set up your monitoring system roughly along the lines suggested in Chapter 1. However, if you now switch to monitoring in mono, you'll hear a clear louder–softer alternation in the chord levels.

The fact that the mono balance shift draws more of your attention to what's going on in the center of the stereo image can be a blessing in disguise, though—after all, the most meaningful sounds in the mix usually reside there. Mono listening also forces you to work harder to achieve clarity for each of the sounds in your mix, because you can't make things more audible just by shifting them to less cluttered areas of the stereo field. As Geoff Emerick puts it, "Mixing in stereo [is] the easy way out."[8]

However, you can definitely get some unpleasant surprises when switching to mono if your stereo mix is sneakily harboring any phase or polarity problems between its left and right channels. A common source of these kinds of nasties is stereo recordings that have used spaced microphone configurations—drums, room ambience, vocal/instrumental ensembles, and acoustic piano are often captured in this way, for instance. If any part of the sound source isn't the same distance from both mics, its sound will arrive at one mic earlier than at the other. In stereo this time-delay between the two channels actually reinforces the stereo positioning of the instrument for the listener, but in mono

FIGURE 2.4
Stereo recordings made with spaced-pair microphone techniques are a common reason for mono-compatibility problems.

the same time delay will cause comb filtering, the frequency-cancellation effect we encountered several times in Chapter 1, which can seriously damage the sound's perceived timbre. In addition to real stereo recordings, simulated stereo reverb and echo effects can also suffer from comb filtering in mono, because they combine delayed versions of the same basic sound by their very nature.

If any signal appears in both stereo channels, but with inverted waveform polarity (i.e., with the waveform shape inverted) in one of them, then that can also cause a serious mismatch between the balance of the stereo and mono renditions of your mix. Fortunately, most normal recorded sounds don't have much in the way of out-of-polarity stereo components—the most common exception being stereo recordings where a rogue cable or signal processor has inadvertently flipped the polarity of just one of the mics. However, a lot of synthesizer patches deliberately do incorporate one-sided polarity inversion, because that gives them a subjectively impressive artificial wideness in the stereo image. (Some dedicated stereo-widening mix effects work along similar lines too.) The drawback of this widening trick, though, is that out-of-polarity stereo components simply cancel themselves out when a stereo mix is summed to mono, which can lead to unwelcome surprises for any important synth lines in your arrangement—I've heard electronica mixes where the main hook effectively vanished in mono, so consider yourself warned!

FIGURE 2.5
Two common real-world speaker setups that prevent any meaningful stereo imaging reaching the listener: hi-fi speakers placed artistically in a CD rack (*left*) and speakers hidden away in high corners of shops and restaurants (*right*).

2.1.4 Is Mono Still Relevant?

But surely now that recordings, downloads, and broadcasts are all routinely in stereo, mono is only really relevant to a few grunting Neanderthals, right? Wrong. Although the majority of music is indeed now produced, sold, and transmitted in stereo, you might as well kiss goodbye to most of the stereo information in your mix once it's been through the end user's reception or playback equipment. Even for the most earnestly bearded hi-fi enthusiast, stereo imaging will still crumple into the nearest speaker if he shuffles out of the sweet spot and, frankly, I can't actually recall ever having seen a home hi-fi system with anything worth calling a sweet spot anyway. Hands up—how many of you have acoustic treatment in your living room? Exactly. And how many of you have had the speakers banished by your flatmates to that corner under the side table, behind the plant? You're lucky to hear the lyrics from most hi-fis, let alone any stereo.

Less fancy domestic playback systems fare even less well. The left speaker of your teenager's hi-fi minisystem is far too busy holding the wardrobe door open to worry about stereo imaging, even if it weren't accidentally wired up with inverted polarity. And it stands to reason that your web PC's speakers have to huddle together at one side of the monitor to leave room on the desk for the phone. Furthermore, watching any TV from the sweet spot between its two built-in speakers is a surefire route to eyestrain. Maybe you can save the proper stereo experience for the kitchen boombox, building a custom shoulder harness to keep you securely between the built-in minispeakers while you peel those spuds and heat up the fryer? Perhaps not. (Are you even sure both speakers are working? My kitchen radio recently presented me with a version of The Beatles' "Yesterday" minus the strings.) To be fair, earbuds are still something of a safe haven for stereo, but only if you resist the urge to share one of them with the person sitting next to you and you're not one of the millions of listeners worldwide who, like Brian Wilson, suffer from one-sided hearing impairment.

Move outside the home and the situation is no less dire. Consider the lengths to which shop and restaurant owners seem to go to avoid any risk of stereo imaging interrupting your retail experience. "No matter what anyone says," says Allen Sides, "if you're in a bar you're going to hear one speaker."[9] Car stereo? I only know of one extortionately priced sports car that actually puts the driver an equal distance from the two speakers, so anyone else wanting a true stereo experience in the car will have to run the gauntlet of a potentially eye-watering encounter with the gear shift. But at least in these cases you have different speakers playing different things. Many other systems don't even make a token stab at stereo! Things such as telephone hold systems, band/club PAs, and shopping-center announcement systems all typically sum their inputs to mono before feeding any speakers. FM radio receivers also often automatically sum the audio to mono in order to improve reception in the presence of a weak transmitted signal. With all this in mind, checking the mono-compatibility of your mix remains just as indispensable as it was 50 years ago, unless you're unfazed by the thought of your audience being treated to a hideously imbalanced and comb-filtered version of your latest track during their weekly supermarket sweep.

2.1.5 Modern Auratone Substitutes

Allow me to recap here. If you want to reap the greatest benefits of the Auratone in your mixing environment, then you need to monitor in mono from one (yes, just the one!) small, unported, single-driver speaker. A lot of people still harbor the misguided view that the Auratone's primary feature was that it sounded horrible (hence its common "Horrortone" and "Awfultone" nicknames); in other words that it simply allowed you to anticipate the effects of worst-case scenario playback. However, this myth doesn't really stand up to much scrutiny when you consider that the Auratone's portless cabinet is unusual in small speakers these days, and that it also delivers unusually low distortion. Clearly, a single Auratone won't impress any of your clients with its pathetic honky sound, but it's nonetheless a brilliant specialist tool that ruthlessly spotlights midrange balance and mono-compatibility issues in your mix.

As with your main nearfield system, you just need to learn to use an Auratone for what it excels at while relying on your other monitoring equipment to make up for its weaknesses.

Finding a modern speaker to fulfil the Auratone role isn't that easy, because most designs have a port, two drivers, or a single driver that is too small. Moreover, as with nearfield monitors, the product line-up available on the market changes so rapidly that any specific suggestions I might make here in print would likely be out of date by the time you read them. So if you'd like my recommendations for affordable Auratone substitutes, check out this chapter's web resource page, which I keep updated with the latest info.

> An Auratone won't impress any clients with its pathetic honky sound, but it's nonetheless a brilliant specialist tool that ruthlessly spotlights midrange balance and mono-compatibility issues in your mix.

An ancillary issue is how to feed the speaker in mono. Thankfully, increasing numbers of fairly affordable mixers and monitor controllers now incorporate the required mono and speaker-selection buttons, but if your own system is lacking these then there are lots of workarounds. For example, software mixing systems will usually let you route a mono version of the mix to a spare audio interface output, so you can feed your Auratone from that output and then switch between speakers from within the software. If you're listening via a hardware mixer, then route the mix through a spare pair of channels (or a dedicated stereo channel) and feed the Auratone using a mono prefader auxiliary send. (Just make sure that these channels don't also route back into your main mix buss or you'll get a potentially speaker-busting feedback howlaround!) The mixer's internal send routing should take care of the mono summing for you, and when you want to hear the Auratone you just mute the mix buss and crank up the relevant auxiliary send controls. Neither of these workarounds will help, though, if you're working on a small system with a lone pair of stereo output sockets. Fortunately, in such cases there's often still an additional stereo headphone output that can be pressed into service if you're handy with a soldering iron. Figure 2.7 shows how you can put together a cable that sums the two channels of a headphone output to mono for this purpose.

FIGURE 2.6
A modern homage to the Auratone 5C: Avantone's Mix Cube.

One other practical consideration is where physically to place your Auratone-substitute. As with your nearfield speakers, I'd ideally try to put it on something solid, away from reflective surfaces that might introduce unacceptable comb-filtering effects, but other than that I think you have quite a bit of leeway, given that stereo imaging isn't an issue. Anywhere that's within a couple of meters of your listening position should be fine, as long as the speaker's pointing towards you. I had mine perched atop my left-hand nearfield monitor for years, for instance, and that worked fine for mixing purposes. That said, I found that my hearing tried to compensate for the off-center small-speaker positioning during longer critical listening sessions, such that when I switched back to my nearfields the stereo image would temporarily feel lopsided until my ears had taken a minute to readjust, so I've since moved my Auratone-substitute directly above my DAW's display screen (on its own separate stand) to avoid this.

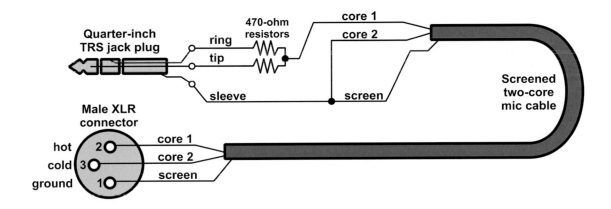

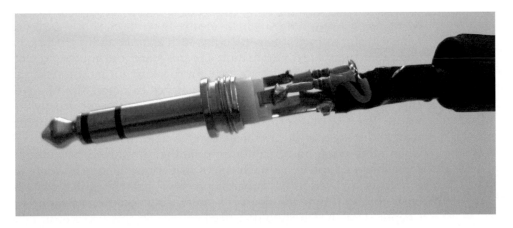

FIGURE 2.7
Schematic for creating a lead that will feed mono to an Auratone-substitute from a stereo headphone socket (*above*) and an example of the finished product (*below*).

2.2 HEADPHONES

Since the Sony Walkman first arrived on the scene back in 1979, headphone listening has become so widespread that you'd have to be pretty cavalier to sign off a mix without checking how it translates for such an enormous group of consumers. Therefore, the second important supplementary monitoring system I'd suggest using as a matter of course is headphones. "When [Phil Ramone and I] would mix, I'd mix and get everything done, and before he OK'ed a mix the last thing he did was listen on headphones," recalls Al Schmitt.[10] Clearly there are limits to how much low-frequency output headphones can deliver, but your Auratone should already have alerted you to any problems that might arise from bass-light playback, so that shouldn't be news to you. What's more important about headphones is that almost all of them transmit each side of your mix exclusively to one ear, whereas with speakers the two stereo channels are always heard to some extent by both ears. For this reason the stereo image is much wider on headphones (the image encompassing an angle of 180 degrees, where speakers only cover around 60 degrees), and any sound at the stereo extremes feels disconnected from the rest of the mix as a result. Whether the image stretching that headphone listeners experience helps or hinders your particular mix will depend on your own preferences and the expectations of the musical genre you're working in, but if you don't check your mix on headphones, then none of your decisions will be properly informed.

But the usefulness of headphones for monitoring in the small studio isn't limited to telling you how a sizeable chunk of the public will perceive your mix. Headphones also serve an additional practical purpose, because their sound isn't affected nearly as much by your monitoring environment. In the first instance, background noise from computer fans, nearby traffic, or next door's line-dancing class are reduced in level relative to the mix signal when using headphones, which means that fewer subtle technical problems will slip through the net. Things such as brief audio glitches, bad edits, and momentary fader-automation slip-ups stand out much more clearly on headphones. "Headphones are great for checking for mouth ticks and pops," says Ed Boyer, for instance.[11] It's also easier to detect the onset of digital clipping on stereo files—the left and right channels will usually clip at different times, so the sprinkles of digital distortion appear characteristically right at the edges of the 180-degree stereo field. The lack of room reflections when monitoring on headphones is another advantage, because it serves as something of a safety net in the event that comb filtering or room-resonance problems are seriously jeopardizing your balance judgments. If you've been able to follow the nearfield-monitoring recommendations I made in Chapter 1, then you shouldn't need to rely on this precaution heavily, but if you're working on an unfamiliar system or have only limited control over your own monitoring environment, then a pair of familiar-sounding headphones can be a godsend.

FREQUENCY-RESPONSE CORRECTION PLUG-INS FOR HEADPHONES

A reasonably flat and well-extended frequency response is as much of an asset for studio headphones as it is for nearfield monitors, so some software manufacturers now offer special plug-ins designed to smooth the spectral non-linearities of real-world headphone hardware. The basic idea is that the developer carefully measures the sound characteristics of lots of well-known headphones in their lab, and then creates special EQ presets to compensate for their frequency–response lumps and bumps. All you need to do is tell the plug-in which headphones you're using, and it then applies the corresponding preset to try to flatten the monitored frequency response. In my own experience, though, I've found that real-world sonic variations between headphones of the same make and model (on account of manufacturing tolerances and the unpredictability of everyday wear and tear) so undermine the effectiveness of the preset-based approach that I reckon you'd get similar usability benefits from just correcting obvious frequency imbalances by ear using your DAW's bundled EQ plug-in. So save your money and spend it on better monitoring hardware, which I think is much more likely to improve your mixing power in the long run.

To be fair, there are a few companies who will actually measure your specific headphones and generate a customized correction curve for that unique pair. However, that won't turn bad headphones into good ones, and I already know several pairs of top-of-the-range studio headphones that I'd consider more than adequately "flat" for mixing purposes straight out of the box, so the considerable extra cost of additional bespoke frequency contouring just seems like overkill to me.

If your budget's tight, then you'll be pleased to know that the audio quality required of a set of headphones in this context isn't tremendous, and £50 ($75) should give you all the fidelity you need, but do still try to get studio (rather than hi-fi) models if you can. In fact, the kinds of headphones that most recording musicians use while overdubbing are usually more than up to the mark. That said, there are actually compelling reasons why most small-studio owners may find it worthwhile to spend more money here if they can. The first is that top-of-the-range studio monitoring headphones are capable of doing many of the jobs you'd expect of speakers, which is great for situations where you only have limited access to decent nearfields. You might be one of 20 students jostling for time in your college's single studio; perhaps you can't afford to set up your own decent nearfield system yet and can only check things periodically on a mate's rig; maybe you've been threatened with slow agonizing death if you wake that baby one more time. Regardless of the reasons, the ability to get a measure of real mixing done on headphones can dramatically increase the amount of time you have available for improving your production's sonics.

Clearly, the peculiarities of the stereo picture will present some additional panning and balance difficulties whenever you work primarily on headphones, plus even the best headphones won't give you a proper impression of your low end. On the flipside, though, the absence of room acoustics problems means that you may find that audio quality and tone decisions across the rest of the spectrum actually turn out to be more dependable. In my experience, the problems when working with excellent headphones are by no means insurmountable as long as you still have at least some access to an Auratone and a pair of full-range nearfields—whoever's they happen to be. As a matter of fact, given that my personal favorite top-end headphones all retail for under £350 ($500), I almost always recommend one of those sets as a first monitoring system for those starting out. My honest opinion is that unless you've set aside at least £1000 ($1500) for nearfield monitor speakers, then you'll get more reliable mixes from a pair of top-drawer headphones—particularly if you're able to hijack someone else's nearfield system briefly from time to time to check the low end and stereo image. "For mixing, I can get the mix to maybe 70 percent of the end result [on headphones]," comments Andy Selby, "and I then tweak things sitting down in front of some speakers, which is what I did with my Josh Groban mixes."[12] (For my latest headphone product recommendations, check out this chapter's web resources page.)

So why bother with nearfields at all if headphones can provide much of the same information? As in a lot of studio situations, a big reason for the outlay is speed. It's quicker to do the bulk of your mixing work on a listening system that gives you the whole frequency response in one shot. Although

FIGURE 2.8
Top-of-the-range headphones, such as the Beyerdynamic DT880 Pro (*left*) and Sennheiser HD650 (*right*) shown here, are a good investment for most small-studio operators, especially when budget is limited.

it's perfectly feasible to produce commercial-quality mixes without nearfields of your own, it involves a good deal of extra legwork to sort out the bass end in particular when headphones are the main workhorse, as we'll discuss further in Chapter 3. General mix balancing on headphones also typically takes more time, because the wide stereo image and overexposed mix details can be misleading—common pitfalls are balancing lead parts too low and misjudging delay/reverb effects levels. To keep yourself on the straight and narrow you'll have to rely more heavily on an Auratone substitute while processing your tracks, and you'll also need to compare your work much more frequently with competing commercial productions. All of this only really makes sense in the long term if you've got a whole lot more time than money. Or to put it another way, a cheap razor blade will cut a blade of grass beautifully, but I still don't begrudge the extra expense of a lawnmower.

EMULATING LOUDSPEAKER MONITORING USING HEADPHONES

For at least a quarter century, audio equipment manufacturers have been marketing hardware and software gizmos claiming to artificially simulate the loudspeaker listening experience for the benefit of headphone users. Naturally, I've followed such developments with considerable interest. After all, who wouldn't want to complement (or even replace) their bulky, expensive, and neighbor-antagonizing loudspeakers with the go-anywhere convenience of an equally capable headphone-based system? Unfortunately, almost every system I've tried (and I've tried plenty of them over the years!) is fatally compromised by one or other of two thorny implementation issues.

Firstly, the way we humans hear is powerfully affected by what we see. If the emulation determines the perceived speaker positions in relation to the headphones, the moment you turn your head the speakers will seem to follow you around the room, and this substantially undermines the spatial realism. Some systems do manage to avoid this stumbling block, though, by tracking your physical head movements in some way (usually with a little motion sensor strapped to the headphones themselves), and then adjusting the simulated loudspeaker positions in real time to compensate, thereby locking the speakers to specific physical locations, as you'd naturally expect.

The second issue is that a vital part of what allows us to identify the locations of sound sources is the shape of our ears—both the internal ear canal and the pinnae (also known as "those waggly bits on the sides of your head"). This shape effectively imposes different frequency–balance anomalies on sounds arriving from different directions, and we all instinctively learn in childhood how to interpret these tonal shifts to deduce the locations of sounds. Only if a headphone-based loudspeaker emulation algorithm accurately mimics this physiological effect can it conjure up a loudspeaker illusion that's convincing enough to cope with serious mix work. What complicates matters, though, is that everyone's ear shape is different, as personalized as a fingerprint, so the only way to guarantee a usable loudspeaker simulation for any individual user is to create a measurement of the unique acoustic

(continued)

properties of their ears (something called a Head Related Transfer Function, or HRTF) and build the processing algorithm around that. Almost all currently available loudspeaker-emulation products try to side-step this requirement by offering a choice of averaged "everyman" HRTFs as presets, which may give reasonable results if you're lucky enough to find one that closely matches your own hearing. However, there's no way for less experienced engineers to get a second opinion about whether the simulated sound is suitable for mixing, because no two people will perceive an emulation based on a given preset HRTF in the same way. For my own part, I've tried more than a dozen such systems and have yet to find a single preset HRTF that worked tolerably well for me. (Maybe I just have mutant ears . . .)

At time of writing, I know of only one company who have surmounted both these technical hurdles in any remotely affordable product: Smyth Research. I tested their SVS technology in real project-studio mixing situations for more than a year, and honestly found it to be a worthy match for my own nearfield monitoring system in terms of pure mixing power. While Smyth Research's proof of concept hardware is far from cheap, it's actually less expensive than the nearfield monitoring setup I suggest in Chapter 1, so I think it deserves serious consideration for small-studio use, especially where silent working and system portability are important concerns.

2.3 GROTBOXES

I explained in Section 2.1.5 that although the Auratone has a reputation for sounding nasty, it's not actually a good reference point when it comes to lo-fi playback, so to round out your supplementary monitoring options it makes sense to have at least one set of "grotboxes," unashamedly cheap mass-market speakers that let you hear how your mix holds up in the face of serious play-back abuse. It's not difficult to find such speakers—try the little ported active ones that you can feed from a PC's headphone output or wirelessly over Bluetooth, for example, or perhaps the speakers of a bargain-basement boombox. Don't be afraid to drive them fairly loud either, as this'll only emphasize the poorly controlled distortion and resonance artifacts, and it's the impact of these that you want to evaluate.

One advantage of using a pair of separate PC-style speakers is that you can put them right alongside each other without much in the way of a gap between them. This is how a lot of people end up using such speakers anyway, and it's also roughly how the speakers are spaced in small boomboxes. The effect of placing the speakers like this is to remove any useful stereo information while at the same time avoiding centrally placed sounds from increasing in level as you'd expect when summing a stereo mix signal to mono electrically. This is about the worst-case scenario for the clarity of central signals, so it's a real acid test of lead-vocal level and intelligibility. If you can hear your melody and lyrics clearly under these circumstances, then that's about the best guarantee you have that they'll make it through to the widest possible audience. I also find it handy to put grotbox speakers outside your studio room if possible and to listen to them off-axis from a good distance away. Again, this is a typical mass-market listening experience, and forces you to work even harder to keep the mix clear when broad-based commercial appeal is the aim.

Despite what you might think, grotboxes can frequently be found in the most high-spec of commercial mix rooms. Take Bob Clearmountain's pair of compact Apple computer speakers, for example: "Those are actually my favorites! I have them placed right next to each other at the side of the room on top of a rack. There's just no hype with the little ones—they're so close together that everything is almost mono, and I get a really clear perspective of how the overall thing sounds. . . . They're also good for setting vocal levels."[13] Michael Brauer uses something pretty similar too: "Most of my mixing is done listening to my little Sony boombox. It sits behind me . . . about four-and-a-half feet up from my ears and six feet back, and I listen at a medium low level. Because the speakers of the boombox are relatively close to each other, I essentially listen in mono. The Sony is like a magnifying glass, it tells me whether my mix sucks or not."[14]

FIGURE 2.9
A pair of cheap speakers placed side by side like this is the acid test for lead-vocal levels and intelligibility in your mix.

Finally, while some engineers like to refer to several different grotboxes during the mix process (boomboxes, TVs, car stereos, you name it), my view is that if you've already used your nearfields, Auratone substitute, and headphones sensibly, then there's little extra information to be gleaned by using any more than one grotbox system, and the extra time and money could be better used in other ways.

CUT TO THE CHASE

- A reasonable small-studio nearfield monitoring system can give a good overview of the full frequency spectrum and the stereo field, and it is well suited to assessing the effects of mix processing on the quality and tone of your mix. It should also give you a good idea of what hi-fi listeners and cinema audiences will hear, and is most likely to impress your clients. However, it won't provide all the information you need to craft commercial-standard mixes reliably, which is why supplementary monitoring systems are so important.
- The most powerful additional monitoring system for the small studio is exemplified by the classic Auratone 5C Super Sound Cube, a small, unported, single-driver speaker working in mono. This type of speaker highlights the crucial midrange frequencies that reach the largest number of listeners, and is much less susceptible to the comb filtering and acoustics problems that compromise many affordable nearfield systems.

Listening to your mix from a single speaker not only confirms the vital mono-compatibility of your production, but it also makes mix balance judgments far more reliable, especially for important sounds at the center of your stereo image.

■ Headphone monitoring is vital to check how your stereo picture and mix balance reach a large sector of the listening public. By isolating you from your listening environment, headphones can also identify low-level technical anomalies and processing side effects, as well as indicating how some room-acoustics problems may be affecting your nearfield listening. Although comparatively inexpensive headphones will cover all these bases adequately, a top-of-the-range set is highly recommended, because it will usually increase both productivity and mix quality for most small-studio operators, especially for those unable to set up a dependable nearfield system as discussed in Chapter 1.

■ Cheap, low-quality "grotbox" speakers also deserve a place in the mixing studio, as they give you an idea of what to expect in worst-case listening scenarios. Use a pair of such speakers placed very close to each other (or a small all-in-one stereo boombox), and then listen off-axis from a reasonable distance to get the "best" results. Although grotbox monitoring is important, don't obsess about it. You should only need one small system to do this job if you're using your other monitors sensibly.

Assignment

■ Get hold of a proper Auratone-substitute of some kind (you only need one), and set up a convenient method of listening to it in mono so that you get into the habit of using it that way.

■ Buy a pair of good studio headphones, if possible something at the top of the range so that you can do meaningful mix work when loudspeaker listening isn't an option.

■ Find some suitable grotbox speakers.

Web Resources

On this book's companion website you'll find a selection of resources to support this chapter, including:

■ the StereoTest and MonoBalanceShift audio test files, along with some audio demonstrations showing the damaging consequences of poor mono-compatibility;

■ my most up-to-date personal recommendations for Auratone substitutes and mixing headphones;

■ my full reviews of the Smyth Realiser system and a variety of other specialist processors designed to enhance the headphone-monitoring experience.

 www.cambridge-mt.com/ms-ch2.htm

CHAPTER 3
Low-End Damage Limitation

It's the low end of the audio spectrum that presents the toughest mixing challenges when you're working under budgetary constraints, and in my experience it's this aspect of small-studio mixes that most often falls short when compared with professional productions. As I see it, the reason for this trend is that the only real low-frequency monitoring in a typical small studio comes from its nearfields, yet the majority of the small nearfield systems I've encountered in use suffer unacceptably from low-end resonance problems as a result of ported speaker designs and/or room modes. Although there are plenty of ways to reduce resonance effects to workable levels (as discussed in Chapter 1), the reality is that few owners of small studios actually invest the necessary money and effort into this area of their listening environment to achieve passably accurate results.

The glib response would be that these guys only have themselves to blame, but many people's budgets are too small to accommodate even relatively affordable monitor-setup and acoustic-treatment measures. There is also a legion of small-studio users who aren't actually at liberty to dictate their gear choice or acoustic setup for mixing purposes: students using college facilities, budding media composers camping out in a corner of a shared living room, or recording engineers working on location, to give just a few common examples. So it's fortunate that all is by no means lost, even if the low end of your monitoring leaves much to be desired or you're forced to work mostly on headphones. As long as you're willing to learn a few special mixing and monitoring techniques, there are ways to work around the worst low-frequency pitfalls effectively nonetheless and thereby achieve commercial-grade results with some reliability. Indeed, given that I've yet to encounter any project-studio room with a flawless bass response (certainly every studio I've ever mixed in has had at least some residual low-end issues), there's a lot that mixing and monitoring "hacks" can offer, even to those lucky souls with sufficient budget to follow all my recommendations in Chapter 1.

> Even if the low end of your monitoring leaves much to be desired or you're forced to work mostly on headphones, all is by no means lost—as long as you're willing to learn a few special mixing and monitoring techniques.

3.1 COPING WITH CHEAPER PORTED SPEAKERS

My first set of tips is to help those engineers who find themselves lumbered with having to mix through cheaper ported monitors for whatever reason. First, it pays to be aware of where the port's resonant frequency is located, because this knowledge can help you to correctly identify obvious resonances in that region as speaker-design artifacts rather than mix problems. You can also make a note of the pitch of the resonant frequency, which will give you an idea of which bass notes are most likely to suffer irregularity on account of the porting.

You may be able to find out a given speaker's porting frequency from the manufacturer's product specification sheets, but failing that it's straightforward enough to investigate for yourself using the LFSineTones audio file I mentioned in Chapter 1. If you play this back through your monitor system at a medium volume, you should clearly be able to see the "motion blur" of the woofer's cone as it vibrates back and forth, which makes it easy to tell how wide the woofer's excursions are—it usually helps to look at the speaker cone from the side. As the tones begin their march up the frequency response, a ported monitor's woofer will start off showing fairly wide excursions for the lowest frequencies, but these movements will slowly narrow as the tones approach the speaker's porting frequency. The tone closest to the porting frequency will give the narrowest cone excursions, following which the movements will begin to widen out again as the tones continue on their way up the spectrum. Once you know which tone gives the smallest cone excursion, you can easily refer to Table 1.1 (page 000) to find out the porting frequency, both as a figure in Hertz and as a note pitch.

FIGURE 3.1
Blocking your speaker ports with dusters or bits of foam can give you a useful extra perspective on the low end of your mix.

The second tactic that can help you deal with ported monitors is to block their ports, thereby defeating some of the resonance side effects. Almost any material or foam can be pressed into service here as long as it impedes the flow of air in and out of the port opening—a rolled up sock will normally do a grand job. (What do you mean, "wash it first"? You'll destroy the vintage sound!) Although this will certainly make the speaker significantly lighter on low end, the bass reproduction that you do get should actually be more usable for mix-balancing purposes. Bear in mind, though, that there may well be other disadvantages of blocking the port in terms of frequency response ripples and increased distortion (the speaker was designed to be ported, after all), so though you should be able to judge bass balances more easily with socks stuffed in the port holes, other mixing decisions may work better without them. Be aware, though, that some active speaker designs do use the airflow through their ports as part of their ampli-fier cooling system, so it's probably best to avoid running the speakers at high volume for extended periods with their ports blocked, as this might risk overheating.

3.2 AVERAGING THE ROOM

In Chapter 2 we noted that a decent pair of headphones can bail you out to some extent when room acoustics are undermining the fidelity of your main nearfield monitor system. However, their low-frequency abilities in this depart-ment are limited by the size of their drivers, so you still have to rely on your speakers there, which means finding ways to work around any room reso-nances that are confusing the low-end balance.

Although one of the problems with room resonances is that they vary as you move your monitoring position, this is also the key to one useful workaround: if you make a point of listening to the bass response from several different locations, then it's actually possible to average your impressions mentally to some extent. In a room with resonance problems, you'll find that all aspects of the bass balance will vary as you move around. If a particular bass note, for example, appears to be too loud in some places but too quiet in others, then you can hazard a guess that the overall level may be in the right ballpark, whereas a note that remains too loud all around the room probably needs reining in.

If you want to get the best out of "averaging the room" like this, then jot down your reactions on paper as you move around. This technique can really help clarify your thoughts, especially where there are multiple bass instruments contributing to the mix—kick drum and bass guitar, for example. You'll also find that your profi-ciency with this trick will improve as you build up some experience of how specific areas of your monitoring room sound. Indeed, it's not unusual to discover a couple of loca-tions in any given room where the bass response as a whole is a lot more reliable than in your stereo sweet spot. "You have to find out where the standing waves are and avoid those spots," says Stuart White.[1] A valuable additional perspective can also be gained by listen-ing to your mix through the doorway from an adjoining room, as the different acoustic space will filter the sound through a differ-ent set of room modes. "I know what the bottom end is like by turning it up really loud and standing at the top of the staircase outside," says Kevin Savigar,[2] for instance, joining numerous other well-known engineers such as Allen Sides,[3] Joe Chiccarelli,[4] and George Massenburg[5] who mention this little dodge.

> Another valuable set of clues about your bass balance can be gleaned from a spectrum analyzer, and there are now so many decent freeware models that there's no excuse not to use one.

No matter how methodical you are about it, though, averaging the room will never be an exact science. It can provide a lot of useful extra clues about the low-end balance, but you should nonetheless beware of basing drastic mixing decisions on conclusions drawn solely from this source.

3.3 SPECTRUM ANALYSIS AND METERING

Another set of clues about your bass balance, and indeed your mix's overall tonality, can be gleaned from a spectrum analyzer, and there are now so many decent freeware models that there's no excuse not to use one. (I've listed some of my favorites on this chapter's web resources page.) Engineers Joe Chiccarelli[6] and Eric Rosse[7] have both mentioned using spectrum analysis. "I put [a spectrum analyzer] across my stereo buss that lets me know when the bottom end is right," says Chiccarelli. "I'm mainly looking at the balance of the octaves on the bottom end, like if there's too much 30Hz but not enough 50Hz or 80Hz. When you go to a lot of rooms, that's where the problem areas of the control room are." You should try to find an analyzer that provides good resolution in the frequency domain, and it helps to have some control over the metering time response, so that you can switch between slower averaged meter ballistics (which will be better for overall level judgments) and faster peak metering (which will track things such as drum hits more closely).

The most important thing to understand about spectrum analyzers, however, is that each manufacturer's implementation will present the frequency information

FIGURE 3.2
A high-resolution spectrum analyzer, such as the Voxengo SPAN and Izotope Insight displays shown here, has a lot to offer the small-studio mix engineer.

in a slightly different way. This means that you can only really begin interpreting a spectrum analyzer's display usefully once you've built some experience of how your particular model responds to commercial material. There's no sense in trying to line up all the bars on the graph with your school ruler if that's not what the target market expects. Remember also that spectrum analyzers can evaluate individual tracks as well as whole mixes, and they're particularly handy for highlighting if the internal frequency components of a bass part are shifting from note to note—a problem that all too often scuppers small-studio balances.

Don't ignore your recording system's normal level meters either, as these can reveal some undesirable level irregularities on individual instruments. Again, you'll get more information by using a meter that can show both average and peak levels, and fortunately such meters are commonly built into most DAW software these days. (If not, check out this chapter's web resources for some decent freeware options.)

3.4 WATCH THOSE CONES!

A final visual indicator of potential low-frequency concerns is your speaker cones themselves. "You can sometimes tell how much low end you have on an NS10 from the way the woofer moves," says Manny Marroquin.[8] As I mentioned when talking about finding a speaker's porting frequency, the woofer cone excursions become visible at lower frequencies and remain so even at frequencies that are too low either to be delivered effectively from a typical speaker cone or indeed to be perceived by human hearing. Particularly on low-budget productions, the buildup of inaudible subsonic energy is a real hazard, because it can interfere with the correct operation of your channel mix processors, prevent your finished mix from achieving a commercially competitive loudness level, create unpleasant distortion side effects on consumer playback devices, and make the mix translate less well when broadcast.

"Many people want the bass to be really loud," says Dave Pensado. "But if it's too loud the apparent level of your mix will be lower on the radio. If you put in too much bass, every time the [kick] hits the vocal level sounds like its dropping 3dB."[9] Just keeping an eye on your woofer excursions can be a valuable safety check in this respect, because a lot of sounds that don't seem to have low end to them can still incorporate heaps of subsonics. "You've got to make sure that you're not adding subsonic stuff," says Chuck Ainlay. "If I see a lot of excursion on the woofer, then I'll start filtering something. A lot of times it exists in a bass guitar, which can go quite low, but what you're seeing there is a subsonic harmonic. That can be filtered out without hindering the sound of the bass at all."[10]

Other common culprits are vocal "p," "b," and "w" sounds, which can generate subsonic thumps by dint of their hitting the microphone's diaphragm with a blast of air. A performer's movements

FIGURE 3.3
Excellent high-resolution peak/average metering is freely available in the form of the Sonalksis FreeG and HOFA 4U Meter Fader & MS Pan plug-ins.

can also create drafts around the microphone that have a similar effect—close-miked acoustic guitars can be prone to this, for instance. The vibrations of traffic rumble or foot tapping can easily arrive at a microphone via its stand too. And the issues aren't just restricted to mic signals. A lot of snare drum samples have unwanted elements of low-frequency rumble to them because they've been chopped out of a full mix where the tail of a bass note or kick drum hit is present. Aging circuit components in some electronic instruments can sometimes allow an element of direct current (DC) to reach their outputs. This is the ultimate subsonic signal, as it's effectively at 0Hz and offsets the entire waveform, increasing your mix's peak signal levels without any increase in perceived volume.

Even with the low-frequency energy that you actually want in your mix, cone movements can occasionally be a useful indicator. For example, it's not uncommon with both live and sampled kick drums for the lowest-frequency components to be delayed compared to the higher-frequency attack noise, and this often isn't a good thing when you're working on hard-hitting urban, rock, or electronic music—it just sounds less punchy. You may not be able to hear the lowest frequencies of a kick drum on a set of small nearfields, but if the onset of the corresponding cone excursions seems visually to be lagging behind the beat, then it can be a tip-off that your choice of sound may not deliver the goods on bigger systems. Many bass-synth presets also feature an overblown fundamental frequency, and if you're not careful with your MIDI programming, this fundamental can wind up wandering uselessly underneath the audible spectrum. A franticly flapping woofer cone (or hurricane-force gusts from a speaker port) can flag up such problems even though you can't hear them. For an illustration, check out the two ConeFlapper audio examples on this chapter's web resources page (but at reasonably low volume level to avoid any damage to your speakers): ConeFlapperOut gives a section of an R&B-style backing track with tightly controlled low end; ConeFlapperIn, by contrast, has a strong subsonic element to the kick drum, which eats up an extra 3dB of mix headroom and will flap your woofer around like crazy, but otherwise creates no significant change in the sound at all.

> A franticly flapping woofer cone (or hurricane-force gusts from a speaker port) can flag up subsonic problems even when you can't hear them.

Once more, though, I want to sound a note of caution, because you should avoid trying to conclude too much about your mix from woofer wobbles. Some engineers, for example, suggest that resting a fingertip on a small speaker's woofer cone allows you to feel imbalance and unevenness in the bottom octaves of the mix. In my view, the evidence of cone excursions isn't actually much help with these kinds of mix tasks, especially if you're using ported monitors, which reduce their excursions around the porting frequency as mentioned earlier. Even with unported monitors, cone excursions still tend to increase for lower frequencies, because that allows some compensation for the way that the woofer's limited size becomes

HAPTIC DEVICES

If you turn up any comparatively full-range loudspeaker system loud enough, you won't just experience its low frequencies through your ears: you'll also begin to feel the vibrations in your body, either created directly in sympathy with the air movements or indirectly transmitted to you through the physical structure of your listening environment. The lack of this tactile (or "haptic") experience is one of the reasons headphone listening doesn't deliver the same sensation of low end as loudspeakers, so a number of enterprising manufacturers now offer haptic vibration devices (often in wearable form) specifically designed to recreate it silently.

My own experience of such devices suggests that they can indeed improve the reliability of low-end decisions made on headphones, especially when you're trying to judge the attack and sustain characteristics of low-frequency instruments, because by nature the vibrations are unaffected by any air resonances in your monitoring room. In practice, though, specific haptic devices may have their own low-frequency resonances and nonlinearities to deal with (especially if targeted at the gaming market), but even if they responded ideally our brains simply can't discriminate low-end frequency balances from haptic input as critically as from aural information. As such, you shouldn't expect a haptic device to fully compensate for insufficient low frequency reach in any headphone system, or indeed to substitute for a subwoofer in bass-restricted loudspeaker setups. Incomplete as it is, though, the additional mix information you can glean haptically is by no means valueless (particularly among hip-hop and EDM producers for whom the kick-drum's time-domain envelope is almost a matter of life and death), and haptic devices are now becoming so affordable that anyone mixing bass-heavy music styles on a budget would do well to investigate them.

less efficient at transferring lower-frequency vibrations to the air. As with all monitors and monitoring techniques, you have to take from your woofer movements only the information that's of practical use and be merciless in disregarding anything that might mislead you.

3.5 PREEMPTIVE STRIKES AT THE LOW END

Despite what all these workarounds have to offer users of small studios, there will inevitably be some unwelcome degree of guesswork involved when crafting the low end of a mix unless you have at least some access to a reasonably well-behaved, full-range nearfield system. Faced with this uncertainty, then, the canny engineer will employ a certain amount of preemptive processing to avoid any low-end problems that the available monitoring can't adequately detect, and will also deliberately craft the final mix so that it responds well to mastering-style adjustments should aspects of the low-end balance prove, with hindsight, to have been misjudged.

Chief among these preemptive strategies is to restrict the field of battle, as far as low frequencies are concerned—if you simplify the problem, you simplify

FIGURE 3.4
The effect of DC (0Hz) on a mix file's waveform. Notice how the positive waveform peaks are clipping, even though the negative waveform peaks still have headroom to spare.

the solution. In the first instance, this simply means high-pass filtering every track in your mix to remove any unwanted low frequencies. (I'll deal with the specifics of this in Chapter 8.) "[I use] a simple high-pass filter . . . on almost everything," says Phil Tan, "because, apart from the kick drum and the bass, there's generally not much going on below 120 to 150Hz. I have always found filtering below this cleans up unnecessary muddy low-level things."[11] "Before I begin any final mix I try to take out as much low-end energy as I can," adds Rik Simpson. "Anything that doesn't need to be there is taken off, which gives me more headroom. It really makes a massive difference."[12] Just because an instrument is supposed to be contributing low frequencies to your mix, that doesn't mean you shouldn't high-pass filter it, either, because even if the filter barely grazes the audible frequency response, it will still stop troublesome subsonic rubbish from eating away at your final mixdown's headroom. "In general I like to take off rumble from the room things were recorded in," says Serge Tsai. "I like to keep my low end clean."[13]

Beyond this general trash removal, there are advantages for the small-studio engineer in deliberately reducing the number of tracks that include low-end information. "You've got to remember," advises Jack Douglas, "that the stuff that's going to take up the most room in your mix is on the bottom end. If you just let the bass take up that space, you can get out a lot of the low stuff on other tracks—up to around 160Hz—and it will still sound massive."[14] So if, for example, your main bass part is an electric bass, but there's also significant sub-100Hz information coming from your electric-guitar, synth-pad, piano, and Hammond-organ tracks as well, then it's not a bad idea to minimize the low-end contributions of all these secondary parts to reduce the overlap. This not only means that you can fade up the main bass part more within the available headroom, but the sub-100Hz region will also become much easier to control, because you can concentrate your low-frequency processing on just the bass guitar track. You might even split off the sub-100Hz frequency components for separate processing, metering their levels to ensure that they remain rock solid in the mix balance. Or perhaps you might decide to replace those frequencies completely with a dedicated subbass synthesizer part, using your MIDI and synth-programming skills to dictate the level of low end with absolute precision (a classic hip-hop and R&B trick that we'll be covering in Chapter 12).

Restricting any serious low end in your mix to the smallest number of tracks possible has a couple of other important advantages too. First, it helps the monitoring and metering workarounds described earlier to be more effective in practice, simply because what's going on in the bottom octaves is subsequently less complicated to unravel. Second, should you discover post-mixdown that the weightiness of your main bass instruments is out of line with the market competition, then you can usually do a lot more to correct this frequency problem using mastering-style processing, without destroying the tone of other instruments or compromising the clarity of the mix as a whole.

Clearly, hedging your bets at the low end like this must inevitably cramp your sonic style to some extent, and you may sacrifice a little low-end warmth and nuance by working in this way. However, you can't expect to have your cake and eat it too. If you don't have commercial-grade low-frequency monitoring, but nonetheless want the kind of clean, consistent, and powerful low end you hear on commercial records, then you've got to consider the demise of a few comparatively minor sonic niceties a pretty meager price to pay.

CUT TO THE CHASE

- The single biggest mixing challenge in the small studio is getting the low end right, partly because the monitoring tool best suited to tackling it is also the most expensive and the most complicated to set up effectively: the full-range nearfield system I discussed in Chapter 1. However, for those who have only limited access to a decent nearfield monitoring environment or those without the means to remedy the shortcomings of a compromised loudspeaker system, there are a number of workarounds that allow respectable low-end balances to be achieved nonetheless.
- If you're using ported monitors, then knowing the port's resonant frequency can help you mentally compensate for some of its side effects. Blocking the speaker ports can also improve your ability to make reliable decisions about the low end, although this may detract from the speaker's performance in other important areas.
- If you can't adequately tackle resonant modes in your listening environment (and few small studios make sufficient effort to do so), then you'll make better low-end mixing decisions if you "average the room," comparing the mix's bass response from different locations inside and outside your studio monitoring room.
- Good spectrum-analysis and metering software can be very helpful to your understanding of the low frequencies in your mix. However, you need to spend time acclimatizing yourself to the way any particular meter responds before you can get the best from it.

- The visible movements of your speaker cones can warn you of undesirable subsonic information in your mix and can reveal some areas for concern with different bass instruments. Beware of reading too much into cone movements, though, because they can give misleading impressions of low-end balance.
- Haptic devices can offer useful additional information about the low end of the mix for headphone users, but still aren't as good for low-frequency balancing tasks as a full-range speaker system.
- If you're in any doubt about what's going on at the low end of your mix, then try to simplify the problem by eliminating unwanted low-frequency information and reducing the number of tracks that carry significant sub-100Hz energy. This will allow you to maintain better control over the low-frequency spectrum; the monitoring workarounds already suggested will assist you more effectively; and post-mixdown mastering processing will be better able to remedy any low-end balance misjudgments you may inadvertently have made.

Assignment

- If you're using ported monitors, work out their porting frequency and make a note of the pitch it corresponds to.
- Find yourself a level meter that shows both peak and average levels, and also a high-resolution spectrum analyzer. Use them while mixing and referencing so that you get to know how they respond in practice.

Web Resources

On this book's companion website you'll find a selection of resources to support this chapter, including:

- the LFSineTones, ConeFlapperIn, and ConeFlapperOut audio test files, along with some audio demonstrations showing low-end "lag" on kick drums and the effects of systematic high-pass filtering on low-frequency mix clarity.
- my most up-to-date personal recommendations for affordable spectrum-analysis and metering plug-ins, as well as links to further reading on the subject of audio metering in general.

 www.cambridge-mt.com/ms-ch3.htm

CHAPTER 4
From Subjective Impressions to Objective Results

Monitoring is usually the biggest stumbling block in the path of most small-studio operators, but even the most laser-guided listening system can't mix your tracks for you. It'll tell you where you are, but it's still your job to navigate to where you want to be. The basic difficulty is that listening is an inherently subjective activity. The same "mix in progress" will usually elicit a different reaction from every different listener (even from the same listener after a heavy lunch!), and different engineers will all suggest different processing tweaks. Unless you can make your mixing decisions more objective, you'll always lack the confidence to direct your own mixing resources efficiently and successfully.

Now it would be lunacy to assert that entirely subjective mix decisions have no place in the mixing process. Of course they do, and so they should. That's how people will know that a given mix is yours rather than mine, and it's this kind of contrast that's essential for the development of new and exciting music. But no book can teach you creativity, so I'm afraid I can't help you with the truly subjective decisions. Not that this actually matters a great deal, because most small-studio users have more ideas in their heads than they know what to do with, so even if I could give you a magic potion bottle full of inspiration, I'm not sure you'd actually be any better off for drinking it! More to the point, it's the inability to make more objective decisions that really holds back most small-studio productions. You can create the vocal sound that launched a thousand ships, but it won't launch a pedalo unless you can make it fit properly in your mix.

This chapter takes an in-depth look at techniques you can use to make the kinds of down-to-earth decisions that really matter to commercial mixing. Can I hear the lyrics well enough? Do I have too much sub-bass? Where should I pan my overhead mics? How much delay should I use, and how much reverb? Answering these and similar questions is the bulk of what mixing work is about, and unless you can answer them appropriately, even a truly inspired mix will sound wrong when set against commercial tracks from your target market.

4.1 FIGHTING YOUR OWN EARS

Ears are pretty important if you want to hear anything, but they're a nuisance when you're trying to stay objective about your mix. This is because the human auditory system doesn't just transmit sonic vibrations straight from the air to your consciousness; it not only colors the raw sensory input through its own nonlinear response, but it also continually adapts itself in response to incoming sounds—partly in order to extract the maximum information from them, and partly just to shield its sensitive physical components from damage. Although the fact that our hearing works like this is helpful in everyday life, it actually works against you when mixing, casting a long shadow of doubt over every balancing and processing decision you make. If your hearing system is doing its best behind the scenes to balance and clarify your perception of the mix, rather than consistently presenting you with the warts-and-all reality, then it's all too easy to be suckered into signing off a second-rate mix as a finished product. The key to eluding this trap is to understand a bit about how our physiology skews what we hear, so that you can work around or compensate for the perceptual quirks when evaluating your mix.

> If your hearing system is doing its best behind the scenes to balance and clarify your perception of the mix, then it's all too easy to be suckered into signing off a second-rate mix as finished product.

4.1.1 Shock Tactics

One of the main problems is that the ear is very good at compensating for tonal imbalances. In one respect this is useful, because it's one of the reasons why tonal discrepancies between different brands of loudspeaker don't make as big a difference to their usefulness at mixdown as you might expect. Once you've acclimatized to a particular set of speakers, your ear will factor out their unique tonal characteristics to a certain degree. But for every silver lining there's a cloud: even if your mix has an obviously wonky frequency balance, it only takes a few seconds for your hearing system to start compensating for that, thereby hiding the problem from your consciousness.

How can we combat the ear's fickleness? Well, one good tactic is to switch between monitoring systems fairly frequently, because this instantly changes the tonal frame of reference and hence offers a few precious moments of clarity before the hearing system has a chance to recover from the shock and recalibrate itself. "After listening to all sorts of frequencies for a long period," says Russ Elevado, "the Auratone will just put everything flat and my ears kind of just recalibrate themselves."[1] Given that each of your monitoring systems offers a different set of strengths and weakness as far as analyzing your mix is concerned, switching between them is already a good idea anyway, so this is one more reason to make a habit of it. It's no accident that the monitor-selection buttons on most large-format professional studio consoles are usually among the most worn looking. "I can easily switch between 15 different speakers," says Cenzo Townshend, for instance. "I couldn't just mix on NS10s or KRK 9000s . . . you get used to them too much."[2] In fact, because switching monitors is so useful while mixing, I usually recommend that small-studio owners without a suitable monitoring section in their mixer get hold of a dedicated hardware monitor controller for the purpose, because it saves a great deal of time in the long run if you can hop between your monitor systems at the flick of a switch—and it makes getting into the routine of doing it easier too. There are some extremely affordable models on the market, but you should bear in mind that everything you hear will pass through your monitor controller, so it's perhaps a false economy to pinch pennies here.

FIGURE 4.1
An excellent tool for recalibrating your ears. It also makes hot drinks.

A second important shock tactic is taking breaks. According to Bob Clearmountain: "Don't just keep plugging away at something. When you start to question everything that you're doing and you suddenly don't feel sure what's going on, stop for a while . . . I go out with my dog or take a 15-minute catnap, then come back and have a cup of coffee and it's much more obvious if something is wrong."[3] "The tougher the mix, the more breaks you should take," says Russ Elevado. "If it's a dense song, and there's lots of frequencies going on, I'll do 20 minutes, then take a 20 minute or more break. On a difficult song, you can be more productive with more breaks . . . It's definitely good to pace yourself and know that you're not wasting time when you're taking a rest. You can get too focused on the details when you're working for a long stretch. Once you take a nice break, you can come back and are more able to look at the overall picture."[4]

Taking a break doesn't just mean sagging back in your studio chair and scoffing pizza, either—leave your studio room and do something to remind your ears what the real world sounds like. "When I'm mixing, I like having my dog there," says Billy Bush, for instance, "because every three hours she'll come up and go "Hey, I'm bored!" . . . I just take a break, go for a walk, go to Starbucks, get some coffee . . ."[5] If anything you hear in your studio setup is a better reality check for your ears than the sound of your own kettle boiling, then you seriously need to get out more! The longer you work on a track, the longer your breaks may need to be. As Bill Bottrell explains: "Objectivity is everybody's big problem, especially if you've worked a long time on a song . . . But, if something ends up sounding wrong, you will hear that it's wrong if you get away from it for five days and then hear it again."[6]

4.1.2 Breaks and Hearing Fatigue

The benefits of taking breaks, though, extend beyond basic shock tactics. Whatever volume level you mix at and however smooth-sounding your music, a few hours of uninterrupted mixing will still tire out your ears, and your sensitivity to high frequencies in particular will suffer. So if you want to be able to mix all day the way the pros can, then breaks are essential to keep your ears from flagging. Even then it pays to be careful about what you commit to toward the end of a long session. After one grueling (and not particularly successful) mixing project, I came up with a basic rule of thumb that has served me well ever since: Never finalize a mix after supper. I find that by that time of the day my ears simply aren't objective enough to do the job properly. If at all possible, try to "sleep on the mix" before finishing it. "When you come back fresh," says Joe Chiccarelli, "there are always a couple of obvious little things that you've overlooked."[7]

Engineers such as Bob Clearmountain,[8] Mike Clink,[9] Russ Elevado,[10] and Thom Panunzio[11] have all echoed this sentiment in interviews, whereas Joe Zook offers this cautionary tale about a hit single he mixed for One Republic: "When I first mixed "Stop & Stare," I made the drums and the bass massive. It became this huge grunge thing, and it just didn't sound unique or fresh. I'd started at five or six in the evening and I went home late at night being really proud of myself. But when I came back in the next morning, it was like, "Fuck! Where did the song go?" There were all these great sounds, but it didn't feel right . . . I'd become too focused on all the unimportant aspects of the engineering, rather than the music itself. So I started again."[12]

4.1.3 Monitoring Level

The simple issue of how loud you listen while mixing also needs some thought if you're going to get the best results. For one thing (and to get the preaching out of the way as quickly as possible), it's just plain daft to listen at thundering volumes for long periods because there's a real risk of hearing damage. If you reach the end of a day's mixing and find that your hearing's noticeably dulled or there's a high-pitched ringing in your ears, then you've probably damaged your hearing to some degree. Every time you do that, you're eating away at your hearing bit by bit, and there's no way to bring the sensitivity back once it's gone. Most of the highest-profile mix engineers actually spend the majority of their time mixing at low volume. Take Allen Sides, for instance. He stated in an interview that he only listens loud for maybe 20 to 30 seconds at a time and does his most critical balancing at a low enough level that he can easily conduct a normal conversation during playback.[13] So think like a professional; take care of your ears!

> It's just plain daft to listen at thundering volumes, because there's a real risk of hearing damage. Most of the highest-profile mix engineers actually spend the majority of their time mixing at quite low volume.

"Be safety conscious when you go to shows, and monitor at reasonable volumes," advises Kevin Killen. "Do not be afraid to protect your most valuable commodity."[14] Take special care when listening on headphones, because the unnatural listening experience they create makes it easy to monitor dangerously loud. And be aware that alcohol and drugs can tamper with the physiological protection systems that are built into your hearing system, so be doubly cautious if you're under the influence. (Yeah, yeah, of course you don't.)

There's more to setting your monitoring level than avoiding concussion, though, primarily because these pesky ears of ours present us with a different perceived frequency balance depending on how loud we listen. If you fancy researching the psychoacoustics of this effect, then graphs such as the well-known equal-loudness contours shown in Figure 4.2 will clue you up on the details, but for mixing purposes you only really need to understand this quick-and-dirty generalization: you'll hear more of the frequency extremes as you crank up the volume, whereas quiet listening will focus your ear more on the midrange. That's not the only factor at work, though, because our ears actually reduce the dynamic range of what we're hearing at high volumes as a safety measure too, and background noise masks more of the low-level details when you monitor softly.

The most immediate ramification of all this is that you need to understand how your production responds to being played at different volumes. For example, it's easier to decide whether the most important musical elements of a mix are loud enough for hostile mass-market listening conditions when the volume's super-low, whereas any trace of upper-spectrum harshness will usually become painfully apparent at party-grade listening levels. "I switch my monitor levels all the time," remarks Bob Clearmountain. "I tell people I work with 'Just give that monitor pot a spin any time you like!'"[15] As Chuck Ainlay clarifies: "I don't see any way of identifying if the whole mix is really holding up unless you crank it. I don't know how you can check the bottom end and whether the vocals are poking out too much without turning it up quite loud. There are also things that you'll miss by listening loud, so you have to work at low levels too."[16] "I will spend a lot of time listening to certain sections loudly to make sure things have the right impact," adds Andy Wallace, "and also very quietly to make sure the balances are really coming out well."[17]

However, although listening at a wide variety of volume levels will help you keep perspective on how your mix translates across all systems in general, it's equally important for the mix engineer to take an active decision as to which volume levels are of the greatest importance for the musical style in question. For example, a teen pop/rock mix might require serious high-frequency enhancement to retain its clarity in low-level home/car/office listening situations, whereas the same degree of fizz coming full-throttle over a club sound system would all but tear your ears off! Furthermore, it's important to realize that the inherent excitement of loud listening can make even a limp mix seem

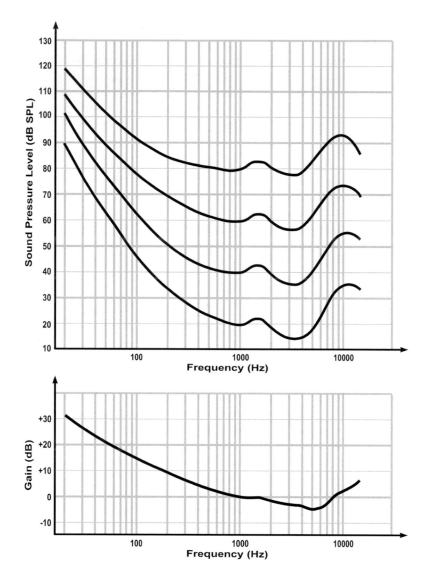

FIGURE 4.2
The top graph here shows four equal-loudness contours. One of the things they show is that our sensitivity to frequency extremes increases with listening level—for instance, moving from the bottom to the top contour (in other words turning the volume up by 60dB) has roughly the same effect on the perceived tonality as the EQ plot shown at the bottom.

powerful. So much so, in fact, that if I find myself reluctant to turn down the monitor volume while mixing, I consider it a sure-fire indication that I need to work harder on the sound! "I do most of my mixing at pretty low level," says Clearmountain. "I get a better perspective that way and I can work longer. If I can get a sense of power out of a mix when it's quiet, then I know that's still gonna be there when I turn it up."[18] Chris Lord-Alge agrees: "I listen really quietly. When you've been doing this job for a while, you tend to listen at such a low volume that in order for it to hit you in the face, you have to really push it."[19]

4.1.4 A Jury of Your Peers

Nervous about playing your mixes to other people? Well, grow up and get over it! I'm not trying to be macho for the sake of it here, it's just that mixing on a professional level is about taking responsibility for your output, and that means playing it unflinchingly for all and sundry. By the time you've worked your way through the mixing process, your own ears will probably be the least objective on the planet about the outcome, which means that it's easy to miss the wood for the trees. There are few reality checks to match another person's dispassionate first impression, so any aspiring mix engineer with even half an ounce of sense should seek out constructive criticism wherever possible.

"You've got to be very careful about losing your objectivity," warns Alan Moulder, recalling his mixing work for Smashing Pumpkins: "It was very handy to have [producer] Flood and [lead singer] Billy Corgan there. They could come in and say, 'That sounds great,' and I'd say, 'Well, so-and-so's not as good,' and one of them would say, 'No, it's fine. I prefer it.' That meant I could just lose myself in the technical world for a while, because these other guys hadn't lost their objectivity."[20]

USING CALIBRATED REFERENCE MONITORING LEVELS

Some engineers recommend maintaining a consistent calibrated reference level for studio monitoring most of the time, such that a given DAW output level will always generate the same loudspeaker listening volume. By minimizing the degree to which loudness-related perceptual effects impact upon your judgment, this approach can help develop a stronger instinct for how different components of a mix should sound—relative to each other, relative to your mix as a whole, and relative to any comparison material. And if you calibrate your mix system appropriately, that instinct can also guide you naturally towards creating mixes that have sensible output levels, without you needing to watch the meters all the time. That said, I feel that loudness-matched referencing against commercial releases is still the most bulletproof method of ratifying mix decisions, and that the metering in any modern DAW is perfectly capable of guiding you towards suitable mixdown levels on its own, so I don't think adhering to a calibrated reference monitoring level is by any means an essential prerequisite for producing marketable small-studio mixes. As such, the main reason to investigate this tactic, in my view, is to improve your "first draft" mixes prior to the mix-referencing stage, thereby speeding up the process of arriving at a finished result, particularly if you are an engineer who, by specializing in a narrower range of musical styles, can most reap the benefits of building a genre-specific sixth sense.

Alan Parsons also advises against going it alone: "I don't think any single engineer, artist, or producer ever achieved anything by working on his own. You learn from other people, and you achieve great things by interacting with other people."[21] Play the mix to your music technology lecturer and watch for winces. Bribe the local DJ to sneak it into his set and find out if it clears the floor. Play it to your pot plant and see if it wilts. The more you get your work out into the public domain, the more informed you'll be about the merits of different mixing techniques, and the quicker your skills will progress.

Clearly you'll want to take some people's comments with a large pinch of salt, because every individual has his or her own opinion about what constitutes good music. Few death metal fans are likely to heap praise on a plastic Eurotrance single, for instance, however stellar the production values. You have to learn how to interpret each criticism, filtering out what you decide is irrelevant and using the remainder to improve your sound. If you can get good at this when you're starting out, it'll stand you in excellent stead for dealing with the comments and revision requests of paying clients further down the line.

Here are a couple of tips that can help with interpreting other people's opinions of your mix. First, the most useful advice tends to come from those listeners who really connect with the sound you're aiming for and seem to like the general approach you've taken with your mix. When that kind of listener says that some aspect of the mix doesn't feel right, then it's usually wise to pay heed. Second, it's not uncommon for people's comments to be apparently nonsensical or contradictory, especially if you're dealing with non-musicians. Although this can be frustrating, it's fair enough. Music isn't an easy thing to describe at the best of times, and much of the language people use to do so is vague. To get around this problem, communicate using music itself. In other words, ask the critic in question to identify commercial mixes that exemplify the kind of attribute that they feel your mix is lacking. Let's say, for example, that one of your listeners wants a "more powerful snare." Get the person to name-check a few professional releases with such a snare sound, and it suddenly becomes easier to work out whether your fader, EQ, compression, or reverb settings need adjusting—or whether that person is, in fact, spouting utter tripe.

4.2 THE ART OF MIX REFERENCING

Although each of the monitoring tactics I've discussed so far in this chapter is important in its own right, I've reserved special attention for the most powerful technique of all: mix referencing. George Massenburg neatly encapsulates what this process is all about: "I wish guys would pick up several of the top 100 CDs—the ones they're trying to emulate—and listen to them, compared with their work. And then tell themselves that their ears are not deceiving them!"[22] Particularly for those starting out, mix referencing against commercial CDs is a dispiriting process, so it's little wonder that so many small-studio users shy away from it. However, this painful trial by fire is also quite simply the best mixing tutor in the world, and the fastest track to commercial-sounding productions.

The fundamental point of mix referencing is that it delivers what your own hearing system and experience cannot: objective decisions. Because we humans are only really equipped to make relative judgments about sound, the only way you can anchor those judgments in reality is by comparing your mix against existing commercial-quality productions. Outside the realms of high-spec professional rooms, mix referencing takes on another role too, because it allows you to compensate somewhat for the skewed listening perspective of unavoidably compromised monitoring systems. In other words, you can start to answer questions such as "Do I have too much 100Hz, or are my speakers telling a big fat lie?"

> Mix referencing against commercial CDs is dispiriting, so it's little wonder that so many small-studio users shy away from it. However, it's quite simply the best mixing tutor in the world.

Although every sensible small-studio operator eventually acknowledges the basic concept of mix referencing, in practice most people working on a small scale fail to extract the maximum benefit from the referencing process. This is a real shame, because conscientious referencing is the ultimate bang-per-buck studio tool. It's the best way of closing the gap between amateur and professional sonics, yet it costs very little money. So let's delve more deeply into the mechanics of the referencing process to demonstrate some of the refinements.

4.2.1 Choosing Your Material

Easily the biggest mistake you can make when mix referencing is to use inappropriate comparison material, and the only decent way to avoid this pitfall is to give the selection process a proper slice of your time. This means vetting suitable tracks from your record collection as methodically as you can, on a variety of listening systems, and then comparing them side by side to knock out the lesser contenders. Sounds like a monumental drag, doesn't it? I agree, but let me tell you a little story.

When I first started mixing, I filled up a CD-R (Remember those?) for referencing purposes with a couple of dozen commercial tracks, selecting my favorites pretty much from memory—much as most small-studio users tend to. Not long afterward, I began to notice that other commercial mixes seemed to sound better than my reference material, so I decided to take a more methodical selection approach and systematically trawled through my entire record collection, making an effort to ratify each track preference on at least three listening systems. (Sad, I know, but I'm a bit of a geek.) After much compiling and recompiling of short lists, I discovered that only one track from my original disc remained among my newly minted top 60. By putting my mixes side by side with lower-grade reference material, I'd effectively been compromising the quality of my own output because I'd not had to work as hard to achieve comparable results.

FIGURE 4.3
Both these records have classic singles on them, but you may find that the mix sonics are better on other tracks from each album.

In a lot of cases, I'd instinctively selected my favorite song off an album, rather than the most useful mix reference from a sonic perspective. In other cases, less high-profile records by a given mix engineer actually sounded better than the mega-famous example I'd initially chosen—Andy Wallace's mix of Nirvana's "Smells Like Teen Spirit" made way for his mix of Rage Against The Machine's "Fistful of Steel," for example. The moral of this tale is that your instincts will play tricks on you if you give them half a chance. The whole point of mix references is that they should allow you to foil such perceptual skullduggery, but they'll only do this properly if you factor out the vagaries of your own hearing system during the selection process. I'm not suggesting that everyone take as tortuous a route as I did (it's probably not healthy), but the more ruthless your shootout procedure, the more powerful a tool your mix references are likely to become.

4.2.2 What Makes a Good Reference?

Of course, you can be as anal about selecting reference tracks as you like, but that won't help you if you don't have an idea of what you're listening for. Clearly, your overall judgment of the sound quality is an important factor when selecting mix reference tracks, and it's right that this judgment should be a subjective one to a certain extent—it's your unique idea of what sounds good that will give your mixes their recognizable character, after all. However, it's nonetheless sensible from a commercial perspective to make the effort to

FIGURE 4.4
If you need suggestions for good-sounding records, there are a hundred lists to be found on web forums. Check out www.soundonsound.com and www.gearslutz.com in particular.

understand and evaluate productions that have achieved an elevated status among the professional community in terms of sheer audio quality.

Any engineer you ask will usually be more than happy to offer recommendations in this regard, and Internet forums such as those at www.soundonsound. com and www.gearslutz.com are regularly host to impassioned "best-sounding album in the Universe ever!!!!" exchanges that can provide food for thought. Another extremely well-regarded source of suggested reference tracks can be found on the website of top mastering engineer Bob Katz. His "Honor Roll" page at www.digido.com lists dozens of albums in all styles. That said, if you browse around his site it quickly becomes clear that Katz is a leading figure in the fight against overloud "hypercompressed" CDs. So without engaging in that whole debate myself here, you do need to bear this in mind.

Beyond just seeking out high-quality sonics, it's good sense to go for reference tracks that relate to the kinds of musical styles you work with, because production values can vary dramatically between genres. "You want to sound contemporary and current," explains Jimmy Douglass, "but you can't know what that is unless you listen to the records that the audience is digging at the moment."[23] "You have to know who your target audience is," adds Anthony Kilhoffer, "and what your peers are sounding like. If it's a Christina Aguilera song, yeah man, that's got to be *bright*! But if it's a Jay-Z song, he doesn't want to hear that on his vocal."[24] There's also the issue of what era of record production you want to emulate. Approaches to the use of compression and reverb, for example, have changed a great deal over time, and tastes continually shift regarding overall mix tonality. However, while you may wish to weed out reference tracks if they don't conform to the genres you work in, it's important to resist the powerful urge to eliminate tracks purely on grounds of musical taste. It's not that quality songs aren't important in a general sense, but it's the sonics of a production that make it a good mix reference, rather than the songwriting. So while I personally remain unconvinced of the musical value of the Pink, Craig David, and Puddle of Mudd tracks on my own reference CD, they've all proven their worth repeatedly for my mixing work.

> It's important to resist the powerful urge to eliminate reference tracks purely on grounds of musical taste. It's the sonics of a production that make it a good mix reference, rather than the songwriting. So while I personally remain unconvinced of the musical value of the Pink, Craig David, and Puddle of Mudd tracks on my own reference CD, they've all proven their worth repeatedly for my mixing work.

Another crucial thing to realize is that referencing isn't about trying to "copy" the sound of another track. "Reference mixes aren't there to copy," says Dave Pensado, "because you're not given the same information . . . It should be for inspiration, just to make sure you're able to compete with what your contemporaries are doing."[25] By the same token, it's rare that you'd want to reference your whole mix from a single commercial release, because different records will be useful for validating different mix decisions. For example, Skunk Anansie's "Infidelity (Only You)," Britney Spears's "Toxic," and Dr Dre's "Housewife" are all tracks I regularly use to help me mix bass instruments. By contrast, I might use Cascada's "Hold Your Hands Up," Gabrielle's "Independence Day," Norah Jones's "Sunrise," or Paolo Nutini's "New Shoes" for exposed vocals. Other tracks can be useful as "endstop markers," such as Natalie Imbruglia's "Torn," where the vocal air and sibilance are pushed too far for my liking, or Eminem's "Square Dance" where the kick drum's low end teeters on the brink of excess. If a vocal in my mix approaches "Torn" or my kick drum rumbles as much as on "Square Dance," then it means that I have to rein in my mix processing somewhere—and that I should probably also take a break to comb the Chihuahua or water the triffids or something.

A/B SWITCHING PLUG-INS

It's perfectly possible in most DAW systems to set up instantaneous A/B switching between your mix and a reference track within the mix project itself, but there are now also dedicated switching plug-ins such as Sample Magic's MagicAB, Melda's MCompare, and Mastering The Mix's Reference designed to streamline this process. However, while I think anything that makes it easier for small-studio engineers to get into the habit of regular mix-referencing is great in principle, the fact is that I don't actually use any of these tools myself. For my own part I prefer to bounce my mix out as a WAV file and then import it into a separate dedicated DAW project for referencing purposes. So why go to this extra trouble? Well, the first reason is simply a question of implementation: convenient as these switcher plug-ins seem at first glance, I usually end up finding them restrictive in one way or other during the referencing process. Maybe it's fiddly to set up which time-region of each reference I want to compare my mix to, or it's not possible to apply plug-ins to the references for analysis or tone-matching purposes. There's a more important issue, though, and that's

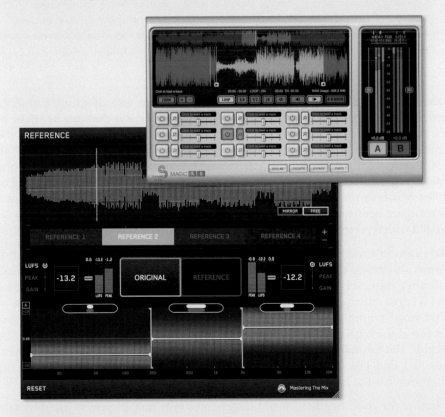

FIGURE 4.5
Sample Magic's MagicAB and Mastering The Mix's Reference make it easy to compare your mix in progress with commercial reference tracks.

(continued)

psychological. By referencing in a separate DAW project like this, I can't immediately react to what I'm hearing by tweaking my mix settings, which makes it easier for me to maintain a disciplined referencing approach. So, for example, because I can't brighten my hi-hat straight away when I hear that it's duller than on one of my references, it makes me more likely to check this mix judgment in relation to my other references, and indeed to confirm whether the decision holds for different listening volumes and different monitoring systems too. The mix judgments I get out of the end of that process are more reliable and objective, and they help make my workflow more clearly directed, with less tail-chasing. Now I'm not saying it's impossible to reference conscientiously using a switcher plug-in directly within your mix project, but I just know from experience that I don't have that much mental discipline!

However, there can be a practical problem with using a separate DAW project for referencing, if you're switching between your mix and the reference tracks using the DAW's Mute buttons. Some DAWs automatically optimize their use of the available CPU power by completely deactivating plug-ins on muted tracks, which means that some plug-ins can take a fraction of a second to recover their proper sound when their host channel is unmuted. It's usually possible to work around this issue one way or another in most cases, but if you can't work out how to do this in your software, then that would be a good reason to use a dedicated switcher plug-in instead.

Reference material is particularly important where the use of room ambience and reverb is concerned, as this is one area of a production that can be tricky to get right. For instance, Outkast's "Bowtie," Solomon Burke's "Other Side of the Coin," Kanye West's "Breath In, Breathe Out," and Keane's "She Has No Time" all take different approaches in this respect, and having access to all of them among my reference tracks helps me make more suitable use of effects across a range of styles. Mix-buss processing is also easy to overdo without some good examples to compare against, and audience expectations of mix tonality can only really be judged with representative tracks from the chosen target market on hand.

"You can do a mix where all the balances are perfect," says Andy Johns, "but it means shit unless the overall sound—the bottom and the top—is going 'hallo'! If it ain't there, see you later . . . You have to have something to compare it to, so I'll constantly be playing CDs."[26] Once more, don't neglect the usefulness of "end-stop" examples here. The tonality of Madonna's "Sorry" works like this for me at the high end, for instance, whereas Pussy Cat Dolls' "Taking over the World" and Eminem's "Square Dance" provide a similar service at low frequencies. If I find I've trespassed beyond these extremes, then I'm pretty sure my afternoon Kit-Kat is overdue.

Whether for good or ill, a lot of budget-conscious mix engineers are now taking responsibility for mastering their own productions these days. If you find yourself in this position, then reference tracks can play an important role here too, especially when it comes to the contentious subject of loudness processing. Just the activity of deciding which commercial tracks deliver an

FIGURE 4.6
The Honor Roll of Dynamic Recordings at www.digido.com has a lot of good suggestions for reference material, including albums such as AC/DC's *Back in Black* and Sting's *Brand New Day*.

appropriate tradeoff between loudness enhancement and its detrimental side effects will be of benefit to you, because in the process you'll make up your own mind where you stand in the ongoing "loudness wars" debate and can choose your own processing methods accordingly. It's even more important with mastering decisions, though, that you make sure you can really hear what you're doing, so don't make any hard-and-fast decisions while whittling down your shortlisted tracks until you've scrutinized them all on a representative range of monitoring systems.

4.2.3 Getting the Best out of Your References

Getting the best out of your mix references isn't just about selecting the right tracks, though. The way you go about making your comparisons is also a vital part of the equation. Because your hearing system is so quick to compensate for the differences between mixes, the quicker you can switch

---

DATA-COMPRESSED FORMATS

Now that so many musicians store their record collections in "virtual" form on their computers and mobile devices, it's tempting to use those for quick-access mix referencing. I'd advise against this, though. The problem is that the stored music files typically use heavy digital data-compression routines to pack the maximum amount of music into the device's onboard storage space. Although the side effects of this data compression are small enough not to concern the majority of the listening public, they still lower the quality bar significantly enough to compromise the effectiveness of your mix referencing. Most audio streamed over the internet is of even lower quality, in order to minimize data-transmission bandwidth, so that's rarely a viable alternative either. The bottom line is that using lower-quality versions of reference tracks gives you an easier target to aim for, so you won't work as hard and the competitiveness of your mixes will suffer.

---

between a reference track and your own work-in-progress, the more revealing the contrast is likely to be. The ear's adaptive capabilities are also the reason why I tend to edit out just the highlights of my reference tracks, because that makes the contrasts between each snippet much more starkly apparent. Even a short five-second intro can heavily skew your perception of how the following full-band entry sounds, for example.

Just as important as instantaneous monitor switching is the ability to loudness-match reference tracks with your mix, as subjective volume differences between

FIGURE 4.7
Two possible mix-tonality references: Madonna's "Sorry" (from *Confessions on a Dance Floor*) has a lot of top end, whereas Pussy Cat Dolls' "Taking over the World" (from *Doll Domination*) has a lot of low end.

the tracks will confuse the evaluation process: it's human nature that anything that sounds louder tends also to sound better, irrespective of whether it actually is. That said, there is a big practical difficulty facing small studios in this respect: if you're referencing against the kinds of super-loud modern productions where the side effects of extreme mastering processing significantly impact on the overall mix sonics, it can be almost impossible to judge whether your own unmastered mix is actually delivering the right balance or tone.

In such circumstances, I fall back on two main workarounds. The first technique is to seek out unmastered mixes of relevant commercial releases or else mastered tracks where the loudness processing has been kept fairly subtle. The former are as rare as hen's teeth as far as most people are concerned, but the latter are slightly easier to find—in Bob Katz's Honor Roll, for instance. The other workaround is to apply loudness processing to your mix while referencing, in order to give you a better idea of, as Kevin Davis puts it, "what the mix sounds like smashed up and brickwalled from mastering."[27] (I'll delve into the practicalities of this kind of processing in Chapter 19.)

BEYOND MIXING: OTHER ADVANTAGES OF REFERENCE MATERIAL

As if improving your mixes weren't a big enough reason for building up a decent selection of reference tracks, the benefits actually reach far beyond mixing. First of all, the simple process of selecting the audio is excellent ear training and helps familiarize you with the sonic templates of different styles. "You have to understand the parameters of the music of the day, of the type of genre that you're doing," says Jimmy Douglass. "And also, more importantly than that, it becomes about selling your own self into the vibe of the music."[28] Then there's the fact that you'll inevitably become familiar with how your reference material sounds on a large number of different systems, and this means that you can start to judge new listening environments in relation to this body of experience—a lifesaver if you regularly do serious work in unfamiliar studios or on location. As John Leckie notes, "What I do whenever I go into a strange control room anywhere in the world [is] put my favorite CDs on, and I soon know exactly what the room is doing."[29] Mike Stavrou stresses the importance of prompt action here: "Use your reference CD before you get used to the coloration of any new room and its monitors. When you first walk into a new control room, this is the moment your ears are most sensitive to the room's acoustic anomalies. After four days you will acclimatize and in so doing will slowly become deaf to the idiosyncrasies and 'nasties' of the room."[30]

Trusted test tracks can also sort the wheat from the chaff when you're auditioning monitor speakers, or indeed any other audio playback component. This is where the Bob Katz Honor Roll really scores for me. Many of these albums will reveal small playback distortions much more starkly than the overprocessed masters that infest the mainstream charts.

4.3 EVERYONE NEEDS A SAFETY NET

So far in this book I've described how to set up sensible nearfield and supplementary monitoring systems so you can hear what you're doing, and I've suggested many ways to improve the reliability of your mix decisions. If you can apply these ideas in your own small studio, then you'll have already taken the biggest step toward achieving competitive, professional-sounding mixes 99 percent of the time. However, unless you've sacrificed your eternal soul in exchange for supernatural arrangement and mix processing skills, there's no getting away from the fact that you're human, and humans have off days. No one, but no one, escapes the occasional handbagging from Lady Luck—not even the top names in the industry. This is why it's common practice for the professionals to hedge their bets when delivering mixes to their clients.

The most flexible way you can do this is to make notes of all the equipment routings and mix settings used, so that the mix can be recalled in its totality if changes are requested at a later date. However, this kind of recall has always been a bit of a minefield in analog setups, because no mix ever seems to come back the same second time round, no matter how well-documented the settings. Your favorite buss compressor might have blown a valve in the interim. Your new studio assistant might not line up the tape machine quite the same way the previous guy did. A roadie might have drunkenly widdled down the back of the console. It's the entertainment industry, for heaven's sake—anything could happen!

> Unless you've sacrificed your eternal soul for supernatural skills, there's no getting away from the fact that humans have off days. No one, but no one, escapes the occasional handbagging from Lady Luck—not even the top names in the industry.

To make mix recall easier, quicker, and more reliable, many mix engineers have now migrated to working entirely "in the box" on a computer DAW mixing system. Although this approach can be much more dependable if you've got serious resources to throw at maintaining the hardware and software, small DAW-based studios often turn out to be no more reliable for recall purposes than their analog counterparts. This is partly because of simple hardware faults and data corruption, but also because of every studio owner's irresistible urge to upgrade—upgrade a plug-in and it might not load into "legacy" projects any longer; upgrade your DAW and the sound of the EQ processing may have been tweaked behind the scenes; upgrade your operating system and you might as well try recreating the mix on a pocket calculator. So while a full mix recall can provide a good solution to mix problems that are discovered in the short term, you'd be loony to bank on that for revisions further down the line. A much more reliable way of managing this kind of longer-term fence sitting is to create alternate versions of your mix—something most pros do as a matter of course. "I'll print as many mixes as needed, depending on how difficult the artist is to please," says Allen Sides. "I just do not want to have to do a mix again."[31]

4.3.1 Alternate Balances

Once everyone concerned is happy with the general sound of a given mix, revision requests are usually a question of altering the balance, and easily the most common part people want to tweak with hindsight is the lead-vocal level. "It's a hard balance," explains Spike Stent. "Record companies always want the vocal louder than God, but you need to keep the power in the track as well."[32] Because lead vocal levels are so hard to get right, most mix engineers will print a "vocal-up" mix for safety's sake, but there's nothing to say you can't print multiple versions with different vocal levels as extra insurance if you're generally less confident with this aspect of your mix. Bear in mind as well that you could later edit between the different vocal-balance versions if there were just the odd phrase that needed tweaking, so you get a lot of extra flexibility this way.

Bass instruments are also frequently the subject of postmix prevarication, partly because there are often such large differences between the low-end presentation of different playback systems, so alternate versions are commonplace

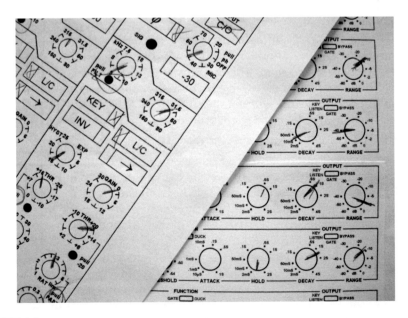

FIGURE 4.8
No matter how carefully you try to save and document your work, mix recall will always be a bit hit and miss—even on computer systems.

here too. "We'll always do a bass version of some sort," says Chuck Ainlay, "whether it just be a 'bass up' or a 'bass and bass drum up' or just 'bass drum up.' When you get to mastering, sometimes there's an element that EQ doesn't necessarily fix, so they might use a different version instead—like that 'bass drum up' version—to add a little bit of punch."[33] Different bass-balance versions are even more relevant in low-budget setups, because acoustic treatment can't ever completely even out typical small-studio room-mode problems. If you carefully restrict your low end to just a couple of tracks in the arrangement, and then do a few alternate mix versions with different balances of these tracks, you should have ample ammo to handle hidden low-frequency gremlins that may surface in the future.

But vocals and bass instruments aren't the only tracks that might warrant alternate mix versions. Any track level about which you're undecided in your final mix can benefit from its own little safety net. "I'll just go through and pick some instruments that somebody might have questioned, 'is that loud enough?'" says Jon Gass, who usually ends up doing roughly a dozen versions of each song as a result.[34] Given that an alternate version will only take a few minutes at most to do, I'd normally err on the side of doing more alternates rather than fewer, but it all comes down in the end to how confident you are of your mix decisions versus how much time you have to spend on speculative ass-saving precautions.

4.3.2 Vocal-out and Solo-out Mixes

There's another type of alternate mix that is frequently requested on solidly practical grounds: a "vocal-out" mix, sometimes also known as a "TV mix" or "public appearance (PA) mix." This is a mix with the lead vocal removed, and it's designed primarily for use in live performance situations where an artist is required to perform a song without the support of a live backing band—either the band is not available or the restrictions of a television/video shoot preclude the use of the band's live sound.

However, the vocal-out mix can also serve a number of other useful functions at the eleventh hour, because it allows you the option of completely replacing the vocals—perhaps the artist's manager has only realized the vocal performance is sub-par now that the mix is otherwise complete, or the band has sacked its singer, or current affairs have rendered the lyrics out of date or tasteless. You can also chop sections of the vocal-out mix into the main mix if you need to delete a stray expletive or even generate a full PG-rated version from scratch. "The instrumental comes in handy sometimes for editing out cuss words and things like that," says Jerry Finn.[35] (That said, it's usually better to create any "clean" version of the lyrics at mixdown time, so you can be sure to catch things such as backing vocals too. It also often makes more musical sense to reverse, process, or replace offending words, rather than simply muting them.)

If there are points in your track where an important instrumental solo takes over the lead role from the vocal part, there's a good argument in such cases for including a mix version without this too. Some singers are also featured instrumentalists as well, so you might want to drop two whole parts for the TV mix so that they can play and sing live in the way they're used to doing—think Alicia Keys or Ed Sheeran, for example.

4.3.3 Instrumentals and *a capellas*

An instrumental mix, completely without vocals, is another sensible fallback, again because it's not uncommon for a singer (or their band/management) to want to rehash their vocal performances and arrangements if they feel that the mixed production isn't quite on the money. But there's another good reason to keep an instrumental backup—because it allows the vibe of the track to be used in more low-key contexts, such as in the background of television and radio shows. This kind of usage may not seem very rock and roll, but it's nonetheless an exploitation of the artist's copyrights, which can generate royalty payments, and extra income is something that draws few complaints from most of the jobbing musicians I know.

> If you find you've completely screwed up your vocal levels, you can sometimes recombine instrumental and *a capella* mixes to give you a "get out of jail free" card.

The inverse of an instrumental version is what is usually called an *a capella*—a solo mix of all the vocals, complete with their mix processing and effects. Indeed, if you find you've completely screwed up your vocal levels, you can sometimes recombine instrumental and *a capella* mixes to give you a "get out of jail free" card. However, the main reason to keep an *a capella* these days is in case anyone wants to nab the vocals for a remix in another style. Again, this might not be what anyone has in mind at the time of the original mixdown, but that's no reason not to plan for it given that a successful club remix could bring in a pile of royalties several years from now.

4.3.4 Mix Stems

Mix engineers have been merrily creating all these kinds of alternate versions for decades, but more recently a further future-proofing tactic has begun to gain currency for music work: mixing to stems. Roughly speaking, this means submixing related sounds in your mix to a limited number of channels and then recording their outputs to separate files alongside your main mix. For a typical band recording, say, you might expect to create stems for drums, percussion, bass, guitars, keyboards, pads, backing vocals, lead vocals, and effects returns. This procedure allows you to reconstitute your final mix by adding them back together. "I have always printed stems, so I can easily do alterations to the mix later on," says Tom Elmhirst. "This happened with [Adele's] 'Rolling In The Deep,' on which Adele added a small vocal section at the end after my master mix and I added some sub to the bass, creating an unnatural and wicked low end in the chorus."[36] Jimmy Douglass also attests that this is standard practice for many professionals, and because artists have now learned to expect this additional level of flexibility, you may seem behind the times if you can't provide it for them.[37] Rich Costey points out that working from stems even at the mixing stage can offer psychological benefits too: "Sometimes when you tweak stems you're able to work on a macro level that allows you to achieve things that are harder to do when you're knee-deep in 100 tracks."[38]

Although the most obvious advantage of mixing to stems is that you get more retrospective tweakability than is afforded by simple alternate mix versions, that's by no means the end of the story. Stems can extend the options available to club remixers, for instance, by giving them more in-depth access to the sounds on the original record. They also give a lot more editing and layering flexibility to computer games developers if they want to weave the elements of your music around onscreen action. For professional productions, however, much of the enthusiasm for mixing to stems at the moment seems to derive from the possibility that the track might need to be cheaply reworked for

MASTER-BUSS PROCESSING

Some styles of music rely heavily on obvious master-buss processing (most commonly compression) to achieve their characteristic sound, and this is an area of mix technique that can be difficult for the less experienced engineer to get right. Also, when processing a complete mix, the unique personality of the specific model of processor tends to make more of a mark on the overall production sonics than if you had used the same processor on just a few tracks. If you're concerned that this element of your mix may be letting you down, then recording versions of your mix with and without any master-buss processing can be worthwhile. That way you can revisit your buss-processor settings later to apply the benefits of hindsight, or maybe rent some esoteric outboard to see if that might elevate the sound beyond what was possible with standard fare. A mastering engineer may also be able to trump your home-brew processing by dint of more high-end gear and golden ears.

You should consider carefully whether any alternate versions of your mix should include your master-buss processing as well. For example, if you leave a compressor in your master buss while exporting instrumental and a capella mixes, you'll find that you can't just recombine them to recreate the final mix, because the two mix passes will result in different gain-change patterns throughout the song. Mix stems will also fail to reconstitute the mix sound correctly if they pass through master-buss processing, for similar reasons. "I'll turn the buss compressor off," confirms Mick Guzauski, "because of course that compressor reacts to everything."[39]

surround sound or that specific playback elements may be required for an artist's live shows. "I'm an extreme stem guy," remarks Young Guru, for example. "After I've laid down a mix, I'll run each individual track via the board and print it back into a new Pro Tools session with effects, EQ, and everything on it . . . When Jay-Z does a live show with a band, and needs certain tracks because the sound can't be reproduced, I can give him stems of these."[40]

CUT TO THE CHASE

- If you're sensible with monitoring and acoustics in the small studio, you should be able to hear everything you need for making a commercial-quality mix. To use this information effectively, however, you need to learn how to convert your own subjective impressions of what you're hearing into more objective mix decisions.
- Part of the problem with listening objectively is that your ears interact in complex ways with what you're hearing, effectively presenting you with a moving target, so you need to understand and work around this. Handy tricks include switching regularly between monitoring systems, taking frequent breaks, choosing and varying your monitoring levels appropriately, and gathering the opinions of third parties wherever possible.
- By far the most powerful technique available for objectifying your decisions is the process of mix referencing. The more time and effort you invest in carefully selecting high-quality reference materials, the clearer your mixing decisions will become and the more you'll push yourself to improve your engineering chops. Instantaneous switching between your mix and your reference tracks is vital if you want to highlight the differences between them most clearly, and the loudness of reference tracks should also be matched.
- Everyone makes mistakes, and professionals respond to this reality by keeping the most important options open for as long as they can. Although full mix recall is now widely available in principle, the time-honored method of printing alternate mix versions is actually more dependable over the long term. Alternate mixes can provide the mix engineer with greater peace of mind regarding critical balance and processing judgments, and they allow the producer enough flexibility to rearrange or reperform critical lead parts should these require further attention post mixdown. Artists can benefit too, both from additional live performance opportunities and from more widespread exploitation of their copyrights.
- A more recent trend is to print submixed stems from which the mix can be recreated. Mixing to stems can provide more postmixdown tweaking capacity, extra royalty streams, and a certain amount of future proofing if there's a likelihood that you'll have to put together a live-show or surround-sound mix in a hurry or with a minimal budget.

Assignment

- Find a simple way to switch between your available monitoring systems and adjust their levels so that it's easy to make this a habit.
- Create your own reference collection, using only audio sources that use no data compression and evaluating the merits of any potential track on all your monitoring systems.

Web Resources

On this book's companion website you'll find a selection of resources to support this chapter, including:

- my most up-to-date personal recommendations for affordable stereo monitor controllers and loudness-matching software;
- links to several different lists of recommended reference recordings;
- further reading about using calibrated reference monitoring levels.

 www.cambridge-mt.com/ms-ch4.htm

PART 2
Mix Preparation

I know what you're thinking. "Enough of the monitoring stuff already; let's get mixing!" That's a fair comment, but for the benefit of anyone who skipped straight over Part 1, let me be absolutely clear: you've missed out the most important bit of the book! Nothing from here on will help you much unless you can actually hear what you're doing.

Even with that bit of tub thumping out of the way, though, it's not quite time to start flexing the faders yet, because one of the key factors in putting together a professional-sounding mix is preparing the ground properly. A major stumbling block for small-studio owners is that they're usually trying to mix the unmixable—their problems derive at least as much from shoddy mix preparation as from any shortcomings in their processing skills. This aspect of production is rarely touched upon in print, so most amateur engineers underestimate its significance. But don't be fooled: proper mix preparation is fundamental to achieving commercial results in any kind of repeatable way.

> A major stumbling block for small-studio owners is that they're usually trying to mix the unmixable—their problems derive as much from shoddy mix preparation as from any shortcomings in their processing skills.

But if mix prep's so critical, why don't members of the industry's top brass talk about it more? The first answer to this perfectly justified question is that the best mixing engineers tend to work with producers and recording engineers who understand how to track, overdub, arrange, and edit their projects into a state where the mix engineer can concentrate on the creative aspects of mixing without distractions. "My goal is always to give something to a mixer where he can push up the faders and it's reasonably close to the architecture that I want in terms of the arrangement," says Glen Ballard, for example. "If you give them that to start with, it frees them up to really focus on detailing the sounds and making everything sound great. I try not to add to their burden the idea that they've got to arrange the song as well."[1] With the demise of many large studio facilities, fewer small-studio users have now had the benefit of the traditional "tea-boy to tape-op" studio apprenticeship and simply don't realize the amount of time and effort expended on professional productions before mixdown.

Because quality-control procedures are by nature usually quite tedious, it's also understandable that high-profile engineers tend to gloss over them, not least because they often have the luxury of farming out heavy editing and timing/pitch-correction donkey work to specialist DAW operators behind the scenes. "The top guys have an unfair advantage, because we all have assistants," admits Jaycen Joshua.[2] And although shrinking recording budgets are increasingly exposing even the biggest names in mix engineering to questionable home-brew production values, it's their assistants who actually bear the brunt of sorting this out, so the boss won't usually dwell on it for long. "The first thing that happens is that Neil Comber, my assistant engineer, will prepare the session for me, cleaning it up and laying it out in the way I like," explains Cenzo Townshend. "All in all, Neil usually has three hours of work to do before I can actually begin the mix."[3] Many other top-tier engineers have also briefly described their own assistants fulfilling broadly similar roles: Michael Brauer,[4] Chris Lord-Alge,[5] Manny Marroquin,[6] Tony Maserati,[7] and Spike Stent,[8] among others.

Because mix preparation is so important, and a matter of course commercially speaking, Part 2 of this book explores it properly within a small-studio context. My aim is to put you on more of a level footing with the professionals when you actually reach the mixdown itself, because that's almost always half the battle where self-produced records are concerned. The more of this kind of technical work you can get out of the way in advance, the more you'll be able to stay in touch with your vital creative instincts during the mixing process itself.

CHAPTER 5
Essential Groundwork

5.1 STARTING AFRESH

For much of this book I'm discussing mixing as if you're working with other people's musical material, and in such situations it's usually easiest to start with a blank slate as far as mix processing and effects are concerned, rather than working from the client's rough-mix setup—even if you have a mixing system identical to theirs. There's more than just convenience to this, though, as it also allows you to bring some valuable fresh perspective to the production, responding to what's actually important without preconceptions. This is why, even if you're actually recording, producing, and mixing everything yourself, I strongly recommend that you still treat the mixdown stage as a separate task. "I deliberately define the point at which I start mixing," explains Fraser T. Smith. "I think it's helpful for my headspace. . . . I'll clear the decks, clean up the session, consolidate all the tracks, and look at the session purely from a mix point of view. . . . It feels like the session is set in stone, and nothing is going to be moved any more. It's a visual/psychological thing."[1] If you've created or recorded the music yourself, then Rob Kirwan also recommends giving your head time to clear before embarking on mixdown: "When I have also recorded the song, I find it very difficult to mix it straight away. I need 10 days or so away from the record to be able to approach it fresh again, and be more objective about things and also not be emotionally attached to things in the mix."[2]

My favored way of clearly defining the start of mixdown is to prepare a set of unprocessed raw audio files, one for each track in your arrangement, much as you'd do if you were having your music mixed by someone else. Make sure to bounce down the output of any live-running MIDI instruments as audio too. This reduces CPU load (allowing you more power for mix plug-ins), discourages endless tinkering with synth settings during mixing (although you can still rebounce an altered version of the part later if you really need to). "I usually print my MIDI to audio pretty early on, because it stresses the computer if you have five or six MIDI synths open," says John Fields. "At some point, I decide: "No more changes, I'm going to print these. But I'll always have the original MIDI track right next to the audio, in case I need to make a change."[3] Bouncing down programmed parts also avoids an insidious problem with some MIDI instruments where they respond slightly differently with every play-through—there are more than enough mind games to deal with at mixdown without this kind of thing going on. "I will print all my MIDI stuff to audio," says Kevin Savigar, "because sometimes there's some delay with the MIDI, and audio locks a little tighter."[4] "I never run anything live, from a sequencer," agrees Tom Lord-Alge. "I don't want to have to worry about synchronization or issues of sound level."[5] Splitting out the individual sounds from virtual drum instruments and multitimbral samplers is a good idea too, because it gives you maximum mixing flexibility.

If there is some effect that is an essential musical feature of the track (perhaps a pitch corrector, a specific guitar-amp simulator, a synth-style resonant filter, or some unique special effect), by all means leave that in when generating your bounced multitrack files, but bear in mind that it can save time later if you bounce out a clean version as well, just in case the effect needs remodeling to work properly in the final mix context.

5.2 ENHANCING NAVIGATION

One way or another, you should end up with a fresh DAW project file containing a bunch of bounced audio tracks that all start at the same point. The first step is to make it as easy as you can to get around the place, and this is something many small-studio users seriously undervalue. In the words of Jaycen Joshua, "Organization is the key to victory!"[6] It might seem petty, but the last thing you want to be doing when you're mixing is waste time working out where the blazes that second tambourine overdub got to. Or resetting the EQ of the wrong guitar part by mistake. Or editing the synth pad out of the wrong chorus. "The more organized you are, the fewer mistakes you make," confirms Leslie Brathwaite.[7] Creative thoughts are fragile little things, so any kind of delay between thinking them and acting on them while mixing is really bad karma. What was I going to do with that second tambourine now? Oh well. Never liked it anyway.

5.2.1 Organizing Your Tracks

One time-honored way of making your navigation of a project instinctive is to standardize the track layout of the instruments in your arrangement so that you instinctively know where to find your kick drum, bass guitar, lead vocal, or whatever when inspiration strikes. Part of this process involves putting the most important instruments where you can easily access them and submixing any large groups of subsidiary parts to fewer mixer channels so that controlling them en masse is easier. Michael Brauer's views are emblematic of most of the biggest names in mix engineering here: "Whatever the session is, it will be a maximum of 44 tracks. . . . I am

If you can stick fairly closely to some kind of generic session layout with each successive mix project, you'll clear headspace for making the really big mix decisions.

not going to mix 200 tracks on the desk for you, so the best thing for you to do is to give me the stereo blends that you like. . . . I'm happy to mix 16 tracks. The simpler the better. I don't want to have to fight my way through a mix."[8]

Many engineers, including both Lord-Alge brothers, also limit their track count in a similar way, with certain sounds always arriving on the same mixer channels and the most important sounds closest to hand. "If everything is parked in the same place," explains Chris, "all you have to worry about is the song. When you're mixing you want to eliminate all the things that make you think outside of the song."[9] Tom elaborates on his own preferences: "I like to have my drums on tracks 6 to 14. The bass guitar will always be on 15 and 16. Channels 17 to 24 contain the main instruments: guitars, keyboards, or whatever they are. The lead vocal will be on 25 and any additional vocals go after that. Any music that's left over will go on 33 and upward. [Percussion] will go to channels 5 and down. . . . The faders for channels 17 to 24 are close to my left and 25 to 32 are close to my right. Tambourines can live on channel 1—I don't want to have to move over there all the time; I don't need that much exercise! Of course, by centering the most important instruments on the desk I can also remain in the ideal listening position between the monitors for most of the time."[10]

Naturally, the "in the box" DAW mixing systems typical of many small studios make physical ergonomics less of an issue, but the general principle of standardized track setups still holds: if you can stick fairly closely to some kind of generic layout with each successive mix project, you'll clear headspace for making the really big mix decisions. Practically every mixing program now has sophisticated organization systems for your tracks, so get busy with them to create a visual connection between all your drum tracks, say, or to provide easy global control over massed rhythm guitars or backing vocals.

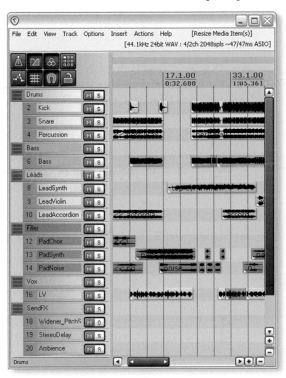

FIGURE 5.1
Laying out your tracks in a logical and consistent manner can seriously speed up your work, as well as freeing up your brain for more creative thinking during the mix.

5.2.2 Colors and Symbols Speak Faster than Words

As part of organizing your tracks for mixing, you should also get into the habit of labeling your tracks sensibly. If you have 15 tracks just labeled "Guitar," you'll forever be tweaking the acoustic instead of the power chords. More mind games; less hair left. So do yourself a favor and try to make the track names mean something. On many DAW platforms, it's also good practice to abbreviate the names as much as you can so that you can still read them when they're truncated to fit across the width of the channel strip in your software mixer.

However, most software users have an even more useful tool for track identification: coloring. Our brains are much quicker at responding to colors than to words (which is why traffic lights use colors, for example), so the more you can color-code your tracks and audio files, the quicker you'll be able to navigate around them. As Robert Orton has noted, "I'll start my mixes by laying out the session in the way I like and color-coding everything, so it's easier to orientate myself and I instinctively know where things are."[11] If all your drum parts are yellow and all your bass tracks red, then you always know which type of audio you're looking at even when you've zoomed right in to edit individual audio waveforms or when your track names are covered by other floating windows. Jochem van der Saag has another useful suggestion: "[I] add colours to remind myself that I need to do something, so I may mark a section in a weird colour so next time I open the window I know I have to fix it."[12]

Graphical symbols are usually quicker to comprehend than text too, and they can be understood even when very small on screen, Therefore, if your DAW gives you the option to add track icons of any type, then these might also be a useful visual aid.

5.2.3 Dividing the Timeline

Mixing typically involves a lot of listening to small sections of your mix, as well as a lot of comparing different song sections against each other, so you can save a lot of aggro by having a clear idea of the musical structure of the production and how it relates to the minutes and seconds of your playback timeline. On dedicated hardware recorders, you'd have had a scrap of paper with track times scribbled next to a list of the song sections, but in the software domain you normally have a much more elegant scheme whereby you can display song sections as bars or markers in the project's main horizontal time ruler. Getting this chore out of the way at the first possible opportunity is especially important if you're working on someone else's production, because it significantly speeds up the process of getting to know the music. Again, naming and coloring these section markers will only help you zip about the place more swiftly.

If there's any work to be done in terms of synchronizing your sequencer's metric grid to that of the music, then that's also well worth sorting out before any proper mixing starts. This kind of thing is usually tedious to do, and trying to tackle it right in the middle of the mix is a recipe for inspiration nosedive.

FIGURE 5.2
Track icons can help you identify tracks much more quickly than text names.

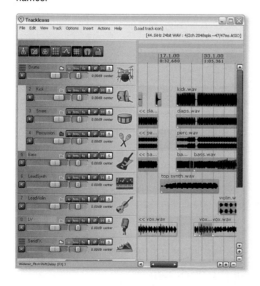

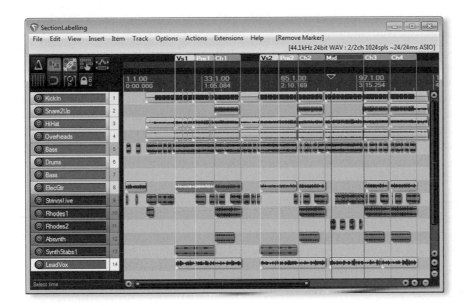

FIGURE 5.3
Use the markers within your own software to name the sections of your song; otherwise it's easy to confuse yourself about which chorus you're actually working on, especially when your view is zoomed a long way in for fine editing or automation work.

5.3 PROJECT RECONNAISSANCE

5.3.1 Spotting Trouble and Hidden Gems

Once you've made sure you can find your way around your mix project as quickly as possible, it's time to start listening through to the tracks individually so you can begin to build a picture of what's there. Even if you recorded the whole production yourself or you're already familiar with a previous rough mix, there's a lot to be gained from listening to at least a snatch of each multitrack file on its own before you properly start work with the mixing. For a start, it's a good opportunity to edit out any silences in your audio files, which not only reduces strain on your computer resources but also typically makes the spread of recorded material across the timeline clearer to see. Plus, you get the opportunity to deal with purely technical issues (such as unwanted background noise, audio glitches, and lumpy audio edits) that might otherwise slip through the net when everything's going at once. "I will always begin a mix with cleaning up the session," says Demacio Castellon, "doing crossfades, making sure there's no headphone bleed or pops. I hate pops! I spend a lot of time making sure everything is smooth and clean."[13]

Another benefit to be gained from checking through the individual tracks before getting down to mixing is that this process often unearths moments of magic that have been all but hidden by competing arrangement layers. Maybe it's a fantastic little string-bend fill from one of your guitarists, a cheeky little ad-lib tucked away at the end of a backing vocal track, or some choice words from the drummer as he spills coffee over himself—basically anything that makes you laugh, smile, boogie, or punch the air. One of the most sure-fire ways to win over most clients is to dramatically unveil forgotten gems like these in your final mix, because they can make the mix sound fresher without you actually having to add anything. In fact, several times I've been congratulated for "putting a new part" into a mix, when all I've done is dredge up one of the client's own buried sounds that had long since faded from their memory. The beauty of this trick when mixing other people's work is that it's low-risk, because anything in the original multitrack files implicitly bears the client's seal of approval, whereas there's nothing like that guarantee for any truly new parts you might add.

Again, do make use of your DAW system's coloring tools to highlight these kinds of special moments, because they're usually scattered thinly across a variety of tracks in most productions and you'll lose track of them all otherwise. There are few activities more soul-destroying mid-mixdown than searching laboriously through 24 tracks of vocals for that one killer vocal ad-lib. (It always seems to be on track 23, and after all that it actually turns out to have the wrong lyrics.)

5.3.2 Multing

The final thing that you can start doing during your project reconnaissance is any multing you think might be necessary. In the context of a DAW mixing system, the word "multing" primarily refers to the procedure of chopping up a single audio file and distributing the pieces across several different tracks, thereby allowing each file section to benefit from its own mix settings. In recent years, most DAWs have also begun allowing you to apply processing directly to audio clips/regions, which provides broadly similar functionality, but without your having to create a blizzard of tracks in the software's arrange

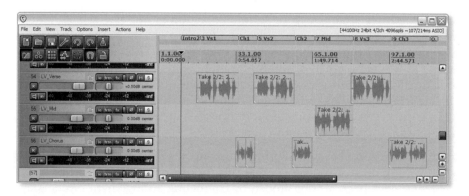

FIGURE 5.4
Multing the lead vocal part, as in this screen grab, is common practice, because it allows you to adapt your mix processing to different sections of the arrangement.

window! Multing is a terrifically powerful ally of the small-studio engineer, especially because modern DAW systems are so powerful you can afford to go on multing till *Pirates of the Caribbean 27* hits the cinemas!

The simplest application is where any instrument plays significantly different parts in different sections of a song. A single acoustic guitar, for example, may well need very different treatments for its verse finger picking and chorus strumming, while a snare-drum close mic might benefit from a change in processing if the drummer switches between regular hits and rimshots during the song, so multing sections to different tracks on both those cases would make sense. Multing vocals is particularly common, given the wide variety of sounds a voice can make (whispering, crooning, rapping, yelling, yodeling . . .). Leslie Brathwaite talks about this in relation to mixing Pharell Williams' "Happy": "The reason to keep the tracks split out is that I'll treat each vocal track slightly differently based on its characteristics . . . Pharell will have sung the hooks a bit more aggressively, because he is singing with the backing vocals behind him. This means that the compression in the verse lead will be a bit different from that on the hook lead."[14]

For similar reasons, backing-vocal parts are a frequent target for multing in my own mixes. A single backing vocal track can easily have several phrases within it that all serve different purposes, especially within dense pop or R&B vocal arrangements: verse "oohs," prechorus ad-libs, chorus harmonies, whispering in the middle section—the list of possibilities goes on. Multing out all these bits and grouping tracks of similar phrases together is often the only way to make sense of things at the mix.

It's also common for important parts such as lead vocals, rhythm loops, or bass instruments to be multed over several tracks so that their processing can be adjusted to meet the demands of a changing backing arrangement—you might, for example, be able to get away with much fuller-sounding vocals during your song's sparse verse texture than during its massive wall-of-guitars chorus. "I use many different EQ and delay settings in many different parts of the song," says Spike Stent, for example. "So when you have one vocal that goes all the way through a track, it may be multitracked three or four times to different EQs and different effects. . . . You will normally find that the verse and chorus vocal sounds are entirely different."[15] Mike Shipley is willing to push this concept to extremes if necessary: "It's just a matter of making the voice sit in the right place, and if it takes ten channels to do it with different EQ for different parts of the song, then that's what it takes."[16]

It's impossible to predict everything you might want to mult until you're actually mixing, so don't be afraid to come back to it at any later stage. However, if you can handle the bulk of obvious mults before the

FIGURE 5.5

A section of a backing-vocal arrangement. Notice how the parts on tracks 2 and 3 are similar only to start with—track 2 then matches tracks 4 and 5, whereas track 3 matches track 1. Multing these parts would make sense here to allow similar parts to be processed together.

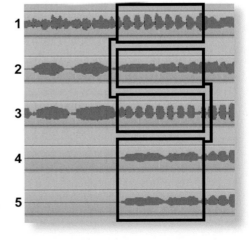

mixdown, it'll not only speed up the process of familiarizing yourself with the recorded material, but will also save unnecessary interruptions of your mixing workflow later on.

5.4 CORRECTION TIME!

If you've followed through all the steps in this chapter, locating the track and time location of anything you hear in the mix should be a no-brainer—which is exactly what you want it to be, because you've got much bigger fish to fry. You should also have a smattering of tracks and audio segments that you've colored to indicate that they need special attention, whether that's corrective processing or just some added emphasis in the mix. The next stage is to attend to any timing and tuning concerns, an activity that warrants close scrutiny in the next chapter.

CUT TO THE CHASE

- Small-studio mix engineers frequently overlook proper mix preparation, leading to a frustrating mixing experience and, in some cases, productions that are effectively unmixable. The only way to get on comparable footing with the professionals is to do your groundwork.
- It's usually best to start a mix with a clean slate, working from raw multi-track audio files in a new DAW project, even if you're engineering and producing everything yourself.
- Speed up your navigation of the mix project as much as you can so that you can concentrate more on critical balancing and processing decisions. Standardize your track layout to make finding the appropriate controls more instinctive, and make as much use of track naming and coloring as you can to avoid confusion. Set up timeline markers to indicate the main mix sections and deal with any synchronization issues as soon as you can.
- Listen through to the individual multitrack files to remove silent regions, fix purely technical problems, and highlight any audio sections that need timing/tuning correction or special emphasis in the mix. While you're at it, mult out any sections of tracks that you suspect might benefit from independent mix settings, but don't be afraid to carry out further multing later on during the mix process if required.

Assignment

- Find a mix that you can work on while progressing through the rest of this book, and create a fresh DAW project file for it. (If you need a selection of raw multitrack material to work with, check out the web resources below.)
- Use colors, icons, markers, and text labels to speed up project navigation and to highlight tracks or audio sections that need special attention.
- Use multing wherever appropriate.

Web Resources

On this book's companion website you'll find a selection of resources to support this chapter, including:

- a special tutorial video to accompany this chapter's assignment, in which I demonstrate how to prepare a simple multitrack session for mixing;
- the "Mixing Secrets" Free Multitrack Download Library, a collection of hundreds of different raw multitrack recordings that you can download for educational use. The projects cover a wide range of musical styles and are all in simple WAV audio format, so can be easily imported into any mainstream DAW system. The library also has a busy Discussion Zone forum where you can compare dozens of different mixes of each multitrack project, and share mix feedback with thousands of other budding mix engineers.
- some audio demonstrations showing the power of multing in practice, along with a number of mix case-studies from the *Sound On Sound* "Mix Rescue" column where multing played an important part in achieving the desired sonic transformation.

www.cambridge-mt.com/ms-ch5.htm

CHAPTER 6

Timing and Tuning Adjustments

Everyone knows that a great live performance has the capability to connect with a listener's emotions, in spite of (or even because of) numerous tuning and timing discrepancies. However, for mainstream productions designed to withstand repeated listening, commercial expectations are now extremely high regarding the tightness of both pitching and groove, therefore much time and energy is routinely expended buffing these aspects of many professional recordings. Everybody's heard songs on the radio where this tinkering feels like it's been taken way too far, but opinions clearly differ widely on what "too far" is. At the end of the day, how much corrective processing you use in your own work remains a personal judgment call. However, I'd urge you to resist any dogmatic ethical stance on the subject, and judge the merits of any such studio jiggery-pokery purely in terms of whether it serves the music. If an ostensibly out-of-tune note encapsulates the emotion of the music, then have the courage to let it lie. If the most emotional vocal take has some distractingly duff pitching, then consider that the general public (and indeed the artist) will probably thank you for tastefully drawing a veil over it.

> If an out-of-tune note encapsulates the emotion of the music, then let it lie. If the most emotional vocal take has some distractingly duff pitching, then the general public (and indeed the artist) will probably thank you for tastefully drawing a veil over it.

Though the audio fidelity of pitch/time-correction processors has improved tremendously during recent years, none of the current crop of algorithms can claim to be completely free from undesirable side effects, and some pundits use that as an argument for rejecting their use out of hand. However, any mix processing can display undesirable side effects, and it's simply the engineer's responsibility to evaluate whether the cure is better than the disease in each specific instance. The slightest pitch-correction artifact on a classical solo piano recording may not be appropriate, whereas even some appalling processing nasties may be a minor concern when tuning one of many wayward background vocals within a busy pop mix.

That said, having heard literally thousands of amateur mixes, it's my firm opinion that most users of small studios don't pay nearly enough attention to timing and tuning touchups, even taking into account the different demands of different styles in this department. This is doubtless partly because these tasks are about as invigorating as watching paint dry, and I can sympathize wholeheartedly there, but it's one of those bullets you've got to bite if you want to compete commercially these days. Those inexperienced with corrective editing also tend to stop short of what's required because they're unable to achieve enough correction without incurring excessively unmusical side effects, and that's the situation I want to try to remedy in this chapter. With a bit of care it's perfectly possible to achieve adequate timing and tuning correction in practically any mix without giving the game away by making things sound unnatural. Certainly, I use a degree of corrective editing on almost every mix I do, but I've never had any complaint about it—in fact, I can count on the fingers of one hand the occasions anyone's even commented on it at all!

6.1 GROOVE AND TIMING

There's a persistent studio myth that tinkering with performance timing will inevitably sap the life out of a production. What this doesn't acknowledge is that there's a big difference between a great natural-sounding groove and a simply sloppy performance. Your drummer can be directly channeling the soul of James Brown, but if the rest of your instruments display the rhythmic panache of Mr. Bean, then you'll still empty the dance floor pretty quickly! True, if you try to turn the drummer into a drum machine, then you'll probably lose more in vibe than you gain in consistency; but if you edit the other parts such that they agree better with the drummer, then the sense of natural rhythmic flow will usually shine through much more strongly.

My top tip for making the best of a production's groove is to work out which single instrument embodies it best, solo it to smooth out any stumbles or areas of inconsistency there if necessary, and then use it as a reference point as you reintroduce the remaining instruments one at a time, tightening the timing of each as you go to taste. In a full mix, it can be all but impossible to work out which combination of instruments is causing some perceived timing unevenness, but if you strip back your arrangement and then progressively rebuild it like this, you can be pretty certain at any stage where to point the finger of blame whenever the groove falters.

For most modern productions, it's usually a no-brainer to start the editing process from the drums, especially if these are MIDI parts or loops, which by their nature flawlessly maintain a consistent rhythmic momentum. But even with live bands, the drummer is typically the player with the best sense of rhythm, so it stands to reason that the drummer's parts will probably offer the best guide. In arrangements without drums, any other rhythm-section part can operate as the main reference point, but it pays in such cases to audition each contender on its own to properly weigh up its appeal before making up your mind. Whatever part you decide to latch on to, though, you should naturally ensure that the quality and consistency of its groove are as good as possible before you start lining anything else up to it. This is the part of the timing-correction process where there's the greatest risk of oversanitizing things, but fortunately there are a few guidelines that help ward off this undesirable outcome.

6.1.1 Timing: A Relative Perception

The most important thing to remember about the perception of timing and groove is that it is entirely relative. You judge the timing of every new rhythmic event based on expectations you've built up from the patterns of events preceding it—not based on how it lines up with any notional metric grid you might have superimposed on those events in your DAW. For example, even if a drummer records his parts along to a click track (and therefore ostensibly in sync with your DAW's metric grid), it's perfectly possible for a drum hit that looks as if it's been played late according to your DAW's bars/beats ruler to actually sound as if it's been played early. This might sound paradoxical, but let me explain.

Let's say you've got a song at 100bpm, and the drummer hits the downbeat of bar 1 perfectly in time with the DAW's grid but then doesn't quite keep up with the click track's tempo—for the sake of example we'll say he sets off at 98bpm. If you're listening to the drummer without the click, the groove of his drumming will sound perfectly fluid as long as he remains true to his 98bpm tempo. However, if toward the end of bar 2 the player notices that he's now roughly

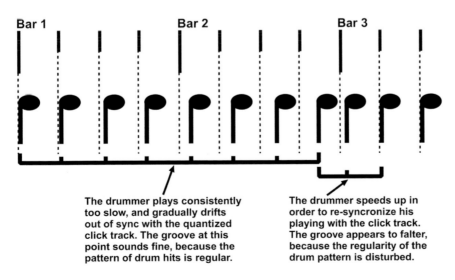

The drummer plays consistently too slow, and gradually drifts out of sync with the quantized click track. The groove at this point sounds fine, because the pattern of drum hits is regular.

The drummer speeds up in order to re-syncronize his playing with the click track. The groove appears to falter, because the regularity of the drum pattern is disturbed.

FIGURE 6.1
The importance of relative timing to groove.

a 32nd note late compared to the click track he's hearing in his headphones, he may begin correcting himself, advancing bar 3's downbeat hit slightly so that it's only a 64th note late compared to the DAW's metric grid. However, although it still looks late in the DAW, what you'll hear if you're listening to the drummer without the click track is that the downbeat hit of bar 3 appears to be rushed compared to the consistent 98bpm groove you've heard so far. (If you're still scratching your head about this, then check out Figure 6.1 for further clarification.)

The clearest lesson to be learned from this is that you can't afford to judge issues of groove just by looking at your sequencer's metric markings; those floppy things on the sides of your head must always claim the final veto. But there are other ramifications. First, you need to understand that you can't properly evaluate the timing of a given bar or beat unless you audition from a point well before it—you'll usually need to allow a run-up of at least two bars for your own internal musical pulse to latch firmly onto the track's groove. Otherwise you won't pick up those small lurching sensations that alert you that there's work to be done. You also have to realize that when you hear a note as being out of time, the best remedy might often be to edit the notes before it. Once you've carried out the edits, you may also have to adjust subsequent notes too, because your earlier edits may have substantially altered the rhythmic expectations of the listener.

It's particularly easy to get yourself into difficulties if your key rhythmic instrument part has been edited together from several different takes or constructed from a few repeated performance sections or loop samples. Even if all the sections in isolation are groovier than a purple velvet suit, you can often find that the first couple of hits after each section boundary sound out of time, because each section has a slightly different relation to the main metric grid—in other words, some are slightly behind the beat and others slightly ahead of the beat. Most people's first instinct in this scenario tends to be to edit the hell out of the apparently out-of-time individual hits, but this rarely yields adequate improvement and usually compromises the overall grooviness into the bargain. A better solution, of course, is to slide the timing of each section as a whole while listening for rhythmic lumpiness across their transitions.

Once you understand and apply these underlying concepts, you should be equipped to get on with timing edits without falling for too many psychological mirages, at which point the quality of your work will largely depend on how much practice you get. The more you do corrective work, the more attuned you become to the subtle stumbling feelings that indicate the presence of an out-of-time note. Fortunately (or not, depending on how you look at it), there is usually no shortage of material to practice on in most low-budget productions, so you should get up to speed fairly quickly. It's worth noting that it can take a little while for you to get yourself sensitized to small timing variations at the start of an editing session, so it's not a bad idea to double-check the first few sections you work on after you've had a chance to warm up.

6.1.2 Tightening the Timing

Once your key instrument is fighting fit and ready to go, you can start to rebuild the rest of the arrangement around it, and I suggest starting this process with those parts that have the most important rhythmic roles. If you've used the drum parts as your initial timing reference, then the bass is often the best part to deal with first, before moving on to any rhythm-section guitars or keyboards. With those parts tightened up a bit, you'll often find that the impetus of the groove is strengthened enough that the remaining tracks don't actually need to be tweaked much at all and may actually help retain more of a natural-sounding vibe.

It's not uncommon in a lot of mainstream productions to layer multiple recordings of the same part in order to create a thicker and more homogenized texture. Rhythm-guitar and backing-vocal parts are most commonly subjected to this treatment. The problem with these kinds of parts from the perspective of timing correction is that once the first layer is edited and in place, it's more difficult to hear timing discrepancies in the second layer. My advice here is to tighten the timing of each layer one at a time, without listening to them together. Once each layer seems to fit the groove in isolation, then you can

WHEN NOTHING GROOVES

For most of this chapter I'm assuming that something in your arrangement has a fairly consistent groove, and for most small-studio arrangements this isn't a bad assumption, given how many tracks these days are constructed from MIDI files or packaged sample loops. However, small-studio band recordings are more than usually prone to the curse of the talentless drummer, and such performances can be difficult to improve with normal corrective edits because the timing is so inconsistent. You can't maintain a groove that isn't there in the first place!

So what to do? Well, probably the least time-consuming tactic is to ditch all but the most passable sections, loop-audition each one to spot-fix as many internal problems as you can, and then cobble together a final take with copy and paste. You'll sacrifice some variety in the player's drum patterns, but at least you'll salvage some semblance of rhythmic momentum. You can sometimes get away with this dodge even if the drummer didn't play to a click.

If there's ostensibly a metric grid around which the drummer flailed hopelessly, then another last-ditch option is to slice the performance into individual hits and quantize them. In a situation where there's no groove to kill, making your drummer sound like a machine can start to seem a lesser evil. However, I'd urge small-studio operators to do their best to avoid this course of action if they value their sanity. First, it'll take you all night; second, a bad drummer's performance will still sound unmusical, even if it's in time; and third, a lot of ropey drummers have a lousily balanced kit, so it'll be a nightmare to mix into the bargain.

combine them all. I find that there are usually then only a handful of minor adjustments still to be made and the whole job's done in half the time.

How much of each part you tighten up and how precise you are about matching it to the groove are also key aesthetic decisions, although I have to say that I prefer to err on the side of more accurate timing where rhythm-section parts are concerned. I've rarely found this to compromise the musicality as long as the original rhythmic reference instrument has a good feel to it. If your reference instrument happens to be a MIDI-programmed part, then extra vigilance may also be called for. "When live elements are being mixed with machine elements, you have to make sure everything is very synced up," recommends Serban Ghenea, "otherwise things like phase incoherence between live and machine tracks can drastically change the sound when you put them together in the mix. . . . It doesn't defeat the purpose [of having live tracks], though. . . . You can still hear the little imperfections that make it human."[1]

Extra precision is also usually a virtue with any double-tracked parts that you intend to place on opposite sides of the stereo field, because the brain is extremely sensitive to timing disparities between similar sounds arriving from different directions—even slight flamming between notes/strums can get distracting. Furthermore, many small-studio engineers who produce otherwise commendably tight backing tracks frequently suffer from a blind spot where the timing of lead vocals is concerned. Because the main singer takes up so much mix real estate in a lot of styles of music, this oversight can really interfere with the way a production swings as a whole, so don't spare the editing scalpel where it's warranted. Don't forget to pay attention to where notes end, either. The tail ends of bass notes in particular can really undermine the momentum of the groove if badly timed, and the same is true for other important rhythmic parts and anything else that is up at the front of the mix.

> Many small-studio engineers who produce otherwise commendably tight backing tracks frequently suffer from a blind spot where the timing of lead vocals is concerned. This oversight can really interfere with the way a production swings as a whole.

While your DAW's metric grid is a pretty unreliable guide if you're trying to refine the groove of your main reference instrument, that instrument's own waveform profile can actually provide a good visual template for editing the other arrangement layers, so you can save some time at the editing stage if you keep the waveform display of the reference instrument close by as you work. I hasten to add, however, that while carrying out preliminary editing "by eye" like this can significantly speed up your work rate (particularly with layered parts), the process of ratifying your editing choices by ear remains utterly indispensable. You should be fully prepared to shift timing away from what looks neat if your ears smell a rat. Er, you know what I mean!

6.2 AUDIO EDITING TECHNIQUES FOR TIMING ADJUSTMENT

To my mind, a significant reason why a lot of small-studio operators stop short of carrying out all the necessary timing adjustments is that their audio-editing chops aren't up to the task. The software isn't really a limitation in this respect, as even extremely affordable DAW applications such as Cockos Reaper now have more than enough editing tools to pull off practically any real-world timing edit. The trick to achieving good results is to know where best to edit, what tools to use, and how to hide any editing anomalies you can't avoid.

On the simplest level, performing a timing edit merely involves cutting out the section of audio containing an offending note and sliding it to an improved position, then subsequently overlapping the audio snippets at each edit point and concealing the transition with a crossfade. A crossfade involves fading down the outgoing audio while simultaneously fading up the incoming audio, which avoids the audible click that can otherwise arise if there happens to be an abrupt step in the signal waveform across the edit point.

6.2.1 Camouflaging Your Edit Points

At the risk of illuminating the blindingly obvious for a moment, the easiest way to keep your edit points from being audible is to place the crossfades in gaps in the audio. Although this may seem a crashingly straightforward observation, what a surprising number of people miss is that small periods of silence are frequently tucked into otherwise seemingly continuous audio, not least those small gaps that typically precede vocal consonants such as "b," "d," "g," "k," "p," and "t." The main editing consideration here is to judge the position and length of each crossfade, but there's rarely any mystery to that in practice.

Obviously noisy sections in an audio signal are equally good spots for edits because the inherently random nature of noise signals tends to obscure any unnatural waveform transitions at the edit point. In these kinds of editing locations you could often use no crossfade at all without any click becoming audible, but I usually put in at least a 2ms crossfade for safety's sake. There are noisy elements in most musical sounds, so there's often a lot of leeway here. For example, vocal parts have breaths and consonants such as "f," "s," "sh," and "ch," acoustic guitar parts have fret slides, and many overdriven electric guitars are so noisy by nature that you can regularly get away with editing them almost anywhere.

When you have to site an edit point in continuous audio, then the best thing is to try to put your crossfades immediately before percussive attack onsets. It might seem like time travel,

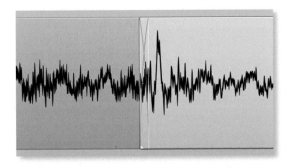

FIGURE 6.2
A premasked edit just before a drum hit.

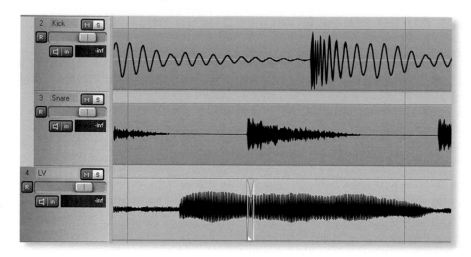

FIGURE 6.3
A masked edit. The snare track can be used to mask even a clumsy edit on the lead vocal track.

but even though the transient only happens after the edit, the later event still effectively conceals (or "masks") the earlier one—a freaky psychological audio effect referred to as backward temporal masking or premasking. Here short crossfades of around 5ms tend to work best. Instruments with a clear percussive attack (acoustic and electric piano, tuned percussion, harp, acoustic guitar) will respond best to these kinds of edits, but even the less well-defined attack characteristics of big-band and orchestral instruments can be used in this way with a reasonable degree of success.

These three basic types of edit (in silence, in noise, and premasked) are already fairly widely used among the small-studio community, and they'll usually handle 80 to 90 percent of the timing edits you need to do. However, you'll not get 100 percent of the way unless you're aware of two other lesser-known editing possibilities. The first is to place an edit point on one instrument such that a powerful transient or noise signal from another instrument masks it (as shown in Figure 6.3), and this hugely expands your options. Take a stereotypical pop/rock recording, for example. Almost anything could be edited underneath snare hits, lower-range instruments could be sliced under the cover of every bass drum, and longer crossfades on higher-range instruments could hide behind cymbals and open hi-hats. When it hits the chorus, the combined masking effect of drums, cymbals, and high-gain electric guitars is likely to be so powerful that you can sneak a 50ms crossfade edit into any background part wherever takes your fancy.

The second power-user edit really comes into its own for time correction of lead vocals, bass parts, and melodic instrument solos, because it gives you the facility to place well-nigh inaudible edits right in the middle of long held notes. The technique relies on the fact that the waveform of any sustained pitch by its very nature repeats regularly. By ensuring that this repeating waveform pattern continues smoothly across the edit point, you can usually get an edit that is somewhere in the range from serviceable to remarkable just by crossfading over a couple of waveform cycles (as shown in Figure 6.4). It has to be said, though, that there is something of an art in selecting the right location for this type of edit. Notes with vibrato can be tricky to deal with, for example, because the pitch variations alter the length of the waveform repetitions. An edit in the middle of a note that is changing tone over time can also stick out like a sore thumb, and this particularly applies to vowel transitions such as diphthongs in vocal parts—out of choice I prefer to put matched-waveform edits in "m" and "n" sounds because their dull closed-mouth tone doesn't vary as widely. Occasionally, though, it's impossible to find a matched-waveform edit point that doesn't sound lumpy, in which case bear in mind that the phasing-like artifacts of a longer crossfade may turn out to be the lesser evil in practice.

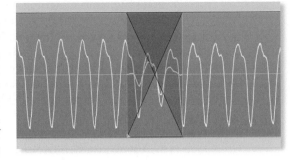

FIGURE 6.4
A matched-waveform edit: essential for exposed edits in sustained pitched notes.

6.2.2 The Role of Time-Stretching

There's one final editing conundrum to be grappled with: whenever you move a note away from its originally recorded position, you create a situation where you have a gap to fill—before or after the edited note, depending on which way you've shifted it along the timeline. If there are already gaps between notes, then one solution is merely to leave a section of silence at the edit point, but this momentary blackout will often sound odd on real-world recordings that have picked up significant levels of background noise, trailing room reverberation, or headphone monitoring spill. In the majority of cases you can just move the edit crossfade somewhere into the body of the note that is too short and then use either a masked edit, a matched-waveform edit, or just a longer crossfade (upward of 100ms) to camouflage the join.

> If you leave a section of silence at the edit point, this momentary blackout will often sound odd on real-world recordings with significant levels of background noise, room reverberation, or headphone monitoring spill.

Sometimes, though, none of these options work out: the arrangement doesn't provide any opportunity for a masked edit, the pitch or tone are too variable for a matched-waveform edit, and the track is too exposed in the mix to justify a longer crossfade's phasey-sounding side effects. This is where dedicated time-stretch processing can save the day by extending the note in question, and the good news is that most DAW systems now have built-in processing that is more than adequate for the purposes of gap-filling fixes.

But if your DAW has got time-stretching built in, why bother with editing at all? The first reason is that even the most state-of-the-art commercially available time-stretching technology I've heard still emits unwanted digital yodeling of one sort or another when it tries to deal with polyphonic audio, transients, or noisy unpitched elements of musical sounds, and as you head further down the food-chain these side effects quickly become worse than the timing problems! However, assuming for the moment that you invest in third-party time-stretching software with acceptably benign processing artifacts, I've found that it's actually slower to work with in a lot of cases, for the following reasons:

- You typically have to isolate polyphonic/noisy sections to avoid excessive processing artifacts, and if any artifacts are unavoidable, then you'll have to fall back on normal audio editing anyway.
- It's often much less easy to line up the waveform in an external editor's display with that of your main rhythmic reference instrument, so you can't readily use a visual guide to your advantage in speeding up the editing process.
- The keyboard shortcut support may be less comprehensive in a third-party application than within your DAW, so there's more laborious mousing around to do.

CHOOSING CROSSFADE TYPES

Some DAW systems give you the option of choosing between different crossfade types, which is useful because this choice can make a big difference. If you select the wrong crossfade type, you'll find that the edited track's overall signal level will increase or decrease unnaturally toward the midpoint of the crossfade. As a general guideline, an equal-power crossfade will usually work best for the majority of edits, whereas an equal-gain crossfade will typically give better results for matched-waveform edits. However, if your particular editing software doesn't label its crossfades in this way, don't lose any sleep over it—just experiment with the different types you have and listen for any level unevenness across the edit point. You'll soon get to know which settings provide the smoothest results.

6.2.3 Finishing Touches

Once you've finished any detailed timing editing for each part of your arrangement, it's an excellent habit to quickly run the song down from the top again before adding in the next part. On the one hand, this provides the opportunity to double-check the remolded audio for unwanted editing artifacts—especially if you listen on headphones, given their greater focus on sonic details. And on the other hand this simple gambit helps combat the loss of perspective anyone will tend to suffer when performing lots of fine edits, making it easier for you to detect subtle medium-term timing drifts that aren't as noticeable when you've got individual notes under the microscope. You may also decide that the overall timing offsets between the tracks would benefit from some adjustment, because the exact relationship between the different tracks will affect the feel of the groove. "An instrument that's playing right on the bar line will feel faster," comments Jack Joseph Puig. "So you find the best place by moving instruments around. I'm talking here about a few milliseconds forward or backward. So within a bar you can have brisk eighth notes or lazy eighth notes, and this is a very important tool when mixing. I always try to find the musicians that are playing more slowly and that are playing more briskly, and I move them around until I get the feel that I want."[2] "I just nudge everything", adds Alex Da Kid. "Nothing ever sits on the 'one' . . . The rhythm of the song is a huge part of what makes me feel good about a song . . . Rhythm is super-important."[3] While you can certainly do this kind of thing by dragging audio regions around, I usually find it's more convenient to use the track's time-offset parameter (or an equivalent utility plug-in, depending on the DAW software you're using), because that way you can experiment with different offsets with no risk of interfering with any edits you've done, and it also makes it easy to reset your changes for comparison purposes.

It has to be said, however, that overall timing decisions like these do also depend to an extent on the balance between the tracks within the environment of the final mix, so while it's worthwhile adjusting overall track alignments like this at the mix-prep stage, you should be prepared to refine these decisions further down the line if you feel the groove isn't flowing well enough. Mike Stavrou has an interesting take on this, from a psychological perspective: "If a player is rushing, you can slow them down by pushing the fader up. Conversely, if a player is slightly behind, you can speed them up by lowering the fader."[4] A similar phenomenon is that the perceived overall tempo of a track can be affected by the levels of the less important beat subdivisions within the groove: fade them up and things drag; fade them down and the tempo often appears to increase.

6.3 TUNING ADJUSTMENT

It is a depressing fact that the sour tuning that blights a lot of low-budget productions is often easily avoidable. An electronic tuner can help a lot here, as long as its internal pitch reference is set correctly by ear; digital MIDI instruments can be completely dependable in terms of pitch, as long as their internal pitch adjustments aren't out of whack; and even a pretty hopeless singer's tuning can be improved if you get the headphone mix right. As far as I can see, though, decent tuning (especially on vocal parts) is increasingly being ignored in small-studio productions, and this puts the onus on the mix engineer to fix things by default, because arriving at a competitive mixdown is otherwise pretty much a pipe dream. "If you don't do it, your records will sound strange," comments Steve Lipson. "People's ears have become used to hearing voices perfectly in tune."[5] The timbre imparted by firm pitch correction has also begun to become a part of the sound of many mainstream chart styles. "Not only do you want it perfectly in tune," comments Ed Boyer, "you also want to hear some of the timbre of pitch correction, because it is part of the sound

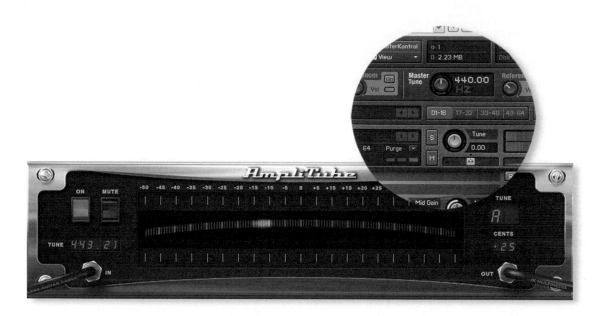

FIGURE 6.5
It's much easier to head off tuning problems while tracking—use a tuner (such as the IK Multimedia Amplitube one shown) for live instruments, and check the Master Tune parameter of any MIDI synths for the best blend.

of modern pop music."[6] It's even happening on rap vocals. "On [Drake's album "Headlines"], I actually tuned the raps," reveals Noah "40" Shebib, for instance. "Drake raps in a very melodic way, which is a conscious decision on his part. I therefore hit it with some Auto-Tune to centre the pitch a little bit."[7]

Now no one is saying that everything has to be absolutely robotic, but it's nonetheless important to comprehend that lax tuning has important mix implications inasmuch as it affects how well your tracks blend. The more the pitching of your arrangement agrees with itself, the easier it is to make everything seem as if it belongs together. If any track's tuning is out of kilter with others in the arrangement, it will stick out uncomfortably and be difficult to balance correctly, even if things aren't far enough off beam to induce lemon-sucking facial expressions. It's for this reason that so many producers talk about tuning drum kits to the key of the track: if the strongest pitched overtones of the drums fit with the track, you can fade the kit up further in the mix before it has trouble blending. Tuning is not just a musical issue; it's an important mixing issue too.

However, although unmixably wayward tuning is the most common ailment in low-budget environments, there's also a significant minority of small-studio practitioners who over-egg the pudding by insensitive use of available pitch-correction tools, and that pratfall is just as capable of hobbling the mix by squeezing all the musical life out of it. "Something perfectly in time, something perfectly in tune, could be perfectly boring," warns Steve Churchyard.[8] Steering a course between these two extremes is the art of tuning correction, and the remainder of this chapter offers advice for maintaining this desirable trajectory.

To begin, there is little excuse these days for inappropriate side effects of pitch correction to be audible in a final mix. Again, the built-in pitch processing within current-generation DAW systems is good enough for a large proportion of applications, and excellent third-party software is comparatively affordable for those who feel their work requires more advanced tools or increased processing fidelity. It's not the tools that are the problem—you've just got to understand how to get the best out of them.

6.3.1 Choosing Your Targets

The first thing you need to know is what pitch-processing algorithms find easy to handle, namely clean sustained monophonic lines. If there's any noise, distortion, breathiness, or rasp to the pitched-note timbre, this will introduce unpredictable waveform elements that can confuse the pitch-tracking code within the algorithm and cause it to misfire in an unmusical way. Note-attack transients or audio crackles also tend to defy the logic of many pitch-adjuster plug-ins, for similar unpredictability reasons. As for polyphonic audio, until recently your tuning options were limited to fairly small global pitch changes in return for moderately nasty artifacts, even when using top-tier correction products; and even now that Celemony's Melodyne has turned the dream of truly polyphonic pitch-adjustment into reality, its processing still exhibits unpalatable side effects in real-world use, especially if you are unwilling to put in a great deal of setup time to ensure the best results.

> If there's any noise, distortion, breathiness, or rasp to the pitched-note timbre, this will introduce unpredictable waveform elements that can confuse the pitch-tracking code, causing it to misfire in an unmusical way.

So in the face of an out-of-tune production, you should think carefully about which tracks to target for correction. For example, even if you only have a couple of out-of-tune guitar overdrive parts, you may well get a better overall sound if you tune the monophonic bass and vocal parts to match them, rather than subjecting these noisy polyphonic sounds to the nasty swirling protestations of overworked digital correction code. Alternatively, if a guitar DI signal is still available, then you might be wise to tune that up instead, subsequently replacing the errant overdriven parts using a digital guitar-amp simulator. By the same token, it's a great deal easier to replace most problematic MIDI synth parts from their source MIDI files, rather than processing bounced audio for pitch reasons. If the keyboard parts were played live, that still doesn't mean you can't replace them with a MIDI-driven part of your own if that's what it takes to reach a commercial-sounding final mix, particularly as many keyboard parts in mainstream productions are often quite simple.

Once you've decided which parts to process, however, you still need to keep your wits about you. Given that pitch correctors don't like unpredictable waveform shapes, it stands to reason that you should do your best to keep noise-like signals out of their way. Mult the guitar's fret-slides to a separate track. Bypass the pitch corrector for the flautist's breaths. Leave the "s" untouched and concentrate your shifting on the "inging in the rain" bit.

6.3.2 The Right Tool for the Job

You can also improve the quality of your correction work by choosing the right tools for the job. On the one hand, you want to achieve the best sound quality, which in most cases translates to finding the processor with the least distasteful side effects. On the other hand, you want to preserve as much of the original musicality as possible, resisting any alluring machine-automated correction facilities where these risk throwing the baby out with the bathwater.

Although real-time pitch-correction plug-ins are now bundled with many DAW systems, and dedicated third-party options are so widely available, there's a lot still to be said for dealing with pitch adjustments using your sequencer's offline editing tools. One definite plus side is that you're probably in less danger of destroying the musicality of a part that way. The main thing that makes a track sound out of tune is if the perceived pitch center of each note is out of line with the context of the track at that moment. As long as the pitch centers feel right, any shorter-term pitch variations mostly just contribute

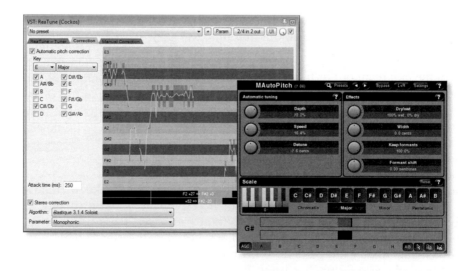

FIGURE 6.6
It's tempting to use real-time pitch correctors like these as a set-and-forget process, but this rarely produces musical or transparent-sounding results.

to a sense of natural phrasing and overall humanity. This is why applying simple pitch offsets to offending notes offline can often sound very transparent.

In addition, offline work encourages you to take more moment-by-moment control over the pitch-correction algorithm itself, adjusting parameter settings to suit each shifted note. For example, although most of the time you want to maintain the length of a pitch-shifted section so that the performance timing doesn't change, it's actually much harder on the digital signal processing (DSP) code, so you may encounter unacceptable audio glitches on occasion. If switching off the time correction (so that pitch and playback speed are always linked, as on a tape-machine or turntable) stops the glitches, you may be able to compensate for the timing change via the time-correction editing techniques mentioned earlier in this chapter and thereby achieve a more successful outcome.

Pitch-correction processors also often apply additional processing to safeguard characteristic instrumental and vocal resonance peaks (called formants), saving your singer from transforming into a chipmunk when you're shifting the pitch upward. For small tuning-correction shifts, though, switching off the formant adjustment won't appreciably affect a sound's character in a lot of cases, and it may again improve the smoothness of the corrected sound.

Different offline pitch shifters do vary a great deal in design, though, and many of their parameters are frankly incomprehensible too, so the main thing is just to use your ears while editing. If you're disappointed with the way the processed sound emerges from its transformation, then take the time to tinker with at least some of the available settings. No pitch processing is perfectly transparent, but that's no reason to let more side effects through the net than absolutely necessary.

6.3.3 Automated and Prescanning Pitch Correctors

There is a whole family of pitch-correction utilities that operate in real time, analyzing and correcting your audio as it happens. Antares were the first to really bring this technology to the masses with their original pioneering Auto-Tune plug-in, but since then many other manufacturers have provided variations on this functionality, and they are now built into many DAW systems. Although these allow you to scrub up a track's tuning in short order using a comparatively small control set, I'd personally advise against falling for the kind of "set and forget" approach they implicitly promote for any important parts in your arrangement. One of the problems is that they tend to process everything passing through them to some extent, so artifact levels on noisy or transient signals can be more of a problem, and expressive nuances such as vibrato are reined in along with pitch-center deviations by the time you've achieved the desired tuning accuracy. However, although it's possible to improve things to a degree with some parameter adjustment, the fundamental problem is that pitch-detection and correction still require musical intelligence, so the only way to make these processors work really well is to manually assist them in achieving a better result.

SPECIALIST SOFTWARE FOR TIMING/TUNING MANIPULATION

In recent years, the power of cutting-edge time/pitch-manipulation software has continued to increase, so if you frequently find yourself with lots of timing and tuning correction to do then such specialist tools may amply repay the investment in terms of time saved. For example, Synchro Arts' Revoice allows you to apply the timing and/or tuning profile of one part to another, which can massively speed up editing multiple backing-vocal layers, while Celemony's most recent versions of Melodyne feature advanced timing-detection features that can greatly reduce the manual editing burden when tightening the groove across an entire multitrack project. That said, the built-in editing functions within most DAWs these days are now so good that the primary reason for choosing more specialized time/pitch-processing software is now convenience, in my view, and even those on the most restricted budgets should be able to get decent-sounding results here as long as they're prepared to put in the manual editing-time required.

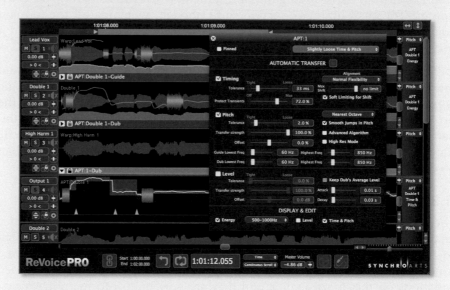

FIGURE 6.7
Syncho Arts's Revoice can apply the timing and/or tuning profile of one part to another.

"With Auto-Tune you really have to work hard to make it do its thing without being heard," explains Steve Bush. "People seem to think it's some kind of quick fix, but it's not. . . . For a start it's not always good at detecting what the correct pitch is, but also even if a whole line is consistently a little out, Auto-Tune will always shift note by note."[9]

The most effective way to improve the musicality of pitch correctors is by taking the guesswork out of the pitch detection, because this is the element of the automatic processing algorithm that most commonly errs. There's usually some way to do this with most plug-ins, whether by using MIDI notes or the DAW's mixer automation system to manually inform the plug-in which pitches are arriving at its inputs. Activating the pitch processing only when needed is a good idea too, as there's no sense in flattening out performance inflections where there's no tuning benefit at all. Beyond these measures, the main thing is to make the effort to match the speed and strength of the plug-in's pitch shifting to the needs of each audio region. To be specific, you should try to keep the correction as slow and gentle as possible to maintain short-term musical nuances, while still shifting the note pitch centers to where they need to be. "I think the way a musician gets to the note is part of their aesthetic," says Matt Wallace. "It's part of their vibe, and sometimes if someone starts flat and sings up to it and then hits that note, the sense of tension and release is really important. I think it's music."[10] Bear in mind that automatic pitch correctors tend to sound best when they don't have to shift anything too far, so you may be able to improve the audio quality by manually dealing with larger shifts offline in the first instance.

Another group of software applications tackle the problem of pitch detection by first scanning the audio in advance in order to analyze the audio data more intelligently offline. This not only leads to fewer pitch-tracking errors (and the facility to correct any that do still occur), but it also allows for more sophisticated user-guided, note-by-note processing. The trailblazer in this field was Celemony's Melodyne, but similar functionality can now be found in competing plug-ins (and even built into some DAW systems), and it's the

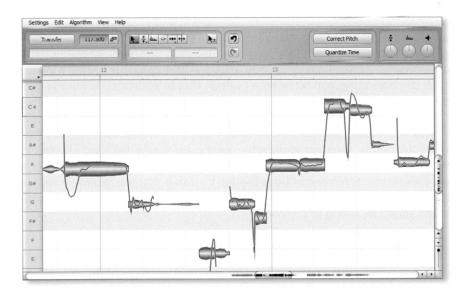

FIGURE 6.8
Pre-scanning pitch-correction software such as Celemony's Melodyne will usually analyze a track's pitch information more intelligently than real-time pitch-correctors.

kind of processing that I would recommend to those who want the fastest route to musical results. In particular, their ability to adjust note pitch-center offsets independently of shorter-term pitch fluctuations is brilliant for delivering sensitive correction only where it's actually beneficial.

Irrespective of the kind of pitch-correction tools you use, the crucial thing to remember is not to trust your eyes more than your ears—just as when editing timing, in fact. "You have to get away from the screen and close your eyes and just use your ears," says Andy Selby. "Watching what happens on the screen really affects how you hear things. You'll be shocked at the emotional and critical difference . . . It is not a drawing exercise. It is about listening. So only go in when your ears hear something that needs changing."[11] Pitch-correction plug-ins have all sorts of meters and graphs telling you what they think they're doing, but it's normal to disagree with them from time to time—for instance, if the underlying pitch-tracking algorithm momentarily slips on a banana skin. "A lot of people do the tuning offline, so to speak, as a standalone operation, as opposed to doing it while listening to the track," complains Joe Barresi. "[I] make the pitch adjustments while the rest of the track is playing. What I'm doing is listening for the pitch cues that the track can give you. It sounds worse if the vocal is in perfect pitch but the track isn't."[12] "The important thing is not whether one track is pitch perfect, but how it sounds in the context of the entire arrangement," confirms Ed Boyer.[13]

> Pitch-correction plug-ins have all sorts of meters and graphs telling you what they think they're doing, but it's normal to disagree with them from time to time if the underlying pitch-tracking algorithm momentarily slips on a banana skin.

Regular breaks during the correction process are very important, suggests Marius de Vries: "It's definitely possible to use the technology to stamp the life out of musical performances. . . . You sit back and listen to it and it's totally flat. In the beginning you often won't admit this to yourself, because you cannot believe that you just wasted a day making something worse, but the fact is there. . . . The only remedy is to keep walking out of the room, and keep walking back in again once you have a clear head."[14] I'd also add that it's worth playing the whole song again from the start once you've finished all your corrections on a given part, listening back with the plug-in window closed (or at least with its visual pitch grid lines hidden), so that you're absolutely sure that things sound right in context over the longer term—whatever any graphical displays might be trying to tell you.

6.4 THE NEED FOR SPEED

Once you get the hang of the general points I've covered in this chapter, decent-quality tuning and timing correction isn't actually that difficult to achieve, even where budgetary restrictions preclude the use of third-party software. However, the reality of small-studio life is that there can be quite a lot of day-to-day audio patchup work. So my last piece of advice is to do everything you can to speed up the process, whether that means investing in dedicated additional software or simply configuring your DAW system's keyboard short-cuts for minimum mouse exercise. As vital as it is to commercial record production, corrective audio editing is never going to be a white-knuckle ride, so at least do your best to make it mercifully brief.

CUT TO THE CHASE

■ Most small-studio productions suffer from insufficient attention to detail where tuning and timing are concerned. Correcting them is tedious, but few budget mixes will compete commercially otherwise.

■ To avoid killing the groove, locate a single instrument (often the drums) as a timing reference, refine its timing if necessary, and then edit sloppily performed parts to agree with that. Rhythm tracks usually benefit from the tightest timing, especially double-tracked parts destined to be panned across the stereo image. Lead vocals can also impact on the groove, so be sure to edit them too if necessary. Don't forget to pay attention to where notes end—especially with bass instruments. To achieve best results with timing edits, listen to at least two bars preceding the section under scrutiny.

■ Simple cut-and-crossfade audio editing is quick and effective for most timing-correction tasks, as long as you learn where to hide your edit points: in silences, in noisy signals, or at premasked and masked locations. Where concealment is impossible, try a matched-waveform edit. Time stretching has useful gap-filling applications, but the bundled processing within most DAW systems is more than adequate for this.

■ Tuning problems are as much a mix problem as a musical problem. The quality of widely available pitch-shifting algorithms is such that there is no need to have unnatural processing side effects on display if you avoid processing noisy, transient-rich, and polyphonic sounds where possible. Focus on lining up the perceived pitch centers of notes, while treating shorter-term musical pitch fluctuations with kid gloves; and adjust the pitch-processing algorithm to suit the audio being processed. If you're using automatic pitch-correction software, provide it with as much human input as you can to retain musicality. Algorithms that work by prescanning the audio usually offer more opportunity for user direction, so they typically give more musical results.

■ If you regularly work with multi-layered parts such as massed backing vocals, then specialist time/pitch-manipulation software may help speed up correction tasks.

■ Although using visual guides can speed up your work, the cardinal rule with any corrective edits is that you should never trust your eyes over your ears. Take frequent breaks while working, and audition the whole part top to tail once you've finished working to gain perspective on the success of your processing decisions. With timing edits, it may prove desirable to apply overall timing offsets to whole tracks at this stage, although these offsets may need refinement later in the mix as the final balance comes together.

Assignment

- Correct any timing or tuning anomalies in your mix project, but take care not to compromise the musicality of the original recordings.
- If you do a lot of tuning/timing-correction work, then invest in specialized software to improve your work rate and sound quality.
- Editing is unavoidably tedious work, so find keyboard commands to speed it up wherever possible.

Web Resources

On this book's companion website you'll find a selection of resources to support this chapter, including:

- a special tutorial video to accompany this chapter's assignment, in which I edit a simple multitrack session's tuning and timing in preparation for mixdown;
- audio demonstrations of the audio-editing and timing/tuning correction methods I've described above;
- links to affordable plug-in tuning-correctors;
- a number of mix case-studies from the *Sound on Sound* "Mix Rescue" column that showcase the practical application of tuning and timing adjustments at mixdown.

 www.cambridge-mt.com/ms-ch6.htm

CHAPTER 7

Comping and Arrangement

Although this isn't a book about recording, there are two specific aspects of the tracking process that I want to discuss briefly in this chapter because they so regularly shoot small-studio mixes in the foot: comping and musical arrangement. Both these factors are important for almost every type of production, regardless of the budget level, yet small-scale projects almost universally undervalue their significance. If your comping and arrangement are good, then mixing is a lot less like hard work; but if they're not, you can be left with an underwhelming record no matter how top-drawer the musical material or your mixdown skills. "I think there's too much dependence on the final mix," says Tony Platt, "and I don't think enough thought and preparation goes into the processes that go on beforehand. . . . The mix should begin the moment you start recording, by moving things around, balancing, blending, and creating sounds. It means you're creating a coherent piece of work rather than recording a whole series of events any old way and then fixing it in the mix."[1] So let's try to head off what tend to be the most common shortcomings.

7.1 COMPING

Comping is very much standard practice in professional productions. It involves recording a given musical part more than once so that the best bits of each take can be edited together to build up the best composite performance. Clearly this is more time-consuming than just capturing a one-take wonder, so it's not something that small-studio operators need do for every part, but any important solo/hook parts will usually benefit from judicious comping. If nothing else, lead vocals should almost always be built up via comping, as I've yet to hear a one-take lead vocal on any amateur mix that couldn't have been improved in this way. Steve Lipson notes, "[Sometimes] you need to take a really close look at things, like when you're comping vocals. Those are situations where you really can't let any mistakes go, and it can really pay off."[2]

Lead vocals should almost always be built up via comping, as I've yet to hear a one-take lead vocal on any amateur mix that couldn't have been improved in this way.

Of course, a comp is only as good as the raw takes from which it's built, and different engineers have different preferences here that range, roughly speaking, between two extremes:

- Recording numerous full "top to tail" takes, and then editing between them after the recording session. The upside of this method is that the musical flow through the part will tend to be more convincing, and you're more likely to maximize the number of unrepeatable little expressive moments you catch for posterity. The downside, however, is that working this way demands much more concentration and endurance from the musician, so in practice you'll only really get superior results with the most talented and experienced practitioners. Less accomplished performers will simply tire themselves out before they nail the more challenging sections and won't be able to maintain enough tonal consistency to allow seamless editing between takes. And if you have to do loads of full-length takes to get all the performance magic you need, then you'll be faced with the mother of all editing jobs to comp them into a final master take. "It takes a long time," said Pascal Gabriel when talking about his work on Dido's "Here with Me." "Out of a two-day writing session we probably spent about three to four hours just comping."[3]
- Recording a single track one phrase at a time, punching in on it repeatedly to create a "patchwork" composite performance. The advantages here are that you concentrate time and effort more efficiently on the sections that need it most; the performer only has to concentrate on perfecting one small section at a time, so it's easier for less experienced musicians; the tone should remain fairly consistent between takes for editing purposes, even as the performer begins to tire; and the comping will effectively be achieved as you go along, so there's no editing to be done later. The disadvantages of this approach, however, are that you lose some of the sense of musical flow between sections of the track, so the final master take can end up feeling slightly disjointed emotionally, and there's also the likelihood that you'll lose nice little performance corners during the real-time patchup process.

Neither of these extremes tend to work particularly well in typical small-studio environments, however, so my recommendation is to chart a course somewhere between them. In most instances, deal with the production one musical section at a time (individual verses, choruses, and so forth) and record a fixed number of section-length takes for each section. If you want to improve any individual take with the odd punch-in, go ahead, but don't get bogged down replacing anything smaller than a whole phrase. I'd say that three takes per section should be considered a minimum if you're going to bother comping at all, and if you're looking for a professional-grade lead-vocal comp, then you'd be well-advised to increase that number to more like eight. Record the takes one after the other until you reach your predetermined limit. By this point your performer will have effectively rehearsed the track several times and will almost certainly be performing better than they were to start with, so now go back through and listen to each take in turn, and rerecord any that don't feel up to the standard of the rest. Do keep an ear out for magic moments within an otherwise lackluster take, though, because you don't want to erase any of those. What I usually find is that a player or singer will only really "inhabit" the musical part properly after a few takes of each section, so I usually end up rerecording at least the first few takes. That said, the freshness of the first take does sometimes harbor a couple of fantastic little inflections, so listen to that particularly carefully before ditching it. "The bits that most people remember," says Pascal Gabriel, "are the little quirky bits you'll find from a take where [Dido] didn't really feel she was confident with the song, but she let rip a bit by accident, and those bits are sometimes worth their weight in gold."[4]

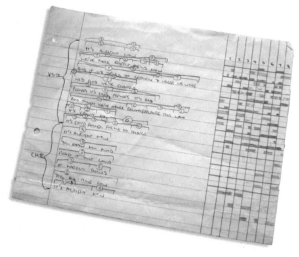

FIGURE 7.1
An example of a real comp sheet for a lead vocal during a song's verse and chorus, and also for a harmony part during the verse. The columns at the right of the page indicate promising sections of the eight vocal takes, whereas final decisions are marked over the lyrics with numbers and brackets.

Once the vocal session has finished, you should be left with a limited number of fairly strong takes from which to choose highlights at your leisure, but at the same time you won't have hours and hours of recordings to toil through. Editing should be straightforward too, because the tone between takes should be quite closely matched. Furthermore, you won't have put performers through the wringer too much, because they've only had to concentrate on a small section at any one time (perhaps over several separate sessions), but the section-length takes will usually lend the comped performance a reasonable sense of musical fluidity.

When it comes to selecting the best bits of your assembled takes, the time-honored route is to draw up a "comp sheet" on lined paper. Each horizontal line of the sheet corresponds to a musical phrase—in the case of an instrument part, you might indicate which is which with bar numbers, whereas in the case of a lead vocal, you'd write the phrase's lyrics on each line. A series of columns down the right-hand side of the sheet can then be used to write comments relating to the quality of each take for any given phrase. What I normally do is listen to each take all the way through, marking any absolutely "must-have" sections, and then try to build as much of the part as possible from those snippets. Only when a phrase doesn't have a standout rendition on one of the takes will I then get into comparing all the individual takes of a given phrase—it might seem more efficient to do phrase-by-phrase comparisons like this straight away, but I find that this tends to encourage more hopping between the takes, so the composite performance doesn't make as much musical sense overall.

VOCAL COMPING: WHAT TO LISTEN FOR

Choosing vocal takes is a subjective pastime, and it can be tricky for those just starting out to get a handle on what to listen for. Mike Clink has some good general-purpose advice: "The number one pointer that I can give is to visualize what you're trying to say. I'm trying to see in my mind if the singer is telling me the same story as what he's written down lyrically. . . . A vocal can't just lay there in a linear manner; it's got to take you up and down, depending on what the lyrics are trying to say . . . If you can look at your lyrics and listen to your vocals and say 'it's telling me a story, and it's taking me on a journey,' then you've accomplished a great performance."[5]

"I'd always look for a vocal performance that had feel over pitch perfection," adds Serban Ghenea. "[Even if it's difficult to correct, it's] worth it because you get to keep the takes that have the best feel, and the best vibe."[6] Ed Boyer agrees: "With digital recording the tuning and timing can be 100 percent, so the focus should be on what kind of vibe and character the recording has."[7] A final perceptive tip for working with rappers comes courtesy of Mark Ronson: "There's the rhythm and the clarity, but there's also a melody in rap. With the really good rappers, even if it's not a song, the melody takes you on a bit of a trip, and I always try to find the little bits in the vocal that have that melody. Sometimes that's almost sing-songy, but not."[8]

Once the comp sheet is completed, the nuts and bolts of the editing shouldn't present many fresh problems if you've already got a grasp of the editing techniques discussed in Chapter 6. However, you should take every precaution to avoid the edits becoming audible or else you'll risk defeating the whole point of doing the comp in the first place—few listeners are likely to find a performance compelling if it sounds like it's been stitched together by Dr. Frankenstein's intern. For this reason, a number of edits may need considerable fine-tuning before they pass muster, and you may have to adjust the level, pitch, or timing of the odd audio slice to get all the pieces of audio to fit together naturally.

Many professional engineers also take this opportunity to clean up undesirable mechanical noises from the vocal/instrumental apparatus: a singer's subtle lip-smack clicks perhaps or over-prominent fret squeaks on an acoustic guitar part. Many small-studio engineers are also tempted to remove breath noises too, but I generally caution against that with lead vocals in particular, as it tends to give the performance a sort of subliminal unnaturalness that undermines its emotional authenticity. If breaths really are too loud, then it'll usually sound more musical if you just rebalance them during the final stages of the mix using fader automation (as discussed in Chapter 19). Background hiss from the recording chain and ambient noise from the recording environment may also be removed at this stage, although again you need to beware unintended consequences. First, such noise may actually be an aesthetically positive attribute in terms of giving the production as a whole more of a live and airy feel, so chopping it out between phrases might actually reduce the appeal of the overall production sound. And, second, the sudden appearance and disappearance of background noise can make otherwise acceptable editing moves distractingly audible, so you need to be extra careful where you place those transitions. Mike Shipley mentioned this issue when discussing his work with Alison Krauss "Alison sings very quietly, and the [Sony] C800 can be a noisy microphone, and quite often there would be noises in the spaces in between when she was singing . . . We spent a long time making sure the cuts weren't too abrupt and sounded musical, for example by cutting on the beat of a bar."[9] In fact, in very exposed arrangements, the most transparent-sounding editing solution may be to copy sections of "clean" background noise to cover over other unwanted between-phase noises, a tactic that's common practice in voiceover work.

7.2 BREATHING LIFE INTO YOUR ARRANGEMENT

If you think of the basic musical material in a production as consisting of lyrics, melodies, and chords, then the arrangement of that musical material constitutes pretty much all the other aspects of the track—things such as the specific instrumentation at different points, the exact distribution of the harmony notes across the pitch range, the particular performance figurations used by each different musician, and the precise rhythmic configuration of the music's groove. If the arrangement works, then it'll play into the hands of the mix engineer, whereas if it fights the music, then there's not much that can counteract that musical ineffectiveness at mixdown. "When you understand arrangement, mixing is easy," says Humberto Gatica. "You know what you're looking for. Otherwise you're fishing."[10] Andy Johns is one of the many other producers who share his opinion: "The way I really learned about music is through mixing, because if the bass part is wrong, how can you hold up the bottom end? So you learn how to make the bass player play the right part so you can actually mix."[11]

7.2.1 Clear Out the Clutter

The first common problem that I hear in small-studio productions is that there are simply too many parts going on all the time. "The fewer tracks you have," says Tony Hoffer, "the fuller [the sound] . . . You can make your bass bigger, the guitars have more room, everything just has more room."[12] Jeff Bhasker agrees: "Keeping it minimal actually makes things sound bigger. Ironically, the more you add, the smaller things sound, because there's no room for anything. If there's a track you can't hear, I'd rather mute it, rather than keep it in there as some kind of background texture or whatever. The biggest thing in the world is the 808 kick drum in [Kanye West's] 'Love Lockdown,' because it's just one thing. When you play that in an arena, it's huge!"[13]

Not only is it tricky to make lots of competing parts clearly audible in the mix, but it also presents a musical difficulty—although a full-textured production

CLASHING TIMBRES

A common problem in less accomplished arrangements is when two sounds clash in some way, usually by operating in the same register. "The aim is to get all the sounds working together so you don't get any nasty surprises at the mixing stage," explains Steve Power, sharpening his scythe. "If two things sound crap together, you probably shouldn't be trying to [fix that] at the mix. If they sound crap, just don't put them together in the first place, because you probably won't rescue them."[17] Eliminating a part completely may not be an option, however, in which case this Fleetwood Mac anecdote may provide some inspiration. "When we were recording *Rumours*," recalls Ken Caillat, "[keyboardist Christine McVie] would ask, 'How does everything sound, Ken?' . . . Sometimes I'd say 'You know Chris, I'm having trouble hearing the keyboard and the guitar.' The first time I said that I didn't really know what I meant, but she said 'Oh, yeah, you're right Ken, we're playing in the same register. Why don't I invert the keyboard down a third and get out of [the guitarist]'s way?' Which is what she did, and it worked brilliantly!"[18]

> Although a full production may appear satisfying at the outset, a lack of light and shade in the arrangement will quickly make it seem bland, and the listener will stop paying attention.

may appear satisfying at the outset, a lack of light and shade in the arrangement will quickly make it seem bland, and the listener will stop paying attention. "I hate records that start in one way and don't change," says Steve Fitzmaurice, "where you can hear the entire record in the first bar."[14]

The first thing that can help is to bear in mind this simple piece of advice from Jack Joseph Puig (echoing an almost identical comment from Wyclef Jean[15]): "You have to consider the fact that the ear can process only three things at once. When you get to the fourth thing, the attention drops away somewhere."[16] So listen to your track and try to isolate which three elements should be the focus of attention at every point in the music. Any part that isn't one of those main components is a candidate for pruning, so see whether you can chop out some sections of it. Let's say that there's an acoustic guitar part strumming away throughout your track; perhaps you should consider taking it out of your verses so that more important drum, piano, and vocal parts can come through more clearly. Save it for adding some extra textural interest during the choruses.

Or perhaps you might edit that guitar part so that it only plays at those moments when one of the other parts doesn't really demand the listener's full attention—perhaps while the singer's taking a breath or the pianist is indulging in some uninspired vamping. This effectively substitutes the guitar for one of your three main parts just for a moment, without robbing attention from them during the bits where they're more important. Most parts in musical arrangements are only playing something interesting some of the time, so if you try to cut out as much of the less interesting stuff as possible, then you will generate some gaps in the texture for other subordinate parts to have their place in the sun. Just this simple arrangement trick can completely transform the sound of a production, because it focuses the listener's ear on what's interesting in the music. There will of course be a limit to how ruthlessly you can perforate the arrangement before the musical flow begins to suffer, but most small-studio producers don't get anywhere near this point before they throw in the towel. As Hugh Padgham observes, "It often takes a lot of effort to have less rather than more. I actually spend more time pruning stuff down than adding things. Doing so can often require a musician to learn or evolve an altogether different part to be played, so what was two tracks is now one track. Every song is different, but I'm always looking for ways to simplify and reduce."[19]

Another thing to consider is whether you might be able to differentiate the sections of your arrangement more effectively by your choice of parts. For example, the first four sections of a common song structure might be intro, verse 1, bridge 1, and chorus 1 (as shown in Figure 7.2), but it's going to be difficult to create any musical development if all of those sections are backed by the same drums, bass guitar, strummed acoustic guitar, and rippling piano figuration. Why not start with acoustic guitar, bass guitar, and drums, but then drop out the acoustic when the lead vocal enters in the chorus? That'll make the vocal appear more interesting and give you the opportunity to differentiate the bridge section by adding the guitar back in again. Then you can leave the piano entry to provide a lift into the chorus. It's ridiculously easy to experiment with your DAW's mute buttons to see what might work here, so there's no excuse for letting the grass grow under your feet. In a similar vein, let's assume that verse 2, bridge 2, and chorus 2 are next in line. It's daft to just repeat the same arrangement, when you can intensify what went before. So maybe let the piano continue from the chorus 1 into verse 2, but drop out the acoustic guitar. The guitar could then join the piano as before to create a thicker bridge 2 texture, and then you might double-track the acoustic guitar or add other parts (such as backing vocals or tambourine) to give another step up into chorus 2.

To put all that in general terms, in a lot of cases in commercial music you want to have enough repetition in the arrangement that the music is easily comprehensible to the general public. But you also want to continually demand renewed attention by varying the arrangement slightly in each section, as well as giving some sense of an emotional buildup through the entire production. That may sound like a bit of a tall order, but it might involve no more effort than poking the odd mute button. "I'll regularly mute the keyboards, or the bass tracks, or the drums, and see how things sound," says Jeff

FIGURE 7.2
A static arrangement can undermine the structure of your song, so try to prune your parts to emphasize each section with some kind of arrangement change. An example of this would be to mute the lighter-shaded parts in the hypothetical arrangement shown here.

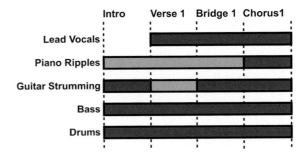

THE DROP CHORUS

FIGURE 7.3
Some albums contain excellent examples of drop choruses, such as Rachel Platten's *Wildfire*, Dido's *No Angel*, and Anastacia's *Not That Kind*.

It's not uncommon to have a double chorus at the end of a typical commercial song's structure in order to really hammer home the main hooks. This is a sound idea in principle, but it does present one arrangement difficulty: if you give the first of these final choruses a really full arrangement to let everyone know it's arrived, then it becomes difficult to sustain that excitement through into the second of the final choruses, because there's no room left in the mix for additional parts. A lot of productions address this problem by using more impassioned vocal parts and ad libs, but I've always felt that there's only so far you can go with that approach, and that there's more mileage to be had in using more imaginative arrangement alternatives instead.

One trick that I just can't resist mentioning is what is often known as a "drop chorus." This is where you wrong-foot the listener at the onset of your song's final choruses by muting loads of the important parts, typically leaving just the lead vocals and some kind of stripped-back rhythm section. In some cases you even get a misleading buildup at the end of the section beforehand, and if you judge that just right, the listener effectively gets two chorus pay-offs at the end of the song, instead of only one. The first of the final choruses uses the element of surprise to make its impact (as well as focusing attention clearly on the lead vocal, which is rarely a bad thing!), whereas the subsequent chorus delivers that all-guns-blazing fullness that everyone was expecting first time around. If you've not consciously noticed any song using a drop chorus before, I encourage you to keep your ears open for it, because it can be extremely effective— Rachel Platten's "Fight Song," Jordin Sparks' "Battlefield," Dido's "Hunter," Fun's "We Are Young,"and Anastacia's "I'm Outta Love" are just some of the hit songs that have made dramatic use of the concept, while many more (such as Sam Smith's "Stay With Me," One Direction's "Drag Me Down," Twenty One Pilots' "Heathens," and Little Mix's "Wings," "Black Magic," and "Shout Out To My Ex") use drop choruses without the "fake" preceding buildup. What's particularly great about a drop chorus from a mix engineer's perspective is that you can often create one entirely at the mixdown stage if you want, without doing any further tracking at all— all you'll usually need is a bit of muting and rebalancing.

Bhasker. "That's how I arrived at having only the drums and vocals after the bridge [of Fun's "We Are Young"], which for me is one of the most magical parts of the record."[20]

Marius de Vries offers another interesting angle: adjusting the musical timeline. "What's often most boring about contemporary songwriting," he points out, "is the way that everything is in bars of four and blocks of eight, which is arguably driven by the way the software is designed. Good creative songwriters will always be searching for ways to jettison a beat in a bar or add another couple of beats before releasing the chorus, for example."[21] Even if you're mixing other people's material, it's surprising how much you can still experiment with these inds of changes using simple editing and time/pitch-manipulation tools.

7.2.2 Adding Detail

The other main shortcoming of ineffective small-studio arrangements is that they don't have enough useful detail in them to sustain interest. Again, there's usually too much repetition occurring. "Why some people's songs kinda just don't go anywhere, is because they just follow that same pattern, so you kinda go brain-dead," says Harmony Samuels.[22] A useful little rule of thumb that can help here is this: if you want something to retain the listener's attention, then avoid playing the same thing more than three times in a row. By the time a given riff reaches its fourth iteration, the listener will start to drift, so if you want to keep people focused, then you need to introduce some variation, usually in the form of some kind of fill. So if your drum part is a one-bar loop and your song works in eight-bar phrases, then it makes sense to perhaps edit the loop a bit during bars 4 and 8 to create fills, providing some variation to maintain awareness of the loop among your audience. How exactly you do the fill is up to you, but remember (as I've already mentioned) that the best location for a fill on any given part is usually where other parts are less interesting. If you feel you want to layer more than one fill simultaneously, then consider putting them in different pitch registers—so combine a bass fill with a snare fill, for example.

If that little "three in a row" test reveals some of your parts to be wanting and a bit of audio editing can't generate anything significantly less repetitive, then it's a signal that you probably want to get other musicians involved—a good performer will intuitively create fills and variations at appropriate points in the arrangement. Even if you're stuck on a Hebridean island with nothing but seagulls for company, that needn't rain on your parade, because you can easily order made-to-measure live overdubs by talented session musicians over the Internet these days—just type "online session musicians" into Google and you'll find plenty of services that can record a wide selection of live instruments over your backing tracks for surprisingly little money.

For seriously chart-oriented music, then it's vitally important to command attention continuously if it's to cater for the attention spans of young, multimedia-bombarded music fans. "Every four to eight bars something has to happen!" comments Rodney Jerkins. "It's very important, because I don't want to bore people . . . I think that's what production should be—it should be about keeping your attention."[23] So every few seconds you might feature a particularly cool backing-vocal lick, a nifty drum fill, a synth hook, a little spot-effect on the lead vocal, or some kind of wacky SFX sample—the possibilities are endless. One other little tip is to consider the bass part of your track not just as a part of the harmony, but more as a second melody line. Brian Malouf says, "Generally, I try to have the bass and vocals be the counterpoint to each other. I look at those two elements as the central melodic components."[24] The more singable you make the bass, the more forward momentum it'll tend to add to the production. The classic example I always think of here is Abba's "Money Money Money," but if you listen to a lot of good arrangements, you'll frequently hear little melodic fragments cropping up all over their bass parts.

> If you're producing seriously chart-oriented music, then it's vitally important to command attention continuously if it's to cater for the attention spans of young, multimedia-bombarded music fans.

CUT TO THE CHASE

- Whatever style of music you're working in, if any of your most important parts are recordings of live performers, you should consider recording several takes and then comping together the best bits. Lead vocals in particular are routinely comped from many different takes in a commercial context. How you go about recording and compiling the takes can have an enormous impact on the quality of the outcome, so don't be afraid to adapt your techniques to suit the particular part and performer you're working with.

- Reducing clutter in an arrangement not only makes it easier to mix, but also helps the production appear more varied to the listener. One general purpose approach is to think in terms of restricting yourself to a maximum of three main points of interest at any one time and then to weed out as many parts as you can that don't make up this main trio—but do bear in mind that the relative appeal of individual instruments can vary from moment to moment, so the identity of the three primary parts may change frequently. Also try to alter the arrangement to give each new musical section its own sound—this will better maintain the listener's interest and support the buildup of the production as a whole.

- Boring arrangements usually suffer from too much repetition, so consider adding some kind of fill if any part plays the same thing more than three times in a row. If you can aim a fill in one part to grab attention from a less interesting moment in another part, then that's a bonus! If you're working in chart-oriented styles, then try to provide some new musical or arrangement diversion every few seconds to keep the listener's attention. Treating the bass line as a second melody is surprisingly effective in improving musical momentum.

- If your song ends with a double chorus, but the second of the choruses seems like it's treading water, experiment with your mute buttons to see whether a drop chorus might bail you out.

Assignment

- If you haven't already comped the most important live recorded part in your chosen mix, then do it now. Compare the comped version to your original one-take wonder to get a sense of why comping is such standard practice on professional productions, especially for lead vocals. Bearing in mind the time-demands of the comping process, decide whether any further parts might warrant similar treatment.
- Go through your arrangement and try to identify which are the three most interesting things happening at any given time; then see how far you can thin out the remaining parts behind them to declutter the sound. Check for sections of your track that share similar arrangements. If you find any such sections, ask yourself whether you might be able to support the music better by altering one of them, either to provide variety for the listener or to enhance the overall buildup of the production.
- Hunt down any musical parts that play more than three times in a row, and see whether a well-placed fill might make them more engaging. Look out for any opportunities to give the bass line some melodic function, and decide whether that helps drive the music along better.

Web Resources

On this book's companion website you'll find a selection of resources to support this chapter, including:

- a special tutorial video to accompany this chapter's assignment, in which I develop a simple multitrack session's arrangement as discussed in this chapter;
- some audio demonstrations showing how careful editing can improve the sound of a mix by reducing background clutter;
- a number of mix case-studies from the *Sound on Sound* "Mix Rescue" column in which I put many of this chapter's editing and arrangement principles into practice.

 www.cambridge-mt.com/ms-ch7.htm

PART 3
Balance

Your mix project is neatly laid out. Timing and tuning gremlins have been banished. The arrangement has been honed. All your faders are pulled down, and you're ready to mix.

NOW WHAT?

That unassuming little question is essentially the crux of what mixing technique is all about, and answering it with conviction at every stage of the process is what foxes a lot of small-studio operators struggling to get commercial-level results. The reason so many low-budget producers can't confidently answer the persistent "now what?" is that they don't have an overarching game plan to provide solid rationale for their processing decisions.

Observing seasoned mix engineers at work is very deceptive in this regard, because they'll often seemingly skip at random between different mixing tasks. In reality they have developed such an intuitive grasp of their own mixing workflow that they are able to respond freely to mix issues as they arise without the risk of losing the underlying plot. As such, I don't think it's very productive for anyone to learn to mix in this

Observing seasoned mix engineers at work is very deceptive, because they'll often seemingly skip at random between different tasks. In reality they have developed such an intuitive grasp of mixing workflow that they can respond freely to mix issues as they arise without losing the underlying plot.

way at the outset, because it's not the apparent randomness that's important; it's the underlying plot!

Therefore, Part 3 of this book unravels the nonlinear real-world mixing procedure into more of a bite-sized, step-by-step process so that's is easier to digest and understand. This inevitably means that I'll be structuring the mix workflow more rigidly than an experienced practitioner would, but by doing this I hope to clarify the thought processes that lie behind even the most intimidating and complex mix techniques. Only once you're thinking logically about the mix process can you be sure you're making decisions for solid reasons— at which point you'll be ready to progress to more a freestyle professional approach with confidence.

Consider the analogy of learning to drive. You don't go straight out on the motorway for your first driving lesson, even though motorway driving is a big part of what most drivers normally do. You first have to learn a variety of potted techniques in isolation: operating the indicators, controlling the clutch, manipulating the steering wheel, swearing at couriers, and so on. It's only once you have the experience to juggle all these tasks subconsciously that you can cruise the fast lane at 80mph while listening to the football commentary and furtively dunking a chicken nugget.

The first task when mixing is finding a balance—at once the simplest concept to express and the most challenging to implement. If you can get every important sound clearly audible and at an appropriate level throughout the duration of the mix, then the battle's mostly won. As John Leckie has noted, "Very often, rough mixes are used on an album. The idea of a rough mix is simply to hear what you've got, so you make your balance so that you can hear everything. And, really, that's all you want out of a finished record—to hear everything. The worst thing is things getting obscured. People will turn the music off when they can't hear everything."[1]

CHAPTER 8
Building the Raw Balance

8.1 ORDER, ORDER!

The order in which you mix the different instruments and sections within your production can have huge consequences for the final sonics, so the first way you can make sense of the mixing process is by working out where to start.

8.1.1 Start with the Most Important Section

Almost every piece of music worth the name has sections that can be considered emotional high points, and a series of drop-downs and buildups connecting them. Your job at the mix is to support this natural ebb and flow (sometimes called the "long-term mix dynamics"), while reserving the greatest musical intensity for the production's most climactic section. This is something that amateur engineers have great difficulty with, the most common complaint being that the verses sound bigger than the choruses. Fortunately, the solution to this problem is disarmingly simple: work on your song sections in order of importance. "I'll start with the louder or bigger section [with] the most instruments . . . the chorus typically, or maybe the bridge" says Neal Avron, for instance. "And then the rest of it is kind of 'unbuilding' for the verses and the intros."[1]

For instance, let's say you're mixing a song with this simple pop structure: verse 1, chorus 1, verse 2, chorus 2, midsection, chorus 3, chorus 4. If you want chorus 4 to be the biggest-sounding section (not an enormously controversial choice!), then mix that first. Throw everything including the kitchen sink at it so that it sounds as breathtaking as you can possibly manage. When you've reached the limits of your mixing capabilities, wipe the sweat from your brow and then see how you can adapt the same mix settings to chorus 3 so that it's just a little less powerful—it should still sound great, but it should also leave room for chorus 4 to sound better. Go back to chorus 2 and chorus 1 with a similar aim. You still want them to sound like choruses, but at the same time you want their impact to be slightly less than any of the following chorus sections.

Once you've defined the subjective importance of each of these choruses, you can set about joining up the dots. Let's say that the midsection of your song

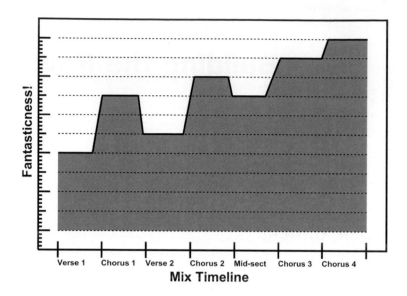

FIGURE 8.1
One possible example of sensible mix dynamics for a typical short song structure. The choruses are more intense than the verses, and each chorus is more intense than the previous one.

is designed to pretty much sustain the energy level of chorus 2; mix that next, perhaps with fairly similar settings as used for the choruses, and then listen how the subjective energy levels change across the section boundaries. If chorus 3 feels like a letdown, then you've pushed the midsection too far. Rein it in a bit. Remember, you can't increase the intensity of chorus 3, otherwise chorus 4 won't sound like a proper climax; and chorus 4 can't get any bigger because you've already maxed out your mixing mojo there. Besides, it's a lot easier to make things sound smaller than bigger anyway. Clearly you'll still want the midsection to sound as good as possible within these limits, but not so good that it undermines the excitement of those final choruses.

Filling in the rest of the blanks in this song structure works just the same: make each section sound as good as you can, but without compromising the musical importance of those sections you've already mixed. Sometimes this means that you end up mixing the beginning of the song last, which stands to reason if everything else is building from there. However, many famous pop songs actually kick off with a modified chorus of some sort, which may not actually be the lowest point in the overall mix dynamics, so mixing your sections in order of importance doesn't mean mixing sections in reverse. It's your responsibility as a mix engineer to tap into the emotional flow of the music in order to decide on the mixing order that will best serve that goal.

Beyond ensuring that a production's climaxes are actually climactic, the hierarchical mixing approach can also highlight very clearly when it's time to use the most underrated of all mix tools: the mute button. If the arrival of our hypothetical chorus 3 is sounding a bit limp, then a good way to give it more of a boost might be to drop the midsection further by removing parts. Maybe the stereo shakers could bite the dust, or some of the guitar layers, and that sax solo was way too "Careless Whisper" anyway. (No-one deserves a repeat of that.) In more extreme cases, you could even cut out some of the core rhythm section, such as the bass or drum parts. Another common target for cuts is verse 1 if its arrangement is currently identical to that of verse 2—without differentiating the two verses in some way it can be difficult to generate any sense of musical momentum through the first part of the song.

At a professional level, there is now an expectation that mix engineers will have this kind of creative input, not only cutting parts but adding new ones too if that's what it takes to achieve results. "Mixers make production decisions all the time," comments Jimmy Douglass, "and I add music to people's records when it's not finished or done correctly. In my work as a mixer, I'm a finisher. That's what I do."[2] Jaycen Joshua confirms this approach: "Once you get into the upper echelons of mix engineers, you are being paid for your taste. You're not hired just to make the rough mix sound better. You are the last line of defense creatively, and you're expected to improve the record. It is your job to change things."[3]

One word of warning, though, when working this way: no matter how much you pare down the introduction of your production to allow room for buildup, it's still important from a commercial perspective that it engages the listener straight away. As Rich Costey explains, "If the intro isn't great, it doesn't matter what happens to the rest of the song. That's something I learned a long time ago: if the song is going to work, the intro has to grab you immediately. The importance of this cannot be overstated."[4] "If you can get someone liking the intro," adds Dave Pensado, "you're halfway home to getting them to like the chorus, and the chorus is where the money lives . . . The intro is like the silverware at a restaurant. If the silverware's dirty, you're not going to eat the meal—and the meal is the chorus."[5]

FIGURE 8.2
A seriously underrated mix tool: the mute button.

8.1.2 Start with the Most Important Instrument

There are some engineers (such as Bob Clearmountain,[6] Chris Lord-Alge,[7] Jack Joseph Puig,[8] and Cenzo Townshend[9]) who like to throw up all the faders at the same time and work on everything simultaneously from the outset. Although there's no arguing with their results, my view is that it's only really possible to pull off this approach successfully if you've got the kind of mixing confidence born of decades of experience. Those starting out are better served, I think, by following the lead of other equally high-profile names (Mike Clink,[10] Jack Douglas,[11] Andy Johns,[12] Tom Lord-Alge,[13] Alan Parsons,[14] and Tony Visconti,[15] to name but a handful) who prefer to build the mix up in stages, adding instruments progressively until the complete mix is happening.

Within this context, the idea of working in order of importance can once again pay dividends, following the lead of Joe Chiccarelli: "It's a matter of finding out what the center of the song is or what makes it tick. Sometimes you build it around the rhythm, sometimes you build it around the vocal."[16] "I'll start by working on the most important instruments," agrees Philippe Zdar. "If the bass drives everything, I'll start with the bass. If it's a riff, I'll work on the riff."[17] The thing to emphasize here is that the sonic importance that you want to be evaluating as much as the musical importance. A common case in point here is lead vocal parts, which despite their central role in delivering the main melody and lyrics, often aren't as important to the sonic signature of the track as the rhythm section parts. The Sugababes track "Red Dress" is one illustration of this idea—although vocal melody and lyrics are the lifeblood of any pop tune, the most important sonic elements in this track are actually the bass and the drums.

Building up your mix one instrument at a time, in order of importance, means that you avoid the all-too-common problem of running out of space for everything. "I always think of mixing as an illusion, and you only have so much space to work with," warns Brian Virtue.[18] If a big vocal sound is important to you, then it's a waste of time to spend hours packing your mix full of Godzilla-grade rhythm guitars before you add that in. When you do fade up the vocal, it'll be fighting for its life against the wall of guitars, and you'll have to trash a lot of the guitar processing you've just slaved over. If every part you fade up is less important sonically than the one before, it's much clearer how much space it can safely occupy in the mix, and therefore how full sounding you can afford to make it before it tramples over more important mix elements.

FIGURE 8.3
Two albums that give lead-vocal sonics extremely high priority, often to the detriment of other backing sounds, are Dido's *Life for Rent* and Rihanna's *Good Girl Gone Bad*.

A couple of high-profile commercial productions that provide particularly extreme examples of this principle in action are Dido's "White Flag" and Rihanna's "Umbrella," both of which give over the majority of their mix space to the drums and vocals while savagely carving away some other background parts. In "White Flag" the acoustic guitars are pretty much reduced to a sprinkling of rhythmic pick noise, whereas the synths that arrive in the chorus of "Umbrella" are so hollowed out that they're more an illusion than anything else. Don't get me wrong, I'm not criticizing these mixes; on the contrary, both of them are extremely bold and admirably effective, and what they demonstrate is that mix processing decisions for any given instrument need to take account of the mix space available to it.

There are some other significant advantages of this methodology from a small-studio DAW user's perspective too. Given how easy it now is to put 15 plug-ins on a track in most DAWs, overprocessing is an ever-present hazard for software users. All plug-ins have side effects, and the less processing you can use to solve a problem, the smoother the final mix is generally going to sound. "The thing I've learned," says Justin Niebank, "is to do as little as is humanly possible, because when the computer has to work that much harder—because you've piled on loads of plug-ins, for example—the sound seems to get smaller and smaller."[19] Because the purpose of a lot of mixdown processing is simply to resolve conflicts between instruments, you'll typically find that each new track you add needs more and more processing to squeeze it into ever smaller pockets in the balance, so later tracks will inevitably suffer more detrimental side effects. By mixing your most important instruments while there's still lots of room to maneuver, they'll end up with less processing, fewer unwanted artifacts, and a more appealing sound.

This way of working can also play into your hands if you're short of processor power, because high-quality plug-ins are usually more CPU-hungry, so you're more likely to use them early on in the mix process when computing power is still fairly thick on the ground. Your most important instruments will therefore be more likely to receive your best-quality plug-ins; and if CPU munch starts restricting your processing options later on, any loss of quality will damage the production sonics less by dint of being relegated to more peripheral parts.

If you're just starting out mixing, then it can actually help to write down a list of all your tracks and then number them in order of importance, as a way of disciplining yourself to stick to the plan. It's not always easy to get off the fence and commit yourself, though, especially if you're mixing someone else's work. The relative importance of different instruments in a mix is heavily genre dependent, and you'll often find that an engineer's choice of what to mix first reflects this—so rock producer Mike Clink[20] talks about building up from the drums and bass, for instance, while pop-centric producer Humberto Gatica[21] concentrates on the vocals straight away. As such it's common sense to look to relevant commercial releases for inspiration or seek advice from any engineer you know who specializes in the appropriate style.

A client's rough mix can also be useful in this respect, as clients will often instinctively balance it heavily in favor of the sounds they want to hear. However, do be aware that some self-recording vocalists would rather hear a copy of *Ultimate Manilow* on repeat than listen to their own recorded voice, so take buried vocals with a pinch of salt. Finally, if during your reconnaissance of the project during the mix prep process you've happened on anything dazzling lurking surreptitiously among the multitracks, you might feel that it's worth reversing some of the established stylistic conventions to make this into a much more prominent feature of the track—it's a risk, but at mixdown it's surprising how often fortune favors the brave.

FIGURE 8.4
If you leave mixing important tracks until the last minute, your CPU may max out before you can give them the quality processing they deserve.

Note that it's entirely normal that the relative importance of certain instruments should vary for different sections of the arrangement, and you need to take account of this when building each new section of the balance. In fact, if you make a point of writing down an instrument ranking for each arrangement section, it can serve as a good indicator as to where multing may be necessary—if the relative importance of any instrument changes dramatically between sections, it's odds on that it'll need multing to allow for different processing strategies in each instance. Lead vocals, acoustic guitars, and pianos, for example, are often of much more sonic importance during sparser verses than during more densely populated choruses.

8.1.3 Time Is Money

One final general concept to bear in mind with mixing is that time is money. In other words, you should try to allocate the majority of your time to those aspects of the production that are most likely to sell your mix: to the client, to the industry, and to the public. Glen Ballard says, "What we sometimes do in the studio is we spend 90 percent of the time on stuff that's worth about 10 percent—the absolute detail of everything that, at the end of the day, probably doesn't matter."[22] In this respect, spending the same time and effort on every track in your mix is usually misguided, because there are only ever a few aspects of any production that carry the main commercial appeal.

> Spending the same time and effort on every track in your mix is usually misguided, because there are only ever a few aspects of any production that carry the main commercial appeal.

What exactly the main hooks in a production are varies a great deal, of course, depending on the style, but it's nonetheless important to be as clear as you can in your mind where the money is. For many mainstream chart styles, the vocals clearly deserve a lion's share of the mixing time (and indeed the mix preparation work), as those elements will command the attention of the widest range of listeners and also carry the main copyright assets: melody and lyrics. It's certainly not at all unusual for vocals to occupy well over half the total production time in pop and R&B, so if you're working in those styles be realistic about what you're up against.

However, in electronic dance styles it may be that the kick, bass, and synth/sample hooks warrant the majority of the attention. Whatever the key tracks in your production, however, the main thing for self-producing small-studio owners is that time is hardly ever on their side, so if you can discipline yourself to abandon some finesse where it's not going to be appreciated, then you'll be able to focus more of your time and energy on polishing those musical features that will really enhance the production's market prospects.

8.2 SIMPLE BALANCING TASKS

Once you've decided which section goes first, you can set up your DAW's transport to loop it for you while you start putting together a basic balance. This isn't a question of merely wiggling faders, though, because there are actually a number of tasks that need attending to at the same time if your level settings are going to have any validity. Stereo panning needs sorting out, for a start, because it's otherwise impossible to judge the effectiveness of the mix in mono (as discussed in Section 2.1.3). Any comb-filtering effects between multimic recordings must also be tackled before fader levels can be set reliably, and low-frequency rubbish is best removed as early in the mix process as possible as well, for the reasons mentioned in Section 3.5.

The simplest type of instrument to balance is one that has been recorded to its own mono audio file, without any simultaneously recorded instruments audible in the background (in other words, without "spill," "bleed," "leakage," or whatever else you might happen to call it). With this instrument all you need to do is remove unwanted low frequencies, adjust the pan control, and set the fader level to where it sounds best. Let's look at these three steps in more detail.

8.2.1 High-Pass Filtering

Cleaning up the low end of the track is a job for your first bona fide mix processor: the high-pass filter. What this does is progressively cut away the low end below a user-defined "cut-off" or "corner" frequency. The increasing extent to which low frequencies are attenuated as you progress down the spectrum below

FIGURE 8.5
High-pass filtering is occasionally provided as a dedicated plug-in (such as Brainworx's Bx_cleansweep, *right*), but it more commonly forms part of a fully featured equalizer (such as Universal Audio's Cambridge Equalizer, *left*).

the cut-off point is usually expressed in decibels per octave, and some more advanced high-pass filter designs give you some control over this. Although you'll find filter slopes of 72dB/octave or more on some filters, these can have problematic audio side effects, so I stick to comparatively moderate filter slopes of 6, 12, and 18dB/octave (sometimes referred to as 1-, 2-, and 3-pole designs, or first-, second-, and third-order filters) for generic mix cleanup work.

Some filters may offer a Resonance, Bandwidth, or Q control that creates a narrow-bandwidth level boost at the cut-off frequency, but this feature is irrelevant to the task in hand so it should be set for the minimum boost and can then be ignored. In fact, as a rule, high-pass filters with resonance controls are more associated with sound-design work than mixing, so they may not be the most transparent-sounding choice for this scenario anyway. You'll normally be better served by the kind of high-pass filter that is provided as part of an equalizer plug-in.

Whatever brand of high-pass filter you use, the key concern is setting its cut-off frequency. Roger Nichols describes a good all-purpose approach here, using the example of an acoustic guitar part: "My rule is: if you can't hear it, you don't need it. Turn on the [filter]. . . . Start raising the frequency slowly until you can hear a change in the sound of the low end of the acoustic guitar. . . . Now reduce the frequency by about 15 percent."[23] Although Nichols targets his filtering primarily at midrange instruments in a professional environment, I think it's better small-studio practice to filter practically everything in the mix in this manner. If you can't rely on your monitoring to inform you whether you're losing important low-frequency information even at your filter's lowest setting, then this is where a spectrum analyzer can provide useful additional insight. And there's another thing I'd add to what Nichols has to say: in my experience the power of a drum's attack can sometimes be softened by high-pass filtering well before the instrument's overall tone changes appreciably. For lower-register instruments such as kicks, snares, and tom-toms in particular, I therefore regularly set the filter's frequency control a little lower and keep my ears open for this easily overlooked processing side effect.

WHAT ABOUT LOW-PASS FILTERING?

Although I suggest high-pass filtering most of your tracks, low-pass filtering (i.e. cutting high frequencies) so systematically holds fewer benefits, because so many instruments have important harmonics and noise components that extend right up the spectrum. That said, there's nothing wrong with low-pass filtering if you really want to kill the high end of an instrument for balancing reasons, and I do resort to it myself from time to time—most commonly with noisy amped instruments or over-fizzy distorted DI signals. However, if you're following this book's step-by-step mix process, I'd leave that until Chapter 11, and even then I'd be as sparing as you can with it, because it's very easy to rob the mix of "air" if you low-pass filter too many things at once.

8.2.2 Panning Mono Recordings

Panning simple mono instrument recordings isn't really rocket science, and although some people wax lyrical about trying to create some kind of "natural" stereo panorama, the reality of mainstream record production is that some panning is carried out for solid technical reasons, whereas the rest is done purely on grounds of the mix engineer's personal taste and with scant regard for any notion of stereo realism. For example, most drum-kit recordings usually spread the instruments widely across the stereo picture, but you'd only get that picture if you were actually seated on the drum stool. "If you're standing in front of a drum kit," says Flood, "you hear it in mono—you don't hear the toms panning neatly from left to right in your ears."[24] Another thing some engineers promote is the idea of LCR panning, whereby most things are panned either centrally or hard left/right. "For many professional mixers, myself included, the pan knobs are rarely anywhere but left, centre, or right," comments Eric "Mixerman" Sarafin. "This isn't to say that there aren't times to soft-pan parts, but in my experience, this should be a rare occurrence in most modern music mixing."[25] But there are plenty of people who use the stereo extremes more sparingly, preferring to spread tracks more evenly across the stereo panorama. "I hate hard-panned L/R," comments DJ Swivel, for instance, when talking about panning layered vocals. "To get the right width I pan them all in pairs: 40/40, 50/50, 70/70, 80/80, 90/90, and OK maybe one 100/100, but this process is ultimately what creates the stereo field . . . I think if something is hard-panned it feels much too isolated."[26] So you'll have to forgive me if I concentrate on the more objectively justifiable reasons for panning here and leave you to make up your own mind about the rest.

From a technical standpoint, there are good reasons for putting the most important musical elements of your mix in the middle of the stereo image, because they'll be the ones that survive best in mono and in situations where only one side of a stereo signal is heard. In most cases, this means that your kick, snare, bass, and any solo lead vocal will normally inhabit the center. This is particularly important for the bass instruments, as transmitting them from the combined surface area of two woofers gets the best low-frequency efficiency out of any given stereo speaker system. (Off-center bass can also cause problems with pressing your music to vinyl as well, if you're concerned with that.)

When any track is panned off-center, a primary consideration is how its balance will change in mono and single-sided stereo. In this respect, panning toward the extremes deserves careful thought, as it leads to a considerable 3dB level drop in mono and might even cause an instrument to disappear completely on the wrong end of a pair of shared earbuds. However, every producer also has to decide how much it matters to them if a panned sound makes the mix feel lopsided, or distracts the listener's attention sideways from the star attractions at the center of the picture. In practice, what most engineers plump for is a compromise called opposition panning: if an important mono instrument is panned to one side, then another similar instrument is panned to the

PANNING LAWS

The way a pan control works for a mono track is by varying the signal level it feeds to the left and right channels of your mix. Pan the track hard left and it'll feed all its signal to the left channel; pan it hard right and it'll feed all its signal to the right channel. Panning the track centrally sends an equal signal level to both channels, although it turns out that there's no universally accepted standard for what that specific signal level should be. One option is for the track to feed only half its signal (i.e. 6dB less than its full level) to each channel, but the principle advantage of this "6dB pan law" is only apparent if you're monitoring your mix in mono: adjusting the pan controls won't change the mono-sum mix balance, because under those circumstances the two 6dB-down signals of a centrally panned track combine to give the track's full signal level. If you're listening in stereo, however, the way phantom images (see Chapter 2) are perceived means that panning a track to the center of the panorama using a 6dB panning law actually makes it appear about 3dB quieter than if you panned it hard to one side. Now, it is possible to compensate for this subjective level-drop in stereo by using a 3dB pan law instead, such that panning moves no longer change the stereo mix balance. However, you can only do that at the cost of upsetting your mono-summed mix balance instead, where a 3dB pan law causes centrally panned tracks to gain 3dB in level. Some mixers even offer a 4.5dB pan law that kind of splits the difference between the 6dB and 3dB pan laws, so that panning affects the mix balance in stereo and mono listening experiences by the same amount, albeit in different directions—in other words, centrally panned tracks gain 1.5dB in mono and lose 1.5dB in stereo relative to hard-panned tracks.

So much for the boring technical details: what pan law is it best to use in practice? Well, in my opinion it rarely matters as long as you set your pan controls fairly early on in the mix process (as I suggest in this chapter), because it then requires no extra effort to address any balance ramifications of the panning as a natural part of refining your mix levels. It's also worth stressing that, whatever pan law you use, center-panned signals are always going to become roughly 3dB louder (compared with edge-panned signals) when you sum a stereo mix to mono, so you still need to check mono compatibility regardless. In short, I think you can simply ignore the subject of pan laws in almost all cases. The only exception that comes to mind is if you want to automate your pan controls to sweep sounds around the stereo image during the mix. (You old stoner . . .) Here, you need to decide whether you want the balance of the dynamically panned sound to remain consistent in mono (a 6dB pan law) or in stereo (a 3dB pan law), or whether you'd prefer to hedge your bets and have a bit of inconsistency in both (a 4.5dB pan law).

other side. This is extremely commonplace with double-tracked vocal and guitar parts (think Destiny's Child and AC/DC respectively—if imagining both those two groups at once isn't too bizarre for you!), but it's important to realize that the same principle can also work well to psychologically separate different instruments that fulfil similar arrangement functions. "If you have a hi-hat and a shaker and you put them on one side, it's going to be very difficult to tell where the hi-hat starts and the shaker ends, but if you pan them onto either side then it becomes much clearer," comments Alan Meyerson.[27] (An example of this concept in action can be heard in the first verse of Tim McGraw's song "Humble And Kind," where acoustic and electric rhythm guitars play different, but complementary, parts on the left and right sides of the image respectively.) Opposition panning provides you with stereo width and interest, but without pulling the overall stereo picture off-center or endangering the musical sense in one-sided stereo playback situations. As such, many panning decisions are more about the width and spacing of opposition-panned instrument groups than about the specific stereo locations of individual sounds.

> Panning toward the extremes deserves careful thought, as it leads to a 3dB level drop in mono, and might even cause an instrument to disappear completely on the wrong end of a pair of shared earbuds.

An extension of the opposition panning idea is to find call-and-response elements within the arrangement and push those towards opposite sides, a technique that often works well for composite guitar riffs, electronica SFX one-shots, or contrasting backing-vocal lines. Bob Clearmountain suggests an even larger-scale application: "I know some mixers [feel] if there's something on the left, you're going to have to have something equal on the right, but sometimes it doesn't have to be simultaneously equal. The verse could have something on the left, and then the B-section might have something answering it on the right."[28]

8.2.3 Setting the Levels

After high-pass filtering and panning, you're ready to set levels. For the first track, all you do is choose a sensible level relative to your mixing system's headroom so you don't risk distortion once the whole arrangement's going. I've had a lot of people ask me to give a target meter reading for this, while others seem to work up a sweat trying to hit some notional "professional standard," but honestly it's not rocket science. The thing is that the mixers in most modern DAWs use a clever DSP calculation method (referred to as "floating-point processing" or "floating-point maths") that effectively provides both an unlimited headroom and a non-existent noise-floor—for all realistic mixing purposes, at least. What this means is that, if you wished, you could merrily overload the level meters on every channel of your mix, and then simply reduce the mixed signal's level to avoid overloads on your DAW's master output, without the mix suffering any distortion or quality loss. That said, it's simple common sense to try to avoid overloading your channels and busses if you can (if only so you can easily export unclipped mix stems from them later on if required, as discussed in Chapter 19), and the lack of noise-floor

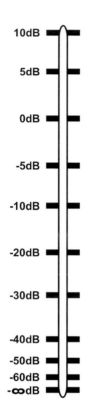

FIGURE 8.6
A typical fader control scale. Fader movements around the unity (0dB) gain position will give smaller and more precise gain changes than fader movements toward the bottom of the level scale.

concerns means that you can afford to leave 24dB or more of headroom on individual channels without any worries. In addition, the level at which you hit your mix processors can make a big sonic difference, but I'll get onto that properly in the next chapter.

While it's natural to reach first for your channel fader when setting initial mix levels, there's also something to be said for leaving the fader at the 0dB (unity gain) position and applying most of the required gain either with the mixer channel's input gain control (if it has one) or with a simple gain plug-in inserted into the channel. There are two reasons for this: first, almost all faders have their best control resolution centered around the unity gain position, so it may prove difficult to make fine level changes if the fader cap's right down at the bottom of its travel; and, second, the fader's upper end stop can be a pain in the neck if you begin the mix too close to it. However, if you'd rather use your faders for initial balancing (perhaps so you can take advantage of a hardware fader controller) there's absolutely nothing wrong with this, and you can always transfer those fader settings to gain plug-ins at a later stage if you run into problems with control range or resolution.

Those starting out with mixing usually underestimate the amount of time and attention they should give to basic level balancing, so resist the temptation to skate over this stage of the mixing process just because it seems so self-evident. Justin Niebank notes, "Balance is way more important than EQ or automation, but people seem to give up half an hour too soon on some mixes. You have to think balance first."[29] Try to zero in on what the level of each new instrument means for the mix as a whole, and be aware that a good balance usually involves some kind of tradeoff: on the one hand, you want the track loud enough that you can hear all the things you really love about the individual sound; on the other hand, you need to keep the track low enough in the mix that it doesn't obscure the merits of more important instruments that are already present.

This is where good monitoring practices can really pay off. Use your main and secondary listening systems to gain a multifaceted view of the situation, paying particular attention to your Auratone substitute, and make appropriate use of the low-frequency workarounds discussed in Chapter 3. Keep your ears fresh by switching monitors frequently, varying your listening levels, taking breaks, and seeking other people's opinions. You might even look to your collection of reference productions for guidance, although I'd sound a note of caution there, because it's easy to get bogged down if you try to do detailed referencing early on in the mix process. In practice, I prefer to play through a handful of relevant reference tracks from time to time to calibrate my ears in general, and then I leave any detailed referencing until the end of the mixdown process when the sonics can be compared with commercial productions on a more equal footing.

SOME TRICKS FOR JUDGING FADER LEVELS

If you can't make your mind up about a particular fader, then here are a few helpful tricks. The first comes courtesy of Gus Dudgeon: "I find the quickest way to decide whether something is actually loud enough or not is to run the mix and turn the signal on and off. If you turn it on and you can hear it, but it's not smack bang in your eye, it's probably where it should be. . . . The best thing to do [if an instrument is too loud] is to take the bloody fader out and start again. Creep it in until you think that's probably where it should be and then try switching it on and off."[30] Mike Stavrou offers another good one: "Instead of focusing all your concentration on the fader being moved, listen to the neighboring instruments instead. While trying to hone the perfect level of the snare drum, for example, do so while listening to the bass drum. . . . You will quite often find a more decisive point of balance this way. That's not surprising, because when you think about it that's exactly how the listener listens—his (or her) mind is elsewhere until you attract his attention."[31]

"It is always easier to tell when something is wrong than when something is right," says Roger Nichols. "While listening to the mix, one at a time change the levels of each instrument in varying combinations. Turn the bass up 1dB. Does the mix sound worse? If the answer is yes, then turn it back down. Turn the bass down 1dB. Does this sound worse? If the answer is yes, then the original level was right for the bass. Now try [the other instruments]. . . . If you can change the level of an instrument in the mix by a tenth or two-tenths of a decibel and you can hear the change that you made, the mix is getting pretty good."[32] (If that "pretty good" feels like a bit of an understatement, just bear in mind it's coming from an eight-time Grammy winner!)

Another nifty hint comes from Roey Izhaki. "Take the fader all the way down. Bring it up gradually until the level seems reasonable. Mark the fader position. Take the fader all the way up (or to a point where the instrument is clearly too loud). Bring it down gradually until the level seems reasonable. Mark the fader position. You should now have two marks that set the limits of a level window. Now set the instrument level within this window based on the importance of the instrument."[33] What's particularly useful about this tip is that the size of the "level window" can also provide advanced warning of tracks where further mix processing is required—as we'll see later.

The level of the lead vocal is probably the most common bone of contention, because there's often a fine line between "loud enough" and "too loud." As Dave Pensado notes: "If it's too loud, the track sounds wimpy. If it's too soft, the vocalist sounds wimpy."[34] One rule of thumb from Bob Power is that the vocal should feel a little too quiet when you turn your monitors up really loud, but it should also feel a little too loud when your monitors are turned down really quiet.[35] Given that the lead vocal usually carries the totality of a song's publishing copyright, i.e. the melody and the lyrics, there's clearly commercial sense in making sure it's clearly audible, but outside chart-centric releases a lower vocal level can actually encourage listeners to turn things up and listen more attentively. "You try to draw the listener in," explains Dave Bottrill. "I learnt that on those early Tool records. You'll notice on *Aenima* sometimes the vocal's quite quiet . . . The idea was that if the listener's thinking "What did he say? What did he say?" then suddenly he's facing into the mix."[36]

Although it's a false economy to rush your initial fader settings, you also shouldn't beat yourself up expecting to get a perfect balance straight away, because it takes a lot more than faders to uphold the balance in mainstream commercial genres. Just try to make the best-informed preliminary decisions you can, so that you minimize the amount of processing required further down the line. And if you stay alert during the balancing process, the faders will start to give you valuable tip-offs about which tracks need plug-in assistance.

8.2.4 Listen to Your Faders!

Faders can't talk, but they can communicate nonetheless; it's only once you learn to tune in to their delicate messages that mix processing actually starts to make any real sense. You need to be sensitive to when any fader setting doesn't seem to feel "stable"—in other words, you can't find a static level for that instrument that achieves a balance you're happy with, because wherever you try to set it, it doesn't feel right. If a fader feels unstable like this, then the track probably needs processing to achieve a good balance. This insight might not seem particularly earth shattering if you already know that a good proportion of your tracks will need processing, but what you can already gain from it is the confidence to leave tracks *unprocessed* if they already balance fine at a static fader level. This is a crucial mixing decision in its own right, and, incidentally, it's one that experienced engineers tend to arrive at much more often than rookies do. "One of the smarter things for a mixer is to know when to leave stuff alone," says Jimmy Douglass.[37] Darryl Swann puts it another way: "Taste your food before you salt it. I know so many guys that feel that they're not earning their money unless they're pushing switches and turning knobs. No, you've got to *listen* to your sound. If it needs work, then plug something in."[38]

> Leave tracks unprocessed if they already balance fine at a static fader level. This is a crucial mixing decision in its own right, and, incidentally, it's one that experienced engineers tend to arrive at much more often than rookies do.

But just spotting an unstable fader is only the first step, because the precise characteristics of that instability are what provide the essential clues you need to choose the best mix fix. The range of conclusions that it's possible to draw is enormous, and we'll be looking at them in more detail over the next six chapters, but for now here are a few commonplace examples to highlight the kinds of things you should be watching out for during the preliminary balancing stage:

- You fade up your drum loop until the kick seems properly balanced, but you can't hear the snare and hi-hat elements well enough at that level. It feels like you need a separate fader for the kick drum, which is a strong clue that you might want to mult the kick hits to a separate track.

BALANCING AGAINST PINK NOISE

One slightly leftfield idea that's been gaining currency in recent years is that if you solo each instrument in your production and match its volume by ear to a fixed level of pink noise (an audio test signal that has equal frequency energy in every octave of the audible spectrum), you can arrive at a set of sensible initial fader levels in a fraction of the time. Opinions differ a little regarding both the mechanics and justifications of the process (I've included links to some detailed tutorials in this chapter's web resources if you're interested), but I won't waste too much space going into it here, because, to be frank, I think it's a massive red herring. As I've tried to demonstrate in this chapter, building a raw mix balance isn't just about setting fader levels. More importantly, it's about developing a detailed knowledge of the raw materials, working out how different tracks interact with each other, and investigating potential balance problems within a mix context—none of which tasks can be carried out effectively by soloing individual channels and smothering them in full-band noise! So even if you find that the pink-noise trick really does reliably deliver sensible initial fader levels, I really don't think it helps you much, because the main point of creating an initial level balance by ear is actually the brain-work it involves. It's this brain-work that takes the lion's share of the time while balancing, not the simple level setting, so it's hardly surprising that the pink-noise trick gives a superficial impression of saving time by comparison. Personally, though, my own earnest attempts to get useful mileage out of the pink-noise trick with a variety of different multitrack projects didn't honestly seem to speed up setting fader levels significantly. Neither was I thrilled with most of the balances I ended up with once the pink noise had been switched off, especially as those fader settings frequently made even less sense in light of subsequent polarity/phase adjustments.

Despite these reservations, though, I will acknowledge a couple of potential ancillary benefits of the pick-noise trick, if you do decide to use it regularly. First, if you set your pink noise to a sensible level (peaking around –15dBFS, say), you can be pretty secure that the mix won't encounter resolution or headroom problems at your DAW's master output. And, second, as you get used to how pink noise sounds, it will begin to act a little like the reference tracks I discussed back in Chapter 4, helping acclimatize your ears to the frequency–response vagaries of your monitoring situation. If you're already happy managing your output resolution/headroom and referencing your monitoring in other ways, though, then I don't reckon it's worth faffing with the pink-noise trick solely on those grounds.

■ You try to set the level for a simple eighth-note rhythm-guitar part, but some chords always feel buried in the mix, whereas others pop out unduly. You're constantly feeling the urge to tweak the fader. What your mix is trying to tell you is that the difference in level between the loudest and softest chords is too large—or, to translate that into tech-talk, the dynamic range is too wide. You therefore need processing such as that in Chapters 9 and 10 to control that problem before the fader's going to settle down to a static balance.

- By the time you've faded up your acoustic guitar to the point where you can hear all the nice sparkly string details, the booming of the instrument's main sound hole resonance is making the lowest notes sound bloated. What you want is a separate fader to turn down just that section of the frequency range—which is effectively what the tools discussed in Chapters 11 and 12 provide.

The more sensitive you can make yourself to the way in which a fader feels unstable, the easier it'll be to select the right kind of plug-in for each track without hesitation. For this reason, time spent working on initial fader settings is rarely wasted—the first balance may not singe anyone's eyebrows on its own, but it provides a rock-solid foundation for all the more complex mix-processing decisions later on.

8.2.5 Additional Considerations for Stereo Tracks

All the same things that apply to mono recordings also apply to stereo recordings, as far as initial balancing is concerned, but there are a few additional factors as well. The first thing to address is that some systems record stereo in the form of two mono files (sometimes referred to as "split stereo"), on the understanding that you'll use individual mixer channels to pan the files hard left and right. However, most DAW systems now have perfectly capable stereo mixer channels, which are much more useful for mixing purposes because they allow you to conveniently process both sides of the audio in tandem. If you find yourself presented with a split-stereo recording, make it your business to convert it to a single "interleaved" file before you do anything else. If that's not practical, then buss the left and right mono channels to a communal stereo group channel and do all your balancing and processing there instead.

That said, on occasion you may wish to offset the whole stereo image to one side, narrow the spread, or reverse the sides. In these situations, some control over the panning of the left-channel and right-channel audio streams can be handy. If you're bussing individual split-stereo channels to a stereo group, then this is no problem. However, things can be trickier if you're working with an interleaved stereo file on a dedicated stereo mixer channel—if that channel sports anything like a pan control, then it probably only adjusts the relative levels of the two audio streams, which is of limited benefit. A few software platforms (Steinberg Cubase, for instance) do actually offer the necessary additional panning controls among the standard channel facilities, but if you can't find any in your own DAW, then there's usually an alternative in the form of a bundled stereo utility plug-in. Still no joy? There are plenty of suitable third-party plug-ins around, and you can find links to my favorites in this chapter's web resources.

The main concern with stereo files, though, is if phase or polarity issues are causing the sound to change dramatically when you listen back in mono.

The sound of any stereo file will always lose something in the transition to mono, so it's unrealistic to expect no change at all, but you do have to decide whether any changes are going to seriously compromise the effectiveness of

CORRECTING SUBTLE STEREO BALANCE OFFSETS

It can occasionally be difficult to decide whether a given instrument in a stereo ensemble recording is exactly central in the stereo image, especially in less-than-ideal monitoring conditions—or indeed if your own hearing isn't exactly symmetrical! A special stereo-metering plug-in called a vectorscope can sometimes help here, but my favorite trick is simply to use a plug-in that allows you to swap the left and right stereo channels back and forth: if the instrument you're concerned with changes its left–right position appreciably when you do this channel swap, then it isn't in the center of the recorded image.

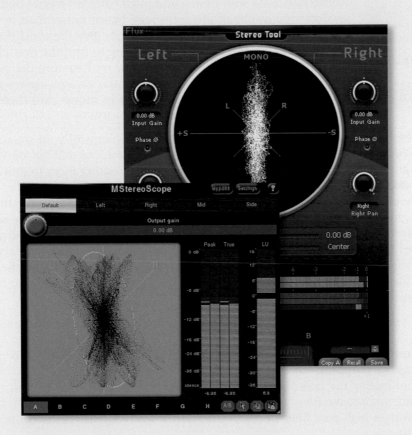

FIGURE 8.7
A couple of good freeware stereo vectorscope plug-ins: Melda's MStereoScope (*left*) and Flux's Stereo Tool (*right*).

the balance. If you conclude that the damage calls for remedial measures, then the first thing to check is that one of the two audio streams hasn't inadvertently been polarity-flipped during recording. It's easy to spot this by eye most of the time by comparing the left and right waveforms, particularly for comparatively close-miked stereo files. Again, if your own DAW system doesn't provide the facilities to correct the problem, check out this chapter's web resources for some third-party plug-in suggestions.

Even once you've corrected any obvious polarity reversals, phase mismatches between signals arriving at the two microphones may also need addressing if comb filtering blights the mono listening experience. Once again, if you can see a clear delay between the waveforms of the two audio streams, then you may be able to minimize the sonic changes in mono by lining things up more closely, although you should be aware that this procedure may undesirably tamper with the spread of the stereo picture as a consequence. As with timing and pitch edits, do ensure that you evaluate the results of this technique by ear rather than by eye, because the tidiest visual match may not give the most mono-compatible outcome. If you're struggling to find the best phase match between the two audio streams, try inverting the polarity of one of them for a moment, shift the streams against each other until the sound disappears most in mono, and then flip the inverted stream back to normal polarity.

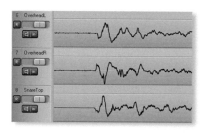

FIGURE 8.8
If your overhead mics have phase and polarity mismatches like this, you'll have real problems with the mono-compatibility of your mix, in addition to a slightly odd stereo picture.

If despite these measures the mono-compatibility problems of your stereo file remain intractable, here's a last-ditch technique that you can still fall back on: pan the two audio streams more centrally, partially summing them to mono in the stereo mix. Obviously this will narrow the stereo image, but it will also partially comb-filter the stereo playback in the same way as the mono rendition, bringing the two tonalities closer together. This may mean that you have some additional processing to do later on to resuscitate the comb-filtered tonality for stereo listeners, but at least the final mix will translate more reliably to a wide audience. If all else fails, you might also consider ditching one side of the stereo file completely, especially if you already have plenty of stereo interest from other tracks in the production.

So to recap quickly, balancing simple mono or stereo files involves these steps:

- routing both sides of a stereo file through a single mixer channel for processing purposes;
- optimizing the polarity and phase relationships between the left and right sides of a stereo file;
- using a high-pass filter to remove unwanted low end within the context of the mix;
- setting the instrument's position/spread in the stereo field;
- balancing the instrument against those already in the mix.

> **GROUPING TRACKS FOR COMBINED CONTROL**
>
> It's not just ensembles that have been multimiked that can benefit from being routed through a communal group channel at mixdown. Any group of parts that you might want to balance, process, or even just visually organize for mix purposes is also fair game, whether that's triple-tracked banjos, dueling prog-rock synth solos, or multiple layers of sleighbell samples. (It also makes it easier to mute the whole lot at once.)

8.3 MORE COMPLEX BALANCING TASKS

8.3.1 Multimiked Instruments

It's not at all uncommon in the studio for a single instrument to be recorded using more than one microphone, the idea being that you can thereby adjust the instrument timbre more flexibly at mixdown. It's especially common for guitar amps to be recorded in this way, for instance. Balancing this kind of recording isn't dissimilar to balancing a stereo file:

- route the individual mic channels to a single mixer channel for processing purposes;
- optimize the polarity and phase relationships between the different microphone signals;
- use a high-pass filter to remove unwanted low end within the context of the mix;
- set each microphone's position in the stereo field;
- balance the instrument against those already in the mix.

The chief difference in this case, though, is that the second step isn't necessarily just about technical problem solving, because it also has enormous potential for creative tonal shaping. The simplest option is to match the polarity and phase of all the mics as closely as possible (using polarity inversion and time shifts as before), such that you can combine their characteristics fairly freely without any real difficulties arising from unwanted comb filtering. Many great electric guitar sounds have been made in this way, so I'm not going to knock it, but there's nonetheless a whole lot more you can do here if you choose to exploit phase-cancellation effects rather than trying to minimize them.

> The simplest option is to match the polarity and phase of all the mics as closely as possible, but there's a whole lot more you can do if you choose to exploit phase-cancellation effects rather than trying to minimize them.

Let's assume for the moment that we're starting from the polarity/phase-matched setup. If you flip the polarity of one of the mics, then rather than adding its characteristics to the composite sound, it'll subtract them. Fading that mic all the way up might unacceptably hollow out the sound of your instrument, but at lower mix levels there will be all sorts of new tonal variations on offer. Likewise, artificially delaying one of the microphone signals a few milliseconds against the others will cause comb filtering between them. Again, you might not want to have the delayed mic too high up in the balance, but at lower levels the milder frequency–response undulations can instantly yield radical tonal changes that would take all day to achieve with traditional mix processors. The best bit about it all is that sculpting your tone with polarity and phase relationships is free of DSP processing artifacts. You'll get big tonal changes with no unwanted side effects—everyone's a winner! Numerous high-profile producers (Ken Caillat,[39] Neil Dorfsman,[40] and Clarke Schleicher,[41] to name just a handful) have professed a fondness for this tone-mangling technique. "When you build a mix," says Jack Douglas, "the real nightmare is when you put something up and the only way you can hear it is by blasting it. . . . If you [sculpt the tone using polarity/phase-relationships] I guarantee you that as soon as you put the sound in the mix, it will be there. . . . Not only that, it won't wipe out everything else in the mix, because it will have such a separate and distinct character."[42]

The option to pan the individual mics of a multimiked instrument to different positions in the stereo field can be another advantage of this recording method. Simple mono recordings have a habit of sounding much narrower than the instrument itself does when you're seated in front of it, so being able to slightly spread the image of a multimiked acoustic guitar or piano, for instance, can help things seem more natural and real, and it can fill up more of the stereo field in a sparse arrangement. In band recordings, double-tracked electric guitar parts can be made to sound bigger and more impressive in stereo in the same way, as well as more homogeneous if the individual guitar images are actually overlapped to some extent.

FIGURE 8.9
The polarity button. It may not be much to look at, but it's absolutely essential at mixdown, for both technical and creative reasons.

You can balance an instrument recording comprising mic and DI signals in much the same way as a straight multimiked recording, because the polarity/phase relationship between the signals creates similar problems and opportunities. However, the extra advantages of a DI recording are that, in the case of electric guitars, it can be reamped to flesh out or even reinvent the sound as required; and that it's pretty much immune to spill from other instruments recorded at the same time, which can be a lifesaver if high levels of spill on the instrument's microphone have rendered that ineffective for adjusting the balance.

8.3.2 Multimiked Ensembles: Preliminaries

The most complicated tracks to work with are multimiked recordings of groups of voices or instruments: singing guitarists, drum kits, choirs, string groups, live band recordings, orchestras, and many other ensembles. The problem in such situations is that every instrument will be picked up by every microphone to some extent, and although it's feasible to polarity/phase-match all these different recorded signals for one of the instruments in the ensemble, it's impossible to do so for every instrument at the same time. Imagine a singing guitarist, for example, recorded with one close mic on the singer and one on the guitar. In both close mics you'll have delayed spill from the off-mic sound source, so if you shift the timing of the guitar slightly so that it lines up with the vocal mic's guitar spill, that'll move the guitar mic's vocal spill further out of alignment with the main vocal signal.

The upshot of this no-win situation is that your goal becomes to turn the unavoidable effects of comb filtering to your advantage. Get it wrong and you can comprehensively trash the sound of the recording, making all the instruments sound hollow, harsh, and washed out. Get it right, though, and you can achieve something truly magical: a production where each instrument is enriched by virtue of the contributions from so many mics, and where the combined spill signals naturally blend the ensemble with very little further processing effort. This is where the initial balance can absolutely make or break a record. Unfortunately, this kind of balancing is by no means an exact science, and every engineer has a slightly different way of tackling it. So let's look at some of the options and discuss the pros and cons in each case.

First of all, it's not unusual to have stereo mic pairs within an ensemble setup, and it's usually worth checking those for polarity and phase issues in isolation before you get into dealing with their interactions with other mics in the rig. So, for example, in a typical multimiked drum kit you'd want to individually check the overhead mic and room mic pairs for stereo spread and mono-compatibility. Likewise, if there are any instruments in the ensemble that have been individually multimiked, then deciding on their initial blend and polarity/phase relationship in isolation is also a good idea. (I usually also route those mics to their own group channel while I'm at it, so I can more easily balance and process them together later on.) In a typical drum set, the snare might have mics both above and below it, for example, or the kick drum might have an inside mic and an outside mic. In both cases, the polarity/phase relationship between the two mics might be adjusted correctively for the best match or creatively to deliver a specific tone. Spike Stent notes, "I'll check whether the kick and snare tracks all line up, and so on. I'm flipping phase all the time. I'm anal about that, because it is essential for getting a really tight mix which sounds big on the radio, cars, laptop, and so on."[43]

Because of all the spill, deciding on a high-pass filter setting for each individual mic becomes more of a gray area, because it depends how much of the low-frequency spill you consider "unwanted" in each case—something that you may only really be able to decide within the context of the whole mix. However, the problem with most normal filters is that, by their very nature, they also adjust the phase response of the filtered signal, which may adversely affect the way the signal phase-cancels against the other mic recordings. My advice is to restrict your use of high-pass filtering here to just removing frequencies well below the range of the lowest instrument; then leave any further filtering until you've had a chance to listen to how all the mics combine in practice (discussed more in Chapter 11).

Panning decisions may be harder to make too, because you need to decide how you pan each instrument's main mic signal in relation to the stereo location of prominent spill from other mics. In this respect, it makes good sense to create a stereo picture that mimics the physical layout of the recording session,

COMPLEX PHASE-MANIPULATION TOOLS

I've already discussed polarity inversion and time shifting as two ways of adjusting the combined tonality of multimiked recordings at mixdown. However, there are also an increasing number of more complex phase-manipulation options such as phase rotators, all-pass filters, and spectral phase adjusters. Technically speaking each operates in a different way under the hood, but in practical terms they all offer the same basic mixdown function: the ability to alter the phase relationship between the different sine-wave components that make up a sound. As with polarity inversion and time shifting, these tools usually have negligible sonic effects on the individual processed track in isolation, but if that track is part of a multimic recording, then the nature of the phase cancellation between the mics will change, and therefore the mixed instrument's timbre. As such, these kinds of complex phase-manipulation tools can be useful for tonal refinement, although I'd personally recommend experimenting with them only once you've exhausted the possibilities of polarity inversion and time shifting.

FIGURE 8.10
A selection of complex phase-manipulators: Solid State Logic's X-Phase (*left*), Voxengo's PHA-979 (*centre*), and Melda's MFreeformPhase (*right*).

because the strongest spill contributions from each instrument should then remain fairly close to the instrument's main mic position in the stereo picture. Where an ensemble recording includes stereo files, matching the images of these with the panning of any mono mic signals is also fairly commonplace. However, these panning tactics may conflict with the technical panning considerations I talked about in Section 8.2.2, so there must often be a tradeoff here. For example, a lot of people record drums in such a way that the snare image in the stereo overhead microphones is off center; but you may nonetheless prefer to pan the snare close mics right down the middle for best mono-compatibility, even if this does mean that the tail of each snare seems to veer off slightly to one side.

8.3.3 Building the Ensemble's Balance and Tone

Once these preliminary jobs have been carried out, mute all the ensemble channels and begin reintroducing each mono mic, stereo mic pair, and multimic combination to the mix in turn. As you do this, use your polarity/phase adjustment tools and fader to search for both the best ensemble tone and the best balance within the context of the rest of the mix. If this sounds like a lot of fiddly work, that's because it is, but it's also a part of the mixing process where miracles can happen, so roll up your sleeves and get your hands dirty!

Broadly speaking, the order in which you tackle the individual mics and instruments at this stage should again reflect their importance sonically, since this helps clarify your tonal decisions. For example, let's say you've adjusted the phase of a newly added instrument mic to give that instrument the best tone you can. If you check its impact on the rest of the ensemble by switching it in and out of the mix and discover that it's deteriorating the sonics of some of the previously added instruments, then you don't need a PhD to work out that some aspect of the newly added instrument's sound may have to be sacrificed for the greater good. Working in this way also partly just makes things easier from a technical perspective, because the mics covering less important instruments are usually mixed at a lower level, which means their spill causes less comb-filtering in combination with the mics of more important instruments, so you have more freedom to adjust the polarity and phase of the newly added instrument for the sake of its own timbre before causing unacceptable sonic damage to other more critical parts. By the same token, if you're able to prioritize any mic signals that have lots of spill, that will also make the mixing process more straightforward, because those signals would have the greatest potential to comb-filter your mix sonics if you left them until later.

There is another factor, however, that may frequently overrule either of these guidelines. If any mic or mic pair has been set up so that it captures a reasonable balance of the whole ensemble, you can often get good results more quickly by using that as the bulk of your ensemble sound: set a level for that first, and then use the remaining signals in the manner of supplementary "spot mics," just for small tone/balance refinements to the main blend. A major advantage of this approach is that the spot mics can usually then be used at lower level, thus reducing the severity of any phase cancellation between their spill contributions and the signal from the dominant overall ensemble mic(s). Indeed, because this method is a little more foolproof in general, it's usually my first recommendation for less experienced small-studio engineers who are grappling with their first experience of a multimiked ensemble—almost always in the form of live multitracked drums. But this approach is by no means just for beginners. Allen Sides,[44] Steve Hodge,[45] and Randy Staub,[46] for example, have all stated a preference for this method. "I'll start with the overheads first and then add the close-in microphones," says Hodge, for instance. "That gives the sound some context."

> If any mic or mic pair captures a reasonable balance of the whole ensemble, you can often get good results more quickly by using that as the bulk of your sound. Indeed, that's usually my first recommendation for less experienced engineers grappling with multitracked drums.

One final issue deserves some discussion for the benefit of those with maverick tendencies. Polarity and phase adjustments provide tremendous freedom for tonal reinvention of multimiked recordings, and in a situation where all the mics are picking up the same instrument, you can pretty much run with that until your legs fall off—no matter what timbre you end up with, you'll still be able to choose whatever level you like for it in the mix. (Do be careful if you're panning those mics to different stereo positions, though, as that can lead to catastrophic mono-compatibility problems.) What's a lot riskier is taking tonal flights of fancy by deliberately emphasizing phase cancellation between different mics in a multimiked ensemble, because this can interfere with your ability to control the most vital thing of all: the ensemble's balance. Imagine the example of multimiked drums again: if you deliberately use the rack-tom close mic to cancel unwanted frequencies in that drum's tone, it'll be of limited use to you if you need to fade up one of the tom fills in the final mix. It's also worth pointing out that any ensemble timbres you lovingly craft with inventive phase-manipulations won't survive for long if instruments or singers move around relative to the mics during performance, so under those circumstances it's best to use spot mics in a broadly additive rather than subtractive role, and I'd stick to polarity adjustment rather than wasting lots of time on in-depth phase tweaks.

There's such a lot to think about when balancing multimiked ensembles that it can be tough keeping an overview in your mind, so allow me to quickly recap the procedure:

- Route stereo files and individual multimiked instruments through single mixer channels for processing purposes, and optimize the internal polarity and phase relationships of each group of files in isolation.
- Use high-pass filtering conservatively, removing only unwanted low frequencies comfortably below the range of the lowest instrument.

- Set the position/spread of each mono mic, stereo mic pair, and multimic combination in the stereo field.
- Introduce each mono mic, stereo mic pair, and multimic combination into the mix in turn, adjusting the polarity and phase of each to achieve the best tone and balance for the ensemble as a whole. It's generally sensible to work through the tracks in order of sonic importance, although any mic signals with lots of spill on them are worth introducing as early as you can, and if the entire ensemble has been captured passably by a single mic or stereo pair, then it may be easiest to build the mix from that.

POLARITY AND PHASE RELATIONSHIPS BETWEEN LAYERED SOUNDS

Although polarity and phase relationships most commonly affect the mixed timbre of multimiked recordings, they can also impact on layered parts that have no acoustic connection at all. The most common culprits here are the kinds of layered drum samples where each drum hit features several sampled layers starting at exactly the same time in relation to each other. In such cases, you can often make big changes to the layered tone by simply inverting the polarity of one of the layers, or by delaying a layer by a couple of milliseconds. The same kind of thing can also crop up when you stack drum loops, when a drum sample is being layered with a live drum kit recording, or when any two similar-sounding one-shot samples are layered together, so it never hurts to experiment with at least your polarity switches if in doubt.

What can complicate things enormously, though, is if the phase relationship between a pair of layered sounds fluctuates, resulting in unpredictable timbral changes from moment to moment. This most commonly occurs where an added drum sample isn't reliably synchronized with a live drummer's playing, but can also occur in many other instances, such as where two live drum performances are combined in a single mix, where a multimiked singer/instrumentalist has moved while performing, where the grooves of stacked drum loops vary, or where several different bass parts are playing in unison. In these cases, editing offers two fairly straightforward (but frequently time-consuming!) solutions: either you can delete layers (or sections of layers) that are interacting with the other, or you can edit their timing note-to-note to maintain a more consistent phase relationship. Equalization can also help, especially with bass instruments, but we'll get to that properly in Chapters 11 and 12.

8.4 CASE STUDY: MULTITRACK DRUMS

To clarify how all these balancing guidelines work in practice, let's walk through the balancing process for an imaginary multitrack drum recording as a kind of mental case study. For the sake of argument, we'll say the recording comprises stereo overhead mics and stereo room mics; close mics above and below the snare drum; close-mics inside and outside the bass drum; and individual close-mics for the hi-hat, two rack toms, and a floor tom. My thought process and workflow might go roughly as follows:

First off, I route and pan the split-stereo overhead and room channels to stereo group channels in my DAW so that I can process both sides of these stereo files at once. The overhead mics had been set up at similar distances from the snare drum, but it's still a bit off-center so I decide to try time-shifting one side of the overheads file to phase-match the snare hits. First I roughly line up the two sides of the split-stereo audio file by eye. Then I flip the polarity of one side and tweak the time shift by ear for greatest cancellation, before returning the polarity to normal. The snare is now closer toward the center of the stereo image and remains more solid sounding in mono. Checking the rest of the overheads sound for mono compatibility reveals a loss of low end in the kick drum spill and a little less brightness in the cymbals. The changes aren't serious enough that I think they'll drastically compromise real-world mono playback. Besides, the stereo width presented by the overhead mics is fairly extreme, and in narrowing that down more toward my own personal preference I also inevitably reduce the tonal change when switching to mono listening. The room mics seem to keep their sound surprisingly well in mono, but they are leaning a little to the left in stereo, so I nudge up the level of the right-hand split-stereo file to compensate.

Because the overheads seem pretty well balanced, I decide to start my balancing with them when the time comes. That said, I'm concerned that the snare drum feels a bit lightweight in the overheads, so I next solo the two snare close mics to hear what they sound like together. The polarity of the bottom mic's waveform is, as expected, inverted compared with that of the top mic, and correcting this immediately fills out the drum's low end—good news. While examining the waveforms, I notice that the bottom mic's waveform is also trailing slightly in the timeline, so I time-shift that to see whether I can achieve any further improvement. There is only a small change, but it's better than nothing. I still feel that the snare's low midrange could be more powerful, so I insert a phase rotator plug-in on the under-snare channel and adjust the amount of rotation by ear, listening for an even fuller sound. This gives me another small but useful improvement.

The kick mics get soloed next, so I turn my nearfields up a bit to hear the low end more clearly. The signals are both in polarity with each other, but there's about 7ms of delay between the inside-mic and outside-mic hits. I try phase matching again, but it doesn't sound as good and unduly emphasizes an undamped low mid-frequency drum resonance. I undo that time shift, and experiment instead with a phase rotator. I get a really useful result here, with more assertive low frequencies and an end to that pitched resonance. The snare and kick mic pairs are also now routed to their own mono group channels, because I don't normally pan the individual mics separately. Inserting a high-pass filter into the kick drum group channel only begins to affect the drum's low-end thump at about 42Hz, so I roll the frequency back to 35Hz for safety's sake and then duplicate that setting to all the other drum channels. The final preparatory stage is to open the

TIME-ALIGNING DRUMKITS AND OTHER ENSEMBLES

I'm frequently asked by small-studio users whether they should time-align their close mics when mixing drums, for instance by delaying their snare close-mic signal so that its snare waveform lines up with the snare waveform within the overhead-mics signal. Well, the first thing to reiterate is that, as discussed in Section 8.3.2, this won't "correct" the phase mismatches inherent in a multimiked drum recording, because (staying with the snare example) delaying the snare close mic in this way won't time-align its hi-hat or rack-tom spill components, so there's certainly no guarantee that it'll provide a more appealing overall kit tonality than if you make phase/polarity adjustments purely by ear. That's not to say that time-aligning close mics like this doesn't give a subjectively different character to a drums mix, because it does, but whether the result appeals is primarily a question of personal taste. For example, drum transients tend to sound slightly sharper on time-aligned recordings and the close-mic signals will usually cohere more naturally with the overheads, and some engineers consider those attributes advantageous; whereas, with unaligned recordings, the slightly smeared transients shift more emphasis onto the body and sustain of the drums in the mix, and the close-mic signals will typically seem to pull the individual drums closer to the listener, which other engineers prefer.

There is one specific technical consideration to bear in mind, though. One of the ways we humans judge the position of a sound source in any reverberant environment is by detecting the direction from which its sound first hits our ears (a psychological phenomenon known as the Precedence Effect), the reasoning being that this first wavefront will have come directly from the sound source without reflecting off anything en route. If you don't time-align your close-mic signals, they mimic this effect in the mix—the snare transient, for instance, will always appear earlier in its own close-mic signal than in the overheads signal, so inevitably provides the first wavefront as far as the listener is concerned. As such, the stereo positioning of the snare drum will be more strongly determined by the panning of the close mic than it would be if the close mics had been time-aligned with the overheads.

The same general principles apply equally to any multimiked ensemble, so whether you delay your close-mic signals to match those of more distant "overall pickup" microphones will likewise impact upon the ensemble's overall timbre, the character of the recorded transients, and the perceived positioning of each ensemble instrument in the front–back and left–right dimensions. What can complicate things further, however, is that many musicians naturally move their instruments while performing, in which case the concept of exact time-alignment goes out the window, and a certain amount of comb-filtering and transient-smearing becomes an integral part of the sound whatever you do with delays at mixdown. If you still wish to delay your close mics under those circumstances (perhaps to avoid bringing a certain instrument too upfront while you're boosting its close-mic level), then you may find in practice that delaying them 5–10ms more than you would expect for simple time-alignment will somewhat mitigate the range of tonal variation triggered by the player's movements.

overhead mics and then fade each of the other four close mics up so that their panning can be matched with what's in the overheads.

Right—time to start balancing! First I mute everything but the overheads group channel and set a level to leave lots of headroom on the mix buss. This production is in an upbeat pop/rock style, so I decide that the snare's probably the most important instrument and add that in next. Fading it in supplements the overheads nicely, adding the snare-drum weight that previously seemed to be missing. Balancing the snare with the cymbals isn't easy, though, because hi-hat spill on the top snare mic is putting the hi-hat out of balance before the snare feels like it's at the right level. I try using more of the under-snare mic, but that detracts from the overall snare timbre, so I just put the snare where I want it and leave the hat too loud for the moment.

The kick drum feels like it's probably the next most important instrument, but when I unmute its group channel, the kick drum's tone becomes rather boxy. Inverting the polarity of its group channel improves things considerably, but repeatedly muting and unmuting the bass drum highlights that its spill is destructively phase canceling with the snare's nice low midrange. I try time shifting both kick drum mics a little while listening for a better snare sound, but by the time the snare is sounding better, the kick drum's low-end power is suffering, and phase rotation of the kick drum's group channel doesn't yield any better outcome. So I go for the best snare sound and try to get a balance for the kick, switching back and forth between my nearfields and Auratone substitute to gain more perspective. Once set, the fader level seems fairly stable most of the time, but occasionally the odd uninspiring downbeat makes me yearn to nudge it up a fraction.

Under normal circumstances, I'd probably turn to the hi-hat mic next, but given the overprominence of that instrument in the balance already, I decide to leave it out completely. That leaves only the toms, which so far feel a bit soft edged and distant compared to the snare. Mixing them in one at a time I check for the most promising polarity of each. For both rack toms, the different polarity settings have different effects on the sound of the ensemble, but it's difficult to decide which I like more. Neither seems to affect the snare unduly, so I choose the settings that seem to give a bit more "snap" to the tom sound. I don't waste time with any more detailed phase matching for tonal reasons, because the tom sound just doesn't feel that important to this particular mix. When it comes to balancing, though, I find I can't push the tom-tom faders up far enough without spill making the cymbal balance overbearing. The floor tom close mic clearly detracts from the snare sound when I first fade it in, but inverting its polarity remedies this situation and otherwise the spill elements picked up by this mic don't seem to cause real problems with any of the other instruments. The main concern while balancing it is that the tom's sympathetic ringing ends up a bit overpowering when the close mic is at a level suitable for the drum fills. As a final touch, I bring up the room mics, but at the low level I choose for them their exact polarity/

phase relationship with the rest of the mics seems to be much of a muchness, so I leave them as they are in that regard. I switch between my nearfields and Auratone substitute a few more times for more perspective, and then I'm ready to add in the next part.

8.4.1 Realistic Expectations

So what has been achieved up to this point in my cloud-cuckoo mix? Well, I've managed to arrive at a reasonable initial balance with a pretty good snare sound. Neither is perfect, by any means, but I've made a lot of important decisions and the sound is already getting better. That's more than enough progress for the moment—after all, I've only just got started! What's absolutely crucial, though, is that the balancing process has already furnished me with the beginnings of a game plan, because I've paid attention to what the instability of certain faders is telling me. So I now know that the kick's low frequencies are balanced too low, and occasional hits are played too quietly; that spill on the snare and tom mics is balanced too high; and that some undamped resonances on the floor tom are balanced too high as well. Notice that I've been able to diagnose all these ailments simply by thinking in terms of balance. Can I hear what I want to hear? If not, then I know I'm going to need further processing. Naturally, the next question is, what processing do I need? The answer to that one is also to be found right there in the balance, but to unveil it you need to learn how to identify exactly what's making each fader unstable—which is the subject of the rest of this part of the book.

> Your first draft mix may be totally up the spout, but don't sweat it. However misguided any of your instinctive mix choices, the cold, hard realities of side-by-side comparison will set you straight at the mix referencing stage.

Clearly, the effects of polarity and phase changes during balancing introduce many very subjective tonal decisions into the process, and no matter how much experience you have of the style you're working in, you'll unavoidably gravitate toward the kinds of sounds you personally prefer. That's perfectly fine—there's room for that in a mix, and it's what makes each engineer's work unique. "After all," says Paul Worley, "the music that really works—the music that really rings the bell and makes people go crazy—is the music that doesn't sound like everything else."[47] Of course, every small-studio user's worry is that their own favorite sounds will make the track less well-suited to its target market and thus less competitive. In some cases this fear will probably be justified, but you shouldn't fret about this possibility unduly at the balancing stage—wait until you have a draft version of the fully mixed sonics that can reasonably be lined up against your mix references. Experience counts for a lot in mixing, and well-honed instincts can quickly create most of a release-ready mix. However, it's important to accept that instincts can be unreliable, so there's no shame in being a little uncertain about everything until you can properly acid-test your work against relevant professional productions. If you're just starting out with mixing, your first draft mix may be totally up the spout, but don't sweat it. However misguided any of your instinctive mix choices, the cold, hard realities of side-by-side comparison will set you straight at the mix referencing stage, and as a result your intuition will be improved next time round.

CUT TO THE CHASE

- The simple act of mixing your musical sections and your instruments in order of importance avoids a lot of the mixing pitfalls most frequently encountered by small-studio users. It also reveals the logic behind many arrangement and mixing decisions and helps you make the best use of limited computer CPU power.
- Different types of instrument recording demand slightly different balancing procedures, but in every case you should set up your mixer so that control is straightforward; optimize polarity and phase relationships; remove any unwanted low end; decide on stereo positioning and width; and set the level.
- Polarity inversion, timing shifts, phase rotation, all-pass filtering, and spectral phase-manipulation are all useful tools for getting the best tone from multimiked and layered instruments, as well as for turning the spill in multimiked ensemble recordings from a curse into a blessing. Be on your guard for situations where phase-relationships vary over time, though, because these can introduce undesirable tonality fluctuations that are tough to remedy with mix processing. With multimiked ensemble recordings, timing shifts on individual microphones may also affect the transient sharpness, stereo imaging, and depth perspective of certain instruments.
- Unwanted low frequencies can usefully be removed from the majority of tracks in a small-studio mix using high-pass filtering. Avoid resonant filters and those with slopes steeper than about 18dB/octave for this purpose. Judge the filter cut-off frequency within the context of the mix, and be especially careful not to rob drum transients of their weight.
- The way different engineers use the stereo field can vary a great deal and largely comes down to personal preference. However, there are some more objective technical issues to consider when panning, such as mono-compatibility, bass playback efficiency, and the danger of nonsensical musical results if only one side of the stereo mix is heard. Your DAW system's panning law may need changing if you want to use dynamic panning moves in your mix. A vectorscope display can be useful for troubleshooting subtle stereo offsets.
- Ensure that you spend quality time with your faders, taking full advantage of all your monitoring systems and techniques to get the clearest view possible of your initial balance. You shouldn't expect a finished mix at this early stage, but if you sensitize yourself to the stability of different fader settings, then that will give you valuable clues about when and how tracks need processing.

Assignment

- Write down all the sections in your production in rank order. Now list the instruments in each of those sections in rank order too. Consider whether any further multing might be sensible in the light of this ranking.
- Build up a balance for the most important section of your production, starting with the most important instrument and then working through your list in order. Deal with routing, phase, high-pass filtering, and panning considerations for each track before trying to decide on the fader level.
- Make a note of any faders that feel unstable.

Web Resources

On this book's companion website you'll find a selection of resources to support this chapter, including:

- a special tutorial video to accompany this chapter's assignment, in which I build a raw balance of a simple multitrack session, using high-pass filtering, panning, polarity/phase manipulation, and fader levels;
- some audio examples illustrating the high-pass filtering and polarity/phase issues discussed in this chapter;
- links to affordable utility plug-ins for high-pass filtering, stereo balancing, polarity/ phase adjustment, and gain setting.

www.cambridge-mt.com/ms-ch8.htm

CHAPTER 9
Compressing for a Reason

Few aspects of mix processing seem to confuse more small-studio owners than compression, despite the fact that this mix process is actually closely related to what the faders do—the only thing that really differentiates a fader from a compressor is that the latter can perform balance changes automatically. It's easy to lose sight of this basic principle when faced with all the technicalities of compressor designs and controls, so this chapter deliberately focuses on that issue while looking at the practical concerns of using compressors at mixdown. Technical bits and bobs will be sprinkled in as they become relevant.

9.1 COMPRESSION WITH TWO CONTROLS

A compressor is effectively a fader that you can program so that it wiggles around in real time. The beauty of it at mixdown is that you can combat undesirable signal-level variations that would otherwise prevent an instrument from maintaining its position in the balance. In other words, a compressor can iron out some of those pesky fader instabilities that we met while performing the initial balance in Chapter 8.

To explain how a compressor can do its job, let's take the example of a lead vocal recording where the singer mumbles some of the words. If you set your fader so that the majority of this vocal part is nicely audible in your mix, the lower-level mumbled words will start playing hide and seek. If, on the other hand, you fade the vocal up so that the mumbled syllables come through, then the rest of the vocal will eat Manhattan! The result is that no single fader setting will give you a good balance, because the difference between the vocal recording's highest and lowest signal levels (the dynamic range) is too large.

Compressors provide a solution to this problem by reducing dynamic range. In other words, compression reduces the level differences between the mumbled and unmumbled words in our hypothetical vocalist's performance, making it easier for you to find a static fader setting that works for all of them. The way the processor does this is to turn down (or "compress") the louder signals in a given audio recording so that they match the quieter signals more closely. All the compressor needs to know to pull off this stunt is which signals you

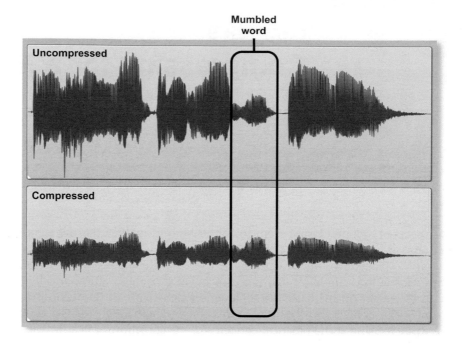

FIGURE 9.1
This example shows how compression can rebalance the relative level of a vocalist's mumbled word by turning down all the other words.

consider to be too loud, and all compressors have a control for specifying this. What can be a little confusing, though, is that this control is implemented on different compressors in quite different ways.

9.1.1 The Many Faces of Threshold and Makeup Gain

There are three different labels you'll commonly see for this first main compressor control:

- *Threshold.* This is the most common control design, but it can seem a bit counterintuitive to start with, because you need to turn the knob down to increase the amount of compression. This is because the control actually specifies the level threshold above which the signal is considered to be too loud. So when the Threshold control is set to maximum, little if anything is considered too loud and precious little compression occurs; as you begin to turn the control down, only the signal peaks are reduced in level; and as the control reaches its minimum setting, all but the softest signals are being sat on.
- *Peak reduction.* A control labeled like this (or sometimes just "Compression") will give you more compression (i.e., more reduction of peak levels) the more you turn it up. Easy peasy!

FIGURE 9.2
Some examples of compressors using each of the three main control types: Peak Reduction (Universal Audio LA3A, *top*); Threshold (Focusrite Midnight Compressor, *center*); and Input Gain (Native Instruments' VC76, *bottom*).

■ *Input gain.* In this case the compressor design has a fixed threshold level above which it will start turning the volume down, so the way you set the amount of compression is simply by varying the input signal's level relative to that threshold. The more you turn up this control, the more the signal exceeds the fixed threshold level, and the more compression you get. Although this sounds a bit like using a Peak Reduction control, the crucial difference is that here the signal's overall level also increases as you add more compression.

I usually recommend that newcomers initially steer clear of compressors with Input Gain controls, because the level hike can easily give the erroneous impression that your processing is improving the sound, even if the amount of compression is way overboard. By contrast, the other two control types tend to make the signal quieter when you compress, so there's less danger of overprocessing. Nevertheless, in the long run it's smart to become comfortable with all three of these common control setups so that you have the widest choice of compressors available to you at mixdown.

It's almost impossible to squish an instrument's dynamic range without simultaneously altering its subjective volume. Most compressors acknowledge this by providing a means of redressing this level change. In most cases, this will simply be a gain control (typically labeled "Makeup Gain" or "Output Gain"), but in some compressors the manufacturers have designed this level-compensation to be automatic. The automatic option seems like a bright idea on the face of it, because it leaves one hand free for beer, but my experience with such designs suggests that they almost always make compressed signals feel louder than uncompressed ones, which again often encourages inexperienced users to over-cook the processing. (Cynics might suggest that such a loudness hike probably helps manufacturers sell more plug-ins too.) Fortunately, many software compressors allow you to choose between manual and automatic gain adjustment here, and I always choose the former if at all possible.

You now understand the theory behind the two most fundamental compressor parameters: threshold and makeup gain. (For the sake of simplicity, I'll refer to them from now on by these names, rather than their variants.) Although many compressors have masses of other controls, you can actually improve your mix balance a tremendous amount without tangling with any of them, so for the moment let's abandon further discussion of technicalities and concentrate on getting the best practical use out of what we already know.

To start with, go back to your initial balance and mute all the channels. Make sure, though, that you leave the fader levels undisturbed at their "best guess so

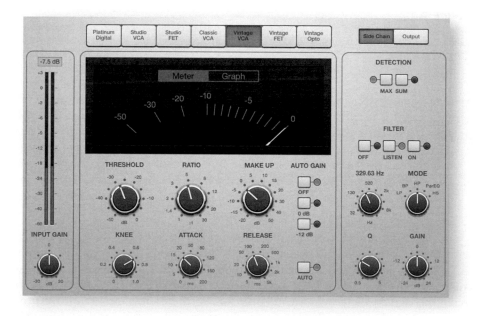

FIGURE 9.3
Apple Logic's built-in compressor is one of many software designs that offer automatic gain compensation (via the Auto Gain switches in this screenshot). For mix purposes, it's usually better to switch it off.

far" settings. Now start reintroducing the instruments in order of importance, but this time concentrate on how compression might be able to improve the solidity of the balance.

9.1.2 Which Tracks Need Compression?

The first thing you'll need to decide for each track you add into the mix is whether it actually needs to be compressed at all, and this is where you can draw on what you've already learned at the initial balancing stage. Remember that the first main aim of mixing is simply to achieve a good balance of the instruments; it's only when you can't find stable fader settings that deliver this goal that you should really consider applying any additional processing. There are lots of possible reasons why a given fader setting can feel unstable, but only some of those problems can be solved with compression, so you need to understand exactly which kind of instability you're looking for in this case.

Automatic gain makeup seems like a bright idea on the face of it, because it leaves one hand free for beer, but it often encourages inexperienced users to overcook the processing.

The simplest clue that you need to compress a track is that you keep wanting to reach over and adjust its fader. So in the vocal example I gave earlier, you might feel you've got the right level for the whole track, but then you find yourself lunging for the fader whenever a mumbled phrase comes along. Because compression can reduce the difference in levels between the clear and mumbled words, it can remedy this problem so that you can leave your fader in peace. To state this idea more generally, compression can be your friend wherever loud moments of a track poke too far out of the balance or wherever quiet details get lost.

If you scrutinize each unstable fader and can say, hand on heart, that none of them suffer from this kind of problem, then no one's going to arrest you for leaving the compression well alone. Randy Staub[1] and Joe Barresi[2] both point out that electric guitars rarely require much compression, for instance. "They're already compressed coming out of the speaker," explains Barresi. "I hardly ever use compression on [electric] guitars . . . If I'm miking the room with some ambient mics I might squeeze the ambient mics, but I never use a microphone in front of a cabinet and compress it. Ever. There's no need." However, although it's vital that you don't assume that compression is needed on any track, there are lots of good reasons why compression is one of the most commonly used effects in mainstream record production. In the first instance, no musician ever maintains their balance perfectly, and though the top session players can get pretty close to that ideal, I've not heard much evidence that those guys hang round small studios very often. In the real world, you have to do the best you can with whatever performers you have on hand, and that invariably involves a certain amount of airbrushing when you're trying to rival the polished performances on typical commercial releases. The drummer may

> ## GAIN MANAGEMENT WITH PLUG-INS
>
> In Chapter 8 I explained that the floating-point maths used in DAW mixers allows you to run your channels and busses at pretty much whatever signal level you want without appreciable sonic degradation, provided that your mixer's master output level remains sensible. There is one important caveat, though: any software that emulates the real-world characteristics of hardware audio equipment may nonetheless respond very differently to different signal levels. As such, getting the best sonic results with plug-ins sometimes involves adjusting how hard you drive them. "I love distortion," says Delbert Bowers, "but overloaded plug-in distortion is the worst type of distortion."[3] There's nothing complicated about this, though: just use a simple gain plug-in to adjust the input level to whatever analogue-inspired plug-in you're using, and if that puts your mix balance out of kilter use a second post-processor gain plug-in to redress the situation.

hit the snare unevenly or may lean on the hi-hat too hard (as in the Chapter 8 case study); the bass player or guitarist may not be able to pick/strum evenly or fret cleanly during a fast riff; and the singer might be lucky to hit the high notes at all, let alone at a consistent level.

The instruments being played can cause just as many problems too. Badly matched components in a drum kit are one of the most frequent culprits, whereas guitars and amplifiers with poorly controlled resonances are also common. Room modes can add further unevenness to the volumes of different instrument pitches. Machine-driven parts aren't immune from problems either, because sloppy MIDI programming can make even the best sampled and synthesized sounds feel lumpy in terms of level. Likewise, there's no end of ways in which a sampler or synthesizer can spoil the balance of a beautifully programmed MIDI part.

But even if there's little remedial work to be done, the demands of many modern music styles are almost impossible to meet without the help of compressors. No bass player I've ever met can play as consistently as is demanded of a mainstream up-tempo pop or rock mix, for example. Lead vocals are seldom mixed without compression either, because singers naturally tend to have a wide dynamic range but most producers choose to narrow this considerably so that the main melody and lyrics remain almost uncannily audible at all times.

9.1.3 Getting Down to Processing: First Steps

Once you've located a track that you think might call for compression, then it's time to choose your weapon. So which compressor should you choose? At the risk of uttering studio heresy, I'd say there are more important things to worry about to start with than the particular model of compressor you use.

As Tony Visconti puts it, "A compressor is a compressor—it doesn't matter whether it's a $5000 one or a $150 one. You have to know how compression works to use it."[4] "Don't obsess," adds Bob Clearmountain. "I hear stories of people who'll try eight different compressors before they decide on the right sound. Really? You're going to waste all that energy and time? They all do pretty much the same thing. They're a little different, of course, but I'm just not that picky!"[5] To be honest, you might as well use whichever one comes to hand, bearing in mind the advice I offered earlier regarding designs with Input Gain controls or automatic gain compensation. It can make things easier if you use one that has a gain-reduction meter of some type, though, as this will show you when and how hard the compressor is working in turning down the levels of louder signals. Gain-reduction displays are typically in the form of VU-style moving-coil meters or LED bar graphs, and sometimes the compressor's normal level meter can be switched to show gain reduction instead. Whichever one you get, it'll usually be calibrated in decibels and you'll see it move whenever compression's happening to show how much the signal level's being turned down. (In case you're still in a quandary about which plug-in to use, check out this chapter's web resources page for some affordable suggestions.)

FIGURE 9.4
When you're learning to use compression, choose a fairly simple design to start with, such as these two excellent freeware VST plug-ins: Melda's MCompressor (*top*) and Klanghelm's MJUC jr. (*bottom*).

Insert your chosen compressor into the channel in question, and if there are any presets available for it, then select something likely looking—again, there's no need to give it too much thought for now, just go with your gut. When it's loaded up, pile on a fair dollop of compression by twisting the Threshold control so that the gain-reduction meter shows at least 6dB of compression occurring on signal peaks, and then adjust the compressor's Makeup Gain control to compensate roughly for any overall level change. Now make use of your different monitoring systems (especially the Auratone-substitute) to help you tweak the fader again, and ask yourself this question: Does the level of that track feel any more stable in the mix?

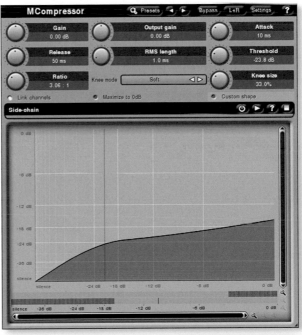

There are a lot of possible answers, so let's look at each in turn. Clearly, if your compression stabilizes the fader, then you've solved your balance problem and the job's done. However, even if you think this is the case, do still try turning the Threshold control back up a little to see whether you

can get away with using less compression. Pushing your channel compressors too hard is a common mistake that can slowly suck the life out of a mix if it's duplicated across all your tracks. "Compression is like this drug that you can't get enough of," says Joe Chiccarelli. "You squish things and it feels great and it sounds exciting, but the next day you come back and it's like the morning after and you're saying, 'Oh God, it's too much!'"[6]

If the balance problems aren't yet solved—you still can't stop fidgeting with your fader—then try rolling the threshold down further to see whether that makes it easier to find a decent fader level. Feel free to completely max out the control if you like, even it if makes the result sound rather unnatural for the moment. The important thing is to keep concentrating on the balance and whether the compression can deliver the static fader level you're after. Again, if you can get the balance you're happy with and you find any side effects of the compression appealing (as they often can be), then consider yourself the champ and turn your attention back to the rest of the instruments in your mix. In the process of experimenting with the threshold, it may become apparent that different settings actually suit different time segments of the track you're working on, in which case you may wish to do some additional multing of the track to implement this effect. Alternatively, many DAWs now offer the facility to apply gain offsets to different audio regions/clips, and that can help improve the sound of the processing. "So that the compressor's working more consistently, I'll do a lot of the donkey work by raising the waveforms with the clip gain," says Tim Palmer. "It makes a massive difference, because the compressor's not screaming for mercy!"[7]

> Pushing your channel compressors too hard is a common mistake that can slowly suck the life out of a mix if it's duplicated across all your tracks.

In the event that you're able to nail down an appropriate balance through heavy compression, you could nonetheless find that the processing isn't doing nice things to the instrument sound in question—perhaps it's making the performance feel lumpy and unmusical or altering the tonality in some unpleasant way. In that case, try switching to a new compressor or compression preset and have another crack. Different compressors and presets can respond very differently for similar settings of our two main compressor controls, and you don't really need to understand why this is to reap the benefits. Compare a few options and choose the one that does the job best. With a little experience, you'll soon build a shortlist of personal favorites for different instruments. Give it a few months, and you'll be sucking your teeth and saying, "That Fairchild's *phat* on glockenspiel!" with the best of them.

9.1.4 When Compression Is Not the Answer

In a lot of cases, though, you won't be able to stabilize the fader fully, no matter which compressor you use or how you set the two main controls. This is the point at which a lot of inexperienced engineers throw in the towel and simply settle for a compromise between dodgy balance and unmusical processing side effects. What you need to realize in this kind of situation is that your mix is trying to tell you that compression isn't the answer, especially if you've already tried a few different compressors or presets. So don't get too hung up on the idea of trying to fix all of an instrument's balance problems with compression, because other types of processing are often required as well. At this stage it's enough just to improve the track's balance in your mix as far as you can without making the sound less likable in a subjective sense.

So let's quickly summarize what we've covered so far. First, concentrate on the balance—can you hear everything you need to hear? When you can't find a static position for a particular fader, then compress (and maybe mult as well) to try to remedy that. When compression does solve your balance problem, then you need to ask yourself a second question: Do I now like the subjective "sound" of my compression? If not, then try a few different compressors or compressor presets. If you still can't find a static fader position that works, then play it safe with the compression until you see what other processing might be able to offer.

SHOULD I COMPRESS INDIVIDUAL TRACKS OR THEIR GROUP CHANNEL?

If you want to control the balance of several vocals/instruments at once, can't you just route them to a common group channel and compress that, instead of painstakingly processing each track separately? Unfortunately not, because the two approaches give different results. While I won't attempt to deny that it's tedious setting up loads of individual per-track compressors, that approach is better able to stabilize level imbalances between them, so I usually find it repays the extra effort whenever I'm faced with mixing project-studio recordings of strings, horns, or backing vocals. By all means use an additional group-channel compressor if you need to control the combined level of the tracks as well, but no matter how hard you push that compression, it won't be able to rein in, say, a single backing vocalist who momentarily sings louder than the rest.

9.2 REFINING COMPRESSION SETTINGS

If that were all there was to using compressors, then you'd be forgiven for wondering why manufacturers bother including any other controls at all. Some classic compressors do admittedly only have two knobs (the Teletronix LA2A, for example), but if you've already taken the opportunity to try out a variety of compression presets on the same sound, you'll have noticed that some of them will work more effectively in evening out the levels of the instrument in question than others, and that's because of the deeper compression parameters in each preset. If you can learn to adjust these parameters for yourself, you can match the compressor's action more closely to the specific dynamic-range characteristics of the input signal, and therefore more effectively achieve the static fader level you're looking for. Besides, once you get some practice with the extra controls, it actually ends up being quicker and easier to set them up from scratch anyway.

Another argument for learning about all the compression controls is that, although the technical *raison d'être* of compression is gain reduction, in the real world compressors usually do more than just reduce gain. They may also change the tone of processed signals quite a lot, even when compressing comparatively little. So if you get into a situation where you like the general tone or attitude that a certain compressor is imparting to your track, but there are no presets that are suitable for the instrument you're processing (or indeed no presets at all), then it's useful to be able to tweak the gain-reduction action manually to suit. That way you can switch between different characterful compressors to find the one that best enhances a given track, while still keeping the balance under control. Let's now introduce some of the more advanced parameters by demonstrating how each allows the compressor to adapt itself to specific tasks.

FIGURE 9.5

The slap-bass peak in the top waveform isn't reined in sufficiently at a 2:1 ratio (middle waveform) and requires much higher-ratio processing at 20:1 to bring it into line.

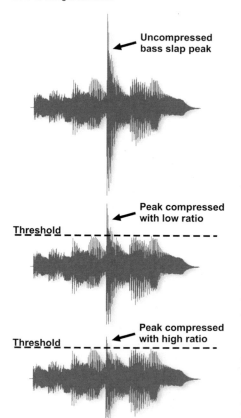

9.2.1 Compression Ratio

As a first example, let's consider a slap-bass part. Now, as everybody knows, the best processing for slap bass is that button labeled "Mute," but let's assume for the moment that this option has been ruled out. This particular slap-bass part balances fine with the rest of the track, except that the odd slap note really takes off and leaps forward out of the mix. The essence of the problem is that you only want to turn down these sporadic signal peaks, but you want to turn them down firmly in order to match the levels of the rest of the bass part. What compressors do is reduce the amount a signal level exceeds the compressor's threshold level, so in this case you want your compressor to put up a proper fight and all but stop the input signal from exceeding the threshold level. That way you can set the threshold level just above the level of the majority of the bass part, and it will then kick in at full force only when the overzealous slap notes hit.

By contrast, imagine an acoustic guitar part where there aren't these kinds of dramatic level spikes, but where the overall dynamic range is still militating against finding a static fader level. In this situation, you want your compressor to act more gently on signals overshooting the threshold so that you can set the threshold just above the level of the softest notes and then subtly squeeze the whole dynamic range down to a more manageable size.

It's a compressor's Ratio control (also sometimes labeled "Slope") that allows it to cope with these two contrasting compression requirements, effectively setting how firmly the compressor reins in signals that overshoot the threshold level. At low ratio settings (something like 1.5:1), the overshoots are nudged politely back toward the threshold, whereas at higher settings (12:1, for instance), they're beaten back by club-wielding thugs! At the highest settings (some compressors offer infinity:1), overshoots are effectively stopped in their tracks, unable to cross the threshold at all. So for our slap bass example, it'll be high ratios you're looking for, whereas for most routine dynamic-range reduction tasks (such as in the acoustic guitar example), the lower ratios (up to about 3:1) will tend to fix balance problems in a more natural-sounding way.

> At low ratio settings (something like 1.5:1) the overshoots are nudged politely back toward the threshold, whereas at higher settings (12:1, for instance), they're beaten back by club-wielding thugs!

When I'm talking about a ratio of 3:1 or whatever, what does that figure actually mean? Well, I could draw you some lovely graphs, but frankly I don't think it'd be a tremendous amount of practical help, because some compressors don't label their Ratio controls and different compressors can react quite differently for the same Ratio setting. Instead of thinking in terms of numbers, a more practical and intuitive approach is simply to use a compressor with a gain-reduction meter so that you can see when and how much the compressor is working as you juggle the Threshold and Ratio controls. In the case of our slap bass, you'd start off with the ratio fairly high, and then find a Threshold setting that caused the gain reduction to kick in only on the slap peaks. Once you'd done this, you'd listen to ascertain whether you'd solved the balance problem, and then adjust the Ratio control accordingly. Still too much slap? (Isn't there always?) Increase the ratio to stamp on the peaks more firmly.

With the guitar example, you might start off with a fairly low ratio (maybe 2:1) and then set up the threshold so that gain reduction was happening for all but the quietest notes. Once the threshold was in roughly the right place, you could then turn back to the Ratio control and tweak it one way or the other to achieve your static fader level. Some quieter notes still too indistinct? Increase

FIGURE 9.6
An acoustic guitar recording (top *waveform*) can be compressed moderately at a 2:1 ratio to add sustain without losing too much of its performance musicality (*middle waveform*). Using a high ratio in this instance (*bottom waveform*) unduly compromises the dynamic nuances of the playing.

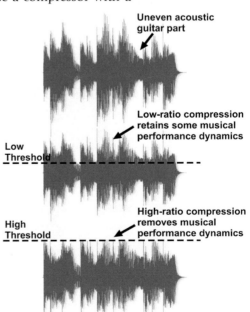

Uneven acoustic guitar part

Low-ratio compression retains some musical performance dynamics

Low Threshold

High-ratio compression removes musical performance dynamics

High Threshold

the ratio to reduce the dynamic range further and see whether this sorts things out. But why not just max out the Ratio control? The danger is that if you turn it up too high, you'll iron out the important performance dynamics that make the part sound musical, leaving it flat-sounding and lifeless, so try to use the lowest ratio that will get the balancing job done.

COMPRESSOR OR LIMITER?

Compressors that are specifically designed to offer very high-ratio compression are often called limiters, so if you find that your particular compressor simply can't muster a high enough ratio to do a particular job, don't be afraid to try a limiter instead. If you do switch to a limiter, though, you'll probably find that it uses an Input Gain control setup, and in some cases the threshold may be set to the digital clipping point for mastering purposes, without any postcompression gain control. This means that you can end up sending your overall signal level hurtling into the stratosphere before you've brought about the gain reduction you require. Fortunately, it's usually easy to add another little gain utility plug-in after the limiter to bring the overall level back down to earth.

FIGURE 9.7
Some good freeware limiters: Sonic Anomaly's Unlimited (*top*), George Yohng's W1 Limiter (*centre*), and Thomas Mundt's LoudMax (*bottom*).

9.2.2 Compressors in Series

But what would you do if that slap-bass part needed not only high-ratio control of the slapped notes but also more general low-ratio dynamic-range reduction? The answer is that you could deal with the problem by chaining more than one compressor in series. This is common in commercial practice, as it lets you dedicate each specific compressor to a different task. Here's Tom Lord-Alge describing a vocal application: "To make a vocal command attention I'll put it through a Teletronix LA3A and maybe pummel it with 20dB of compression, so the meter is pinned down. If the beginnings of the words then have too much attack, I'll put the vocals through an SSL compressor with a really fast attack, to take off or smooth out the extra attack that the LA3A adds."[8]

A second advantage of a multilayered compression approach is that you can use it to achieve industrial-strength dynamic-range reduction without driving any individual processor too hard. "I don't use a great deal of compression from each unit," says Rik Simpson. "I will just tickle it a little bit with each one, and each one adds a different characteristic. I could just add a load of compression from one plug-in, but like this it doesn't sound over-compressed."[9] Indeed, many classic recordings have benefited from this principle—a given signal might first have been moderately compressed on the way to the analog tape recorder to maximize signal-to-noise while tracking; the tape itself may then have compressed the signal slightly too, and further layers of compression would have been applied during mixdown. There's another significant reason why some engineers chain compressors: it allows them to blend the sometimes desirable sonic side effects of several different characterful units. This is particularly common with vocal parts, when you're looking for the most successful combination of tone and level control. However, these coloristic uses of compression won't get you anywhere unless you've mastered the processor's basic mix-balancing properties first.

9.2.3 Attack Time and Release Time: Why They Matter

Here's another example for you. Let's say that we're mixing a song where the strummed acoustic guitar has a nice natural sustain that works really well when it's at the right level in the mix, but you find that you have to turn the fader down whenever the player digs in more during your song's choruses. "Sounds like a job for Captain Compressor!" you cry, but when you actually start dialing in the processing you find that, rather than just addressing the level differences between your song sections, our plucky gain-reducing superhero seems intent on evening out the much shorter-term level differences between the attack-transient and sustain portions of each strum. Although you might be able to sort out your overall balance problem with this compressor, you'll be paying an unacceptable price: the impact of each strum will be softened or the instrument's sustain will be unnaturally overemphasized.

SOFT-KNEE COMPRESSION

Part of what differentiates compressor designs from each other is whether they begin working only the moment the input signal level exceeds the compression threshold (so-called hard-knee designs) or whether in fact they apply small amounts of gain reduction to signals well below the threshold level (soft-knee designs). The main benefit of soft-knee compression is that it makes the onset of gain reduction less obvious and therefore retains slightly more of the natural musical phrasing within the processed part. Quite a few classic studio compressors are by default soft-knee designs, a factor that has contributed to their long-lasting appeal. However, there are many times when the unnatural sound of the compressor working can actually be desirable, so it's not uncommon to find compressors that offer switchable or fully variable hard/soft-knee operation. When using different hard/soft-knee settings in practice, ask yourself the same questions as when trying out different compressors or compression presets: Am I achieving the desired balance, and do I like the processed sound itself?

FIGURE 9.8
A variable Knee control is very useful for adapting a compressor's action to different tasks.

The Attack Time and Release Time controls on a compressor provide a remedy to this ailment, because they determine how quickly the compressor's gain reduction reacts to changes in the input signal level: the former specifies how fast the compressor can react in reducing gain, whereas the latter specifies how fast the gain-reduction resets. The reason the compressor in our example isn't doing the job it's being asked to do is because it's reacting too fast to changes in the signal level; in other words, its attack and release times are too small. Increase them and the compressor will begin reacting more slowly, which means that it's likely to deal with this particular balance problem more efficiently, because it will track longer-term level variations (those between our verse and chorus) rather than short-term ones (those between the individual strum transients and the ringing of the guitar strings between them).

If you look at the scales used on these controls, you may notice that the times are usually expressed in milliseconds, although you do occasionally find microseconds and whole seconds in some cases. However, as with the Ratio control, I wouldn't recommend getting too hung up on exact numbers here, because the figures can only ever be a rough guide to how a specific compressor will actually respond. A much better tactic is to focus on adjusting the controls by ear until you get the best mix balance with the fewest unmusical side effects. If you're using gentle compression, you may find that it helps you hear the effects

of different attack and release times better if you temporarily increase the severity of the compression with
the Threshold and Ratio controls. A compressor's gain-reduction meter can be a useful visual
guide here too, as it'll show you not only how
much compression is being applied, but also
how fast it's changing in response to the track
you're processing.

> If you're using
> gentle compression, you
> may hear the effects of different
> attack and release times better
> if you temporarily increase the
> severity of the compression with
> the Threshold and Ratio
> controls.

9.2.4 Drum Compression: Three Different Settings

The ability to adjust a compressor's attack and release times significantly increases the range of balance problems than can usefully be tackled. To
illustrate this, let's look at another common example: a snare-drum backbeat:

- *Fast attack, fast release.* If you have a fast attack time, then the compressor
  will respond quickly to the fleeting initial drum transient, reducing the gain
  swiftly. If you then set the Release time very fast, the gain reduction will also
  reset very rapidly, well before the drum sound has finished, such that the
  lower-level tail of the drum hit won't be compressed as much. Result: less
  drum transient.
- *Fast attack, slow release.* Partnering your fast attack with a slower release will
  cause a rapid compression onset, but the gain-reduction will then reset
  very little during the drum hit itself, and mostly between the hits. The bal
  ance between the transient and sustain portions of the drum will therefore
  remain pretty much unchanged, and the compressor will primarily just
  make the level of each drum hit appear more consistent. Result: more consistent performance.
- *Slow attack, slow release.* Increasing the attack time will allow some of
  each drum transient to sneak past the compressor before its gain reduction clamps down properly. This effectively increases the level difference between the transient and the rest of the snare sound. Result: less
  drum sustain. (It's worth noting here that, although compression is
  normally associated with reducing dynamic range, in this case it might
  easily increase it.)

Here we can see how having variable attack and release times makes possible
three different balance results (less transient level, more consistent hit level,
and less sustain level), all from the same compressor. This ability to achieve
very different effects is partly what confuses some newcomers to compression,
and it's also one of the reasons why promisingly named compressor presets
don't necessarily do the trick—if your "Rock Snare" preset has been set up to
reduce the drum transient, for example, that won't help if you actually want it
spikier than a hedgehog's mohawk.

AUTOMATIC ATTACK AND RELEASE TIMES

Many of the most musical-sounding compressor designs treat short-term and long-term level variations differently, adjusting their attack or release times to adapt to the musical material. However, in some cases a dedicated automatic mode is provided that disables the manual attack/release time controls, and this will tend to work best with complex sounds that need transparent gain reduction—perhaps lead vocals or acoustic guitars in a sparse acoustic arrangement. However, if you're after well-controlled transient envelope adjustments (such as those demonstrated via the three contrasting snare-processing examples below), any artificial intelligence can actually hamper you in getting predictable results for large gain changes. "I can't let you do that, Dave."

9.2.5 Side Effects of Time Settings

Although thinking in terms of balance answers most questions pertaining to attack and release times, in certain circumstances you may find that your settings produce unwanted processing side effects, and it's as well to be on the lookout for these. The first occurs when you set fast enough attack and release times that the compressor actually begins to react to individual waveform cycles, rather than to the signal's overall level contours. The gain reduction then effectively changes the waveform shape and thus produces distortion, the nature of which will depend on the specific sound being processed and the compressor you happen to be abusing. Bass sounds, with their slow-moving waveforms, are particularly prone to this effect, but delicate acoustic instruments can also present difficulties because they will ruthlessly expose the smallest of distortion artifacts.

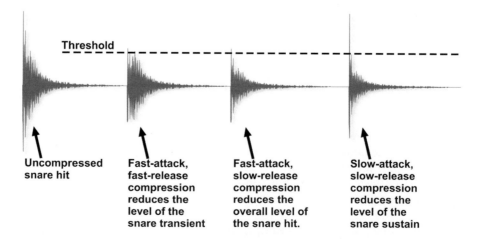

FIGURE 9.9
The effects of three different sets of attack and release times when compressing a snare-drum recording.

Another common problem is with percussive bass sounds, such as kick drums, which can lose bass content if you compress them with attack times under about 50ms. This is because the compressor begins clamping down during the first couple of waveform cycles, something that seems to impact on the lower frequencies more than the higher ones, shifting the tonal balance of the sound. Once you know that this is a danger, it's not that tricky to avoid, but if you're not listening for it, then it's easy to overlook while you're concentrating on wider balance issues.

One final observation is that changing a compressor's attack and release times will affect the amount of gain reduction that's occurring for a given combination of threshold and ratio parameters. For example, a side-stick sound (which comprises a short transient and very little sustain) might completely bypass a compressor with a long attack time, even if its level shoots way over the compressor's threshold. So it's not uncommon to keep adjusting compression threshold and ratio alongside attack and release times to take account of these kinds of changes.

LEVEL DETECTION: PEAK VERSUS AVERAGE

Another way you can adjust the time response of a compressor is by altering the level-detector circuit's reaction to short-term signal peaks. If the compressor is aware of every tiny level spike, it'll catch threshold overshoots reliably, but it will also react unmusically to many normal instrument recordings—our own hearing isn't as sensitive to short signal peaks as to slightly longer-term level variations. A lot of compressors, therefore, average their level-detector readings over time to some extent in order to match the response of the ear more accurately and achieve a smoother compression sound, and some designs even provide a Peak/Average (or Peak/RMS) control of some sort so you can vary the amount of averaging manually. If you do have a separate Peak/Average control, though, bear in mind that it will inevitably interact with the Attack Time and Release Time controls, so you may need to hop back and forth between all three to find the best final setting.

9.3 PARALLEL COMPRESSION

Although most studio users first experiment with compressors as insert effects, they can also be used in a send–return loop setup, such that you end up mixing both processed ("wet") and unprocessed ("dry") versions of the same signal. This technique has been in widespread use for decades within the professional community (where it's sometimes called "New York" compression), but it tends still to be underused by small-studio operators. The main advantage of the parallel setup is that the uncompressed signal retains some of the track's original transients and musical dynamics, even if you choose to pummel the living daylights out of the processed channel. "I'm a big fan of parallel compression," says Joe Barresi. "I'll often mix in noncompressed sound with compressed sound to maintain the transients."[10]

As such, one common application of this approach is to help reduce undesirable side effects wherever fast attack and release times prove necessary for balance purposes. Transient-rich instruments such as drums, tuned percussion, piano, and acoustic guitar frequently benefit from parallel processing for this reason, because that way you can use fast compression to bring up the sustain tails between individual hits/notes (a fairly everyday requirement in a lot of styles) without dulling percussive attack or flattening desirable performance nuances. In fact, so good is parallel processing at concealing the side effects of heavy compression that you can frequently afford to compress with gusto in this scenario, all but eliminating transients from the compressed signal so that it can bolster the uncompressed channel's sustain almost entirely independently. The crucial thing to remember is that the sound of the compressed signal on its own here is irrelevant—it's what it sounds like in combination with the uncompressed signal that counts. "It doesn't matter what it sounds like in solo to me," says Toby Wright. "It can sound like the weirdest thing on earth; as long as it fits and complements the track and what's going on around it, it's there to stay."[11]

Another common reason to employ parallel processing is if you want to overdrive characterful vintage-style compressors in search of juicy tonal and distortion artifacts, because you're free to do so without nearly as much danger of producing a lifeless final sound. In professional circles, this approach most often tends to focus on the most important sonic components of the mix, such as drums and lead vocals, and several parallel compressors might even be blended together. Michael Brauer,[12] Dave Pensado,[13] DJ Swivel,[14] and Spike Stent[15] all talk about using multiple compressors like this on lead vocals, for example. However, this kind of compressor usage is so subjective that there's little real advice I can offer here beyond Joe Meek's famous adage: if it sounds right, it is right!

There is one pitfall to be extremely wary of, however, whenever you adopt a parallel-processing approach, whether with compression or any other signal processor: loudness bias. It's very easy to make processed instruments sound louder when mixing a parallel-processed channel into the mix, and mixing newbies are frequently enticed into overusing parallel compression because the extra loudness makes pretty much anything sound better! My suggestion for overcoming this effect is to group both the uncompressed and compressed signals to a new channel and then use that group channel to set the balance. The other two faders can then be treated as an extension of the compressor's control set rather than being used for mix balancing. In fact, a number of compressor designs have actually begun to incorporate Wet/Dry Mix controls into their user interfaces to sidestep this issue and allow parallel compression to be employed as an insert effect, although I personally find these a bit hit-and-miss in operation, much like the automatic gain-compensation routines mentioned in Section 9.1.1.

COMPRESSING LEAD VOCALS

Although there are no special compression techniques specifically for lead vocals, it's common for their processing to be more in-depth to cope with the difficulties they present. Not only does their naturally wide dynamic range usually need narrowing a great deal, but the listener's everyday experience of natural-sounding vocals makes unnatural side effects from heavy processing difficult to get away with. My main suggestion for getting successful results is to tackle the task in stages:

- Mult sections of the vocal that have very different sounds to different tracks, and use clip/region gain offsets to deal with any obvious level disparities between vocal phrases.
- Use an initial layer of gain reduction to even out overall levels—soft-knee 2:1/3:1 compression with moderate attack and release times is likely to work best here if you're after fairly transparent gain control. Alternatively you might wish to use a multicompressor parallel setup to achieve more tone-coloring options.
- If vocal peaks are still poking out unduly, then follow this general compression with a faster, higher-ratio compression setting, or even some kind of limiter. A soft-knee model will be more appropriate for understated level control, but understatement isn't exactly the name of the game in many modern musical styles. With heavier compression, though, pay particular attention to how the attack and release times affect the singer's consonants. Slower attack settings, for instance, may overemphasize the hard onset transients of "k" and "t" sounds, whereas faster release settings may cause "s," "sh," "f", and "ch" sounds to become overbearing.

It's as well to realize, however, that compression (and indeed any other type of automatic mix processing) will never be able to achieve the inhuman degree of level consistency demanded of chart-ready lead vocals. The missing piece of the puzzle is detailed fader automation: "Once I get [the vocal] to where it's sitting nicely," remarks Bob Horn, for instance, "I go through with little automation pushes and push up the consonants and words just to poke them out a little bit. All that kind of helps it seem real 'in your face'."[16] However, it doesn't make sense to get into detailed automation moves until much later in the mix, so we'll leave a proper examination of that subject for Chapter 19. For the moment, just try to do the best you can without incurring overbearing processing side effects. If you try to nail every last syllable into place with compressors while balancing, you're almost guaranteed to overprocess.

A second thing to be aware of with parallel compression (or indeed when processing the individual tracks of a multimiked recording) is that you can easily run into comb-filtering problems if the compressor delays the processed signal. This is an ever-present danger in digital systems where analog-to-digital conversion and DSP processing both have the potential to incur short latency delays. Fortunately most small-studio DAW users don't run into latency problems too often while mixing because they rely mostly on software plug-in processing and their DAW system will normally have automatic plug-in delay-compensation routines. However, it nonetheless pays to be aware that some plug-ins don't declare their latency to the host software properly, so do keep an ear pricked for the frequency effects of comb-filtering whenever using a parallel processing setup. And even if a plug-in induces no obvious delay, some software nonetheless alters the internal phase relationships between the different frequencies passing through it, so don't be surprised if mixing together compressed and uncompressed channels occasionally alters the combined tone in ways you weren't anticipating.

9.4 BACK TO THE BALANCE

There are clearly a lot of ways that you can refine the way a compressor operates, but none of those refinements will help you at mixdown if you let them distract you from the fundamental questions you need to answer while balancing your mix:

- Is the compression helping the balance?
- Do I like the subjective quality of the compression sound?

If your first compressor choice doesn't give you the balance or the attitude/tone you desire, even once you've tweaked its settings, then switch to another. If you need more clean balancing power or want to use overt compression side effects to your advantage, then try chaining compressors in series or adopting a parallel processing setup. It shouldn't be a huge rigmarole, and there's no need

HYBRID ANALOGUE/DIGITAL MIXING

You'll find plenty of people who'll extol the virtues of analogue hardware for mixing purposes, but it's rarely a very good use of a limited small-studio budget in my view. There's a legion of top professionals working right across the stylistic spectrum who mix entirely in the software domain, so I don't think the lack of analogue hardware is by any means a deal-breaker when it comes to delivering commercial-sounding mixes. Basically, if you can't mix without hardware, then you simply can't mix! In fact, the inconvenience of recalling mix settings for analogue equipment actively militates against many high-level mix engineers using it, simply because their clients demand multiple mix revisions. Moreover, most small-studio engineers who approach me about using analogue mixing hardware haven't yet invested properly in monitoring and acoustic treatment, where a cash injection would improve their mixes much more, in my opinion. If you do decide to incorporate hardware into your system, though, the main things to check are that the send and return signals hit the DAW's A-D/D-A conversion stages at a sensible level, so you're not suffering from clipping or poor noise performance, and that the software correctly compensates for any delay incurred by the send–return loop, otherwise you'll get troublesome comb-filtering artefacts when working with multimiked recordings or parallel-processing configurations.

to despair if you can't achieve perfection. The aim is to make the mix better, not perfect, so don't push your processing too hard at this point. If any balance problem is proving insoluble, then that's probably a sign that simple compression isn't the right tool for the job.

CUT TO THE CHASE

- From a mix perspective, the primary purpose of compression is to achieve a stable balance. If a track balances fine without compression, then don't feel you have to compress it at all. When compressing you need to ask yourself two questions: Is the compression helping the balance; and do I like the subjective sound of the processing?
- The most important controls on a compressor are threshold and makeup gain (or their equivalents), and you should be able to get a long way with compression using just those controls. If they can't achieve a good balance for you without undesirable side effects, then in the first instance just try switching compressor or compressor preset.
- The remaining compression controls allow you to adapt a compressor's gain-reduction characteristics more closely to the needs of the target signal. While this can help you reduce unmusical compression artifacts, it also allows you to take advantage of any desirable sonic side effects of a given compressor in a wider range of situations. Be on the lookout for undesirable distortion or loss of low end when compressing with fast attack and release times.
- There is no rule against using more than one compressor in a single channel. Sometimes different compressors are required to do different balancing jobs, sometimes two compressors can achieve heavy dynamic-range reduction more transparently than one, and sometimes you might want to blend the tonal flavors of two different characterful units.
- Compressing several instruments via their group buss won't give as consistent a balance as compressing each track individually.
- Parallel processing can reduce some of the undesirable side effects of fast compression, especially when heavily processing transient-rich signals such as drums, tuned percussion, piano, and acoustic guitar. It also allows you to drive individual compressors harder in order to exaggerate their unique tonal personality in the mix. Just be careful to avoid phase-related comb-filtering issues and the loudness bias.

Assignment

- Track down all the compressors and limiters on your DAW system so that you know what options are available. If you have little choice, then consider supplementing your selection with third-party plug-ins.
- Check that your DAW system has automatic plug-in delay compensation, and that it is activated.

- Mute all the tracks in your mix, and then rebuild the balance as before, but this time experimenting with compression to see whether it can clear up any of the fader instabilities you identified while initially balancing.
- Make a note of any faders that still feel unstable.

Web Resources

On this book's companion website you'll find a selection of resources to support this chapter, including:

- a special tutorial video to accompany this chapter's assignment, in which I demonstrate how different compression techniques might be used to improve the balance of a simple multitrack session;
- numerous audio examples illustrating the effects (and side effects) of compressor time settings; the characteristics of parallel compression; the tonal variations between different analogue-modeled compressor plug-ins; and the way level automation can complement compression for vocal-balancing purposes;
- links to a wide range of affordable compressor and limiter plug-ins, together with further reading about the most common applications of different classic analogue compressors.

 www.cambridge-mt.com/ms-ch9.htm

CHAPTER 10
Beyond Compression

Compression isn't the only automatic gain-changing effect that has applications at mixdown—it's one of a whole family called "dynamics processes," or simply "dynamics." Other members of this family seem frequently to be ignored in the small studio, but they can also have their uses. This chapter explains some of these potential applications.

10.1 EXPANSION AND GATING

Whereas compression reduces dynamic range by turning down loud sounds, expansion does the opposite—it increases the dynamic range by turning down quiet sounds. There are a number of common applications for expanders when balancing a mix, including the following:

- reducing low-level background noise between vocal or instrument phrases;
- reducing hi-hat spill on a snare close mic;
- reducing the level of a drum sample's sustain tail relative to its transient peaks.

An expander's interface is no more complicated than a compressor's. As before, there's a Threshold control to tell the expander which parts of the signal need to be processed, although in this case it's signals below the threshold that have gain-reduction applied to them. The Ratio control adjusts how vigorously the expander turns things down once signals dive below the threshold level. A ratio of 1:1 causes no expansion at all, whereas a higher ratio of, say, 4:1 will dramatically reduce the levels of subthreshold signals. Some expanders even offer a ratio of infinity:1, which effectively mutes any signal lurking below threshold, in which case the processing is commonly referred to as gating, and some specialized expansion processors called gates offer only this most extreme ratio. (Incidentally, some expanders use different labelling conventions for their ratio controls, so that 4:1 expansion, for example, might be expressed as 1:4 or 0.25:1. Don't worry too much about this, though, because it's usually pretty obvious how the control works in practice once you start cranking it!)

FIGURE 10.1
A well-specified software gate, the Sonalksis SV-719 plug-in.

What can confuse some users initially with expanders and gates is that the Release Time control determines how quickly gain reduction is applied to subthreshold signals, while the Attack Time knob adjusts how fast the gain-reduction resets. However, given that attack transients in the processed track are often what actually cause the gain reduction to reset in these processors, this control legending is actually more intuitive for a lot of people, so it has been adopted as convention.

Using expansion for balancing purposes involves answering the same two questions we posed before with compression: Does the processing solve a balance problem, and has the subjective quality of the processed signal suffered in the process? If you can't get the required result with the basic four controls, then try a different expander or expander preset. If that doesn't improve matters, then just do the best you can with your expander and make a note to tackle the problem again with the more specialist tools we'll explore in later chapters.

EXPANSION: PRE- OR POSTCOMPRESSION?

If you want to use expansion on a track that already has a compressor on it, then in most circumstances it's sensible to insert the expander before the compressor in the processing chain. If the compressor squashes the dynamic range first, it will reduce the level differences between the quieter sounds you want your expander to attenuate and the louder sounds you want left alone, with the result that finding a successful expansion threshold becomes more difficult—or even impossible. However, there are inevitably exceptions to this guideline. For example, if you're using an expander to emphasize a drum sound's attack, you might set a high threshold level just to catch the peak of each hit, but if there's any variation in the peak levels, then you'll get some inconsistency in the transient enhancement. Putting a compressor or limiter before the expander in this case could actually make the expander's operation more reliable.

10.1.1 Stop That Chattering!

There is one practical issue you may have to deal with when using either a high-ratio expander or a gate. If you use fast attack and release times, any signal that hovers around the threshold level can trigger bursts of rapid and extreme gain-reduction fluctuations. At best these will simply sound like ungainly stuttering of the audio, and at worst they'll give rise to a particularly unmusical distortion often called "chattering"—it sounds a bit like whizzing snails in a blender! Although the obvious solution would be to back off the ratio or attack/release times, there are situations where that might not be an option—for example, if you wanted to cut off the unwanted tail of a snare drum sample abruptly.

One fix for gate chattering is to have an additional Hold Time control on your expander that enforces a user-specified delay between a gain-reduction reset (the gate "opening" to let sound through) and the next gain-reduction onset (the gate "closing" to mute the signal). Just a few milliseconds of hold time can be enough to deal with most cases of gate chattering. Another option sometimes provided is something called hysteresis, which effectively creates separate threshold levels for gain reduction and gain reset. A Hysteresis control will usually be calibrated in decibels, but don't pay the markings much heed—just leave the control at 0dB unless gate chattering becomes a problem, and then turn it up only just far enough to save those gastropods from their unpleasant demise.

> If you gate with fast attack and release times, any signal that hovers around the threshold level can trigger a burst of rapid gain-reduction fluctuations called "chattering"—it sounds a bit like whizzing snails in a blender!

10.1.2 Parallel Processing and Range Control

The same kind of parallel processing setup introduced in Section 9.3 with relation to compressors can also extend the possibilities of expansion and gating. Let's say you have a snare drum close-mic recording that catches some important elements of the hi-hat sound between drum hits, but at an unacceptably high level—a fairly common scenario. You might tackle this problem with expansion, setting the threshold so that the gain-reduction resets with every snare hit and then adjusting the ratio to reduce the hi-hat spill to taste. However, this approach will also increase the hi-hat spill's dynamic range, and that might cause a couple of problems. First, the hi-hat might be picked up with a different tone on the snare and hi-hat microphones (or there may be considerable comb-filtering effects between them), so the louder hi-hat hits, which have more spill, will therefore have a different sound. Second, if the snare and hi-hat close-mics are panned to different locations, the hi-hat image may appear to wander between them as the expansion varies the spill levels.

A solution to this problem might be to set up a high-ratio expander or gate as a parallel process instead, completely removing high-hat spill from the

> ## USING GATING TO RESHAPE DRUM ENVELOPES
>
> Almost all Hold Time controls on expanders and gates have a much wider range of delay settings than is necessary to avoid gating chatter, and this opens up a technique that can powerfully remodel the level envelopes of individual drum sounds, especially when working with programmed parts that have little level variation. For example, if you put a high-ratio expander or gate onto a snare drum track and set the threshold as high as possible (while still ensuring that a gain-reset triggers reliably for each hit), then you can subsequently use the expander/gate's Attack, Hold, and Release Time controls to dramatically adjust the level envelope of each hit. You could increase the attack time to remove the initial transient completely, decrease the hold time to shorten the drum's tail, and adjust the release time either for an abrupt drum-machine-style ending to each hit or for a more gradual and natural-sounding decay.

processed channel and then mixing in the resulting "snare-only" signal alongside the unprocessed snare close-mic. By balancing the levels of the processed and unprocessed tracks together, you can freely decide how much hi-hat spill you include in the mix, without altering its dynamic range. Another useful application of this parallel expansion is emphasizing drum transients. Set the expander/gate with a high threshold and fast attack and release times to isolate just a tiny "blip" at the start of each hit, and then mix that back in with the unprocessed sound. Although on its own the isolated transient may sound uglier than the portrait in Keith Richards's attic, that's immaterial as long as it sounds good in combination with the unprocessed track.

You'll remember that some compressors have a Wet/Dry Mix control to achieve parallel processing within a single insert effect, and you'll find the odd expander/gate with one of these too. However, it's much more likely that you'll find a Range control instead, which achieves a broadly similar effect by setting the maximum allowable gain reduction. In our hi-hat spill example, then, you'd first set up the expander/gate using the maximum range setting to completely eliminate the hi-hat spill, and once you'd managed to get it triggering reliably you'd roll back the Range control to achieve the desired level of hi-hat spill.

10.2 TRANSIENT ENHANCERS

We already have at our disposal various means to emphasize a drum sound's attack transient: slow-attack, fast-release compression (as described in Section 9.2.4); fast high-threshold expansion; and parallel or limited-range gating. However, there is often a need for specialist tools to adjust the nature of transients in a mix, not least because digital recording doesn't smooth signal spikes in the way classic analog machines used to. "You spend a lot of time carving in analog to make things poke out," comments Jimmy Douglass, "but in digital everything is poking out and sticking in your face already, so the challenge is to smooth it out and stick it back there."[1] Al Stone complains about the impact of digital recording on drum sounds in particular: "If you get the best signal down on multitrack, you end up balancing your multitrack at the bottom of the faders on the way back, because all the transients remain."[2]

10.2.1 Threshold-Dependent Processing

A number of transient-modification options rely on some kind of level-threshold system to work. There is, for example, a processor called an upward expander (or decompressor) that increases the dynamic range of signals above the threshold level, rather than below it. If you set attack and release times very short, you can adjust the Threshold and Ratio controls to boost just the signal peaks. However, if the peak levels aren't absolutely consistent, upward expansion can exacerbate this irregularity, which is often undesirable.

Another type of processor design uses its threshold level simply to detect transients, and whenever one arrives it triggers a momentary gain boost to emphasize it. You usually get control over the amount of "attack" gain

FIGURE 10.2
Schwa's Dyno plug-in is an example of a threshold-dependent transient processor.

applied, and sometimes also the length and shape of the gain envelope used. Although you could create an identical effect using parallel gating, plug-ins that use triggered gain envelopes often have another trick up their sleeves: the ability to trigger a second independent "sustain" envelope when the input signal level ducks back below the threshold. This isn't something you'd use every day, but it has the capability to increase the sustain of drum samples or repetitive synth parts in a much more dramatic and controllable way than is possible with compression.

10.2.2 Look Mum, No Threshold Control!

Threshold-dependent transient enhancement has an important limitation, though: it only works on transients that are higher in level than the rest of the signal, so adjusting complex transient-rich tracks such as acoustic guitar or piano can therefore be unrewarding. For this reason, a dynamics processor that detects transients in a different way is typically a better bet for balancing them at mixdown. Whether a transient is loud or quiet, it will always by nature involve a rapidly rising signal level, so if a processor looks for this level change, it can detect transients more reliably without the need for any Threshold control. It's then a comparatively simple matter to allow the user to boost (or indeed cut) each transient. (An extension of this idea is for the processor also to detect rapidly falling signal levels, so that you can adjust post-transient sustain levels in the same threshold-independent way.)

Jason Goldstein describes one of his favorite applications for this kind of processing: "[I use it] a lot on acoustic guitars. In cases where they're just playing rhythm and they are too 'plucky,' I can . . . take some of the attack off, without using compression. . . . It's also a lot to do with the issue of apparent loudness. If you make the attack harder, something will sound louder. It will cut through the mix without having to add additional volume."[3] Jack Joseph Puig used a similar effect on Fergie's "Big Girls Don't Cry": "If you were to listen to the acoustic guitar without the plug-in, it would sound kind of lazy and not urgent . . . Instead I wanted to give the guitar an immediate attacking, in-your-face sound, as if it's really digging in and played with fingernails as opposed to the skin of the finger. . . . The brain subconsciously analyses where the front of the note is and what the feel of a record is. If you have attack like this, the recording feels exciting and vibrant, like it's moving and is fast. It's a large part of the way the record feels."[4]

Such processing is a very powerful addition to the dynamics arsenal, and it is often better suited to rebalancing attack and sustain components within individual notes than things such as compression and expansion. You may well not have this type of plug-in built into your DAW, but there is now a good selection of affordable third-party plug-ins available—check out the links in this chapter's web resources for some affordable recommendations. It is worth noting, though, that threshold-independent transient processors

TARGET-ORIENTED DRUM LEVELERS

One of the problems with trying to even out drum parts using compression is that you'll only get complete level consistency by using extreme ratio settings, but those often bring with them unwanted side effects such as transient loss and reshaped decay tails. In recent years, though, a new type of processor has been developed to remedy this. It still makes use of threshold levels to decide which hits to process, but also has a target level parameter. Each time a hit is detected, the plug-in applies a simple gain offset to bring its level closer to the target, and you can also usually set how far towards the target level each hit is shoved, how long each gain offset lasts for, and how quickly the gain resets afterwards. This scheme's primary appeal is the transparency with which it can make your kick or snare sit absolutely solidly in the mix balance, but the creative potential of some implementations also extends well beyond this. You may be able, for example, to set a target well below the detection threshold, such that you can turn down detected hits much more radically than a compressor would be able to do—perhaps even to silence! There may also be a threshold range provided, rather than a simple level, so you can process softer drum hits (such as a snare close-mic's ghost notes) without affecting the levels of the main signal peaks.

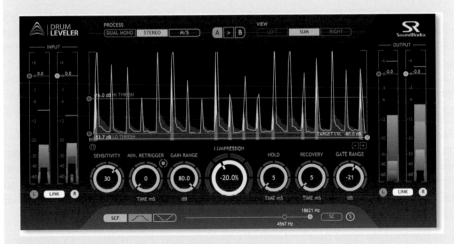

FIGURE 10.3
A target-oriented drum leveler such as Sound Radix Drum Leveler can make drum levels extremely consistent with comparatively few audible side effects.

from different manufacturers seem to respond differently in practice, in terms of exactly how many transients they detect and how they implement the triggered gain changes, so if you have access to more than one such processor on your system, then compare them to determine which design works best in each mix scenario.

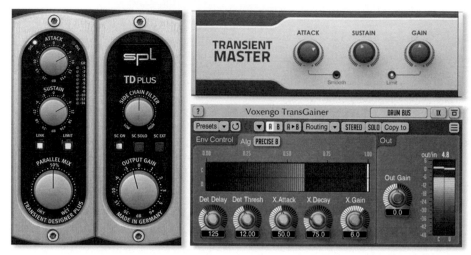

FIGURE 10.4
There are now lots of threshold-independent transient processors on the market, including (*clockwise from left*) SPL's software Transient Designer Plus, Native Instruments' Transient Master, and Voxengo's TransGainer.

10.3 TEMPO-DRIVEN BALANCING

Here's a final useful dynamics trick: triggering gain changes in sync with the song's tempo. So let's say you have a rhythm-guitar part playing eighth notes, and you want to increase the level of the off-beats in the balance. You try doing this with compression, but that messes with the note sustains. What you really need is a way to switch the track's gain between two different levels in a regular pattern. Fortunately, modern DAW systems now provide several fairly simple ways of achieving this as long as the song you're mixing lines up with the software's metric grid.

One method we've already covered is multing. Slice up the offending guitar part and chuck the bits onto two separate tracks so that you can balance them at independent levels. You might need to fade the edited sections in and out carefully to smooth the level transitions, but the results can nonetheless be very effective and there's the added advantage that you can easily create less regular tempo-related alternations, should you have an unusual fondness for capes and dry ice.

Another option is to find a dynamics plug-in that will respond to incoming MIDI notes—some plug-in gates work like this, for example. Once you have this kind of processor, it becomes easy to set up tempo-matched dynamic effects by programming simple metronomic MIDI parts within your DAW software. In our rhythm-guitar example, you might insert a MIDI-driven gate and then trigger that from an off-beat pattern of MIDI notes to let only the off-beat guitar chords through. You could then back off the plug-in's Range control to reinstate the on-the-beat guitar chords at a more appropriate level.

A dedicated tempo-synchronized gating, chopping, or tremolo effect can also give you this kind of dynamics control. Ed Boyer, for instance, talks about using one "to draw an envelope under the shaker, essentially manually gating out some of the in-between sounds that don't add to the musicality of the part."[5] Carlo "Illangelo" Montagnese, on the other hand, used one on some synth

FIGURE 10.5
Programmable tremolo units, such as U-he's Uhbik-T (*left*) and Cableguys's VolumeShaper (*right*) offer time-based balancing options that extend beyond the capabilities of normal level-based dynamics processors.

stabs "to adjust the transients and get the slope the way I wanted it. The original samples did not have quite enough punch."[6] As you can see, there are plenty of different applications for this kind of technique, just as long as you get hold of a processor that's fairly well specified, with a good choice of different level-modulation waveforms. In my experience it's a good idea to try to find a plug-in that also provides a Phase control, so that you can slide the selected waveform shape relative to the beat. This allows you to adjust the tremolo gain changes to suit any instruments that play slightly out of sync with the DAW's metric grid.

LOOKAHEAD AND LATENCY

There are some times with dynamics processing when it's nice to be able to predict the future. For example, if you're using high-threshold gating on a percussive sound, even with a very fast attack time, you can lose some of the sound's initial transient while the gate is still trying to open. To get around this problem, it's handy to get the expander/gate to react slightly before the signal level exceeds the threshold, and some digital processors do provide this option in the form of a Lookahead control. This is usually calibrated in milliseconds, and you only normally need a few milliseconds of lookahead to deal with this kind of problem. Mastering-style limiters commonly have lookahead facilities too, so that they can handle signal peaks more smoothly. While lookahead is undoubtedly useful, however, there is one thing to be careful of: some plug-ins that offer the facility can throw a spanner in the works of your DAW's plug-in delay compensation, so keep your ears open for comb-filtering problems if you're processing in a parallel configuration.

CUT TO THE CHASE

- If insufficient dynamic range is stopping you from achieving a good balance for a given track in your mix, then expansion may be the answer. An expander's concept and controls are similar to that of a compressor, so it shouldn't present much of an additional learning curve. In most situations, use expansion before compression in any chain of effects. If you're using a high-ratio expander or gate with fast time settings, then you may need to use Hold or Hysteresis controls to avoid unwanted distortion or level stuttering. Parallel or limited-range gating may be preferable to moderate expansion in some cases. A gate that has a lookahead facility can improve your results when expanding percussive sounds, but if you're using parallel processing, then make sure that the function doesn't induce latency and cause comb filtering. Gates can also be used to reshape the envelopes of drum hits, especially when you're working with sampled sounds.

- Threshold-dependent transient processors have some uses, but you'll get much more mileage from threshold-independent designs at mixdown. Different brands of transient processor tend to sound quite different, so make a point of comparing any alternatives to find the best match for each specific balancing task.

- Target-oriented drum levelers can provide particularly transparent-sounding balance control.

- If you can align your DAW's metric grid with the tempo of the music you're mixing, then there are various ways you can use tempo-driven dynamics processing to your advantage: via detailed multing, by using a MIDI-driven dynamics plug-in, or by applying a dedicated tempo-synchronized effect.

Assignment

- If you don't have at least one well-specified expander/gate and one threshold-independent transient enhancer in your DAW system, then find third-party plug-ins to fill these gaps in your processing lineup.
- Mute all the tracks in your mix and rebuild the balance again, this time trying out expansion/gating, transient processing, and tempo-driven dynamics in search of a more stable balance.
- Make a note of any faders that still feel unstable.

Web Resources

On this book's companion website you'll find a selection of resources to support this chapter, including:

- a special tutorial video to accompany this chapter's assignment, in which I demonstrate how other dynamics techniques can reach beyond the scope of simple compression when balancing a simple multitrack session;
- some audio examples demonstrating transient-processing and tempo-driven balancing in practice;
- links to a wide range of affordable plug-ins for expansion, gating, transient-processing, and tempo-driven balancing;
- further reading on the many mixing applications of gating.

www.cambridge-mt.com/ms-ch10.htm

CHAPTER 11
Equalizing for a Reason

Speaking in the broadest of terms, Chapters 9 and 10 are concerned with situations in which different time segments of a given track in your mix feel as if they demand different fader settings. It's these time-domain problems that create the kind of fader instability that makes you want to readjust the level setting the whole time, because a fader position that works one moment doesn't work the next. Dynamics processing provides the tools to balance all the audio events in an individual track's timeline relative to each other, which makes it easier to find a single, stable fader setting for that mixer channel.

However, dealing with time-domain issues is only half the story, because there will normally be frequency–domain balance problems to address as well—in other words, situations where an instrument's different frequency regions feel as if they demand different fader settings. The fader instability in this case has a different subjective character to it: you'll struggle to get a satisfactory level setting anywhere in the timeline, because a fader position that works for one frequency region doesn't work for another. This kind of problem requires a whole different set of mix-processing tools, the most common of which is equalization (or EQ for short). Because equalization is so important for mixing, and so many small-studio users seem to get the wrong end of the stick with it, I'm going to dedicate the entirety of Chapter 11 to the subject, before digging into some more left-field frequency–domain tricks in Chapter 12.

11.1 FREQUENCY MASKING AND BALANCE

An equalizer adjusts the levels of different frequency regions relative to each other. As such, it can tackle mixdown frequency–domain imbalances head on, so it's hardly surprising that it's used so much. However, most small-studio operators seem habitually to overuse it, largely because of a fundamental misconception about what it's there for. The root of the problem is that inexperienced engineers primarily use mixdown EQ to try to improve the tone, richness, and general subjective appeal of each of the instruments in their production. Although this is a worthy goal, it's actually a secondary consideration, because EQ has a much more crucial task at mixdown: achieving a stable balance.

To understand why equalization is so vital to creating a good mix balance, you need to know about a psychological phenomenon called "frequency masking" that affects our perception whenever we hear several instruments playing together at once. To put it simply, if one instrument in your mix has lots of energy in a certain frequency region, then your perception will be desensitized to that frequency region of the other instruments. In other words, if you have cymbals thrashing away and filling up the frequency spectrum above 5kHz, you'll perceive this frequency range a lot less well in the lead vocal part—the cymbals will be "masking" the vocal above 5kHz. Although the vocal might sound lovely and bright on its own, the moment the cymbals are added it will appear dull. To retain apparently the same vocal sound against the cymbals, you must either reduce the levels of the cymbal frequencies above 5kHz or exaggerate those frequencies in the vocal sound. "I'm creating space for the vocals," says Noah "40" Shebib, for instance. "I start pulling out the top end of a lot of the tracks to let the vocal actually fill that space."[1]

> EQ presets are of no use whatsoever at mixdown, because the designer of the preset can't possibly predict how masking will affect any given sound in your specific situation.

The ramifications of frequency masking for mixing are enormous. First, it should be clear that EQ presets are of no use whatsoever at mixdown, because the designer of the preset can't possibly predict how masking will affect any given sound in your specific situation—you might just as well ask your Aunt Mavis to set up the controls so they look pretty! For similar reasons, this book contains no handy "pull out and keep" list of recommended frequencies for EQ'ing different instruments, and I make no apologies for that, because my experience from critiquing thousands of real-world small-studio mixes is that this kind of prescriptive advice almost always hampers people's mixing much more than it helps. By the same token, an EQ setting that worked on one mix can't be expected to work on the next. Mick Guzauski: "You shouldn't say, 'Oh, I'm gonna add 5kHz to the vocal and pull out 200Hz, because that's what I did on the last song.' It may not work. Your approach has to be modified somewhat for each song, as each will have different instrumentation, a different singer, a different key, and so on."[2]

The effects of masking also mean that even if each individual instrument in your arrangement sounds good enough to eat on its own, you'll still need some equalization to compensate for frequency masking between the instruments in order to maintain the apparent tone of each one within the final mix. "The more and more I do this, the more I realise how important EQ'ing is," comments Eric Valentine. "Just voicing the instruments in a way where they're not fighting each other . . . so it feels like everything's upfront, nothing's getting buried."[3] What's more, carrying out this necessary processing may make individual sounds a lot less subjectively likeable when soloed, either because certain frequency regions have been exaggerated to overcome masking from less important instruments or because some frequencies have been cut to avoid masking more important instruments. In short, a good mix EQ setting is not necessarily the one that makes the instrument sound best in its own right. In some cases the only way to fit a subsidiary instrument into a crowded mix is if its frequency balance is mangled to the point of absurdity. That's why trying to make every individual instrument sound fantastic is a fool's errand.

Says John Leckie, "Sometimes the drums sound spectacular, but it's not what you want, because you want the guitar to be spectacular, and you can't have spectacular everything—then you wonder why the mix doesn't sound any good, because everything's crowding everything else. When you solo the instruments, everything sounds good, but when it's all put together it's a jumbled-up mess, so something's got to give way."[4] "Or the opposite can happen," says Tchad Blake. "You solo the kick drum and it'll be just awful. But then listen to it with the bass and it can be fantastic."[5]

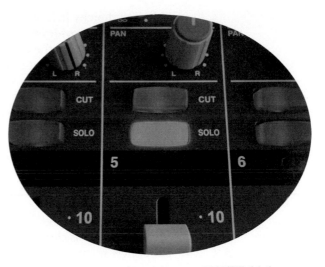

FIGURE 11.1
Don't try to evaluate the success of your mix EQ with the solo button down.

Now I'm not saying that EQ has no role to play in improving the subjective timbre, character, and general splendidness of individual instruments, but this needs to remain well below the balancing role in your list of priorities, otherwise the mix as a whole will never live up to its potential. Ideally, the tone of each instrument should have been optimized at the tracking stage, so if you start mixing one of your own projects and discover a track that's sonically underwhelming (even when it's soloed), then it's often a good indicator that the production's not actually ready to be mixed. Clearly, though, life in the small studio is often far from ideal, and there are usually a few instruments that need extra sparkle, warmth, aggression, moistness, flocculence, or whatever—but EQ often isn't the best tool for adding these qualities. So this chapter is going to concentrate primarily on the balancing role of EQ, and I'll leave it to Chapter 12 to describe some alternative tone-sculpting methods that can deliver serious subjective tonal changes on those occasions where you simply don't like one of the raw sounds you're presented with.

Another lesson to draw from frequency masking is that there's little to be gained by making EQ judgments with the solo button down, because you can't judge the impact of frequency masking between any two instruments unless you listen to them at the same time. Soloing tracks while EQ'ing can be useful for hearing exactly how and where you're altering the frequency balance, but it's imperative that you subsequently validate and refine those EQ decisions within the context of the mix. It's also worth pointing out that frequency masking is weakened between sounds that appear at different locations in the stereo image, so listening in mono will make such masking interactions more audible and thereby help you refine your anti-masking EQ moves. Finally, frequency masking provides another good justification for building up your mix balance by introducing the tracks in order of importance. If each newly added instrument is less important than the previous one, then you can be pretty sure which one needs to capitulate when masking conflicts arise.

11.2 BASIC MIX EQ TOOLS AND TECHNIQUES

With the key balancing role of EQ firmly in our minds, let's go back to the mix balance we've already fortified with dynamics processing. Once again, mute all the channels and rebuild the balance in order of importance. As you introduce each new track, listen carefully in order to answer these two questions:

- Can I find a fader level for the new track that allows me to hear all of its frequency regions as clearly as I want to?
- Is the new instrument leaving the perceived frequency balance of the more important tracks essentially unscathed?

If you can reply in the affirmative both times, then back away from the EQ with your hands in the air. A surprisingly large number of tracks in most productions only require a bit of high-pass filtering to mix, and you should already have attended to that troubleshooting task while you were putting together your first balance. When your answer to either of the questions is no, however, then it's a clue that EQ might be worth a try.

11.2.1 Shelving Filter Basics

An equalizer is effectively a kind of processing buffet, offering a selection of different filter types for users to dish out as they like. Each of these filters can change the frequency balance of a sound in a different way, so the first decision you have to make with any equalizer is which filter types to use. The best one to start off with when balancing a mix is called a "shelf" or "shelving" filter, a broad-brush processor that can change the level of one whole end of the frequency spectrum. The filter comes in two variants: a low shelf, which affects the lower end of the spectrum, and a high shelf, which affects the higher end. Shelves have at least one user-variable parameter, gain, which simply determines the level change applied to the filtered frequency region. The other main attribute of the filter is its "corner frequency," the notional boundary between the processed and unprocessed frequency ranges. Some EQ designs have shelving filters with fixed corner frequencies, but this usually restricts your options too much for critical balancing work, so seek out an equalizer that has a freely variable Frequency control instead.

> A surprisingly large number of tracks in most productions only require a bit of high-pass filtering to mix.

Although a few equalizers take the "processing buffet" concept into all-you-can-eat territory by letting you add new filters indefinitely, most equalizers restrict you to a finite number of simultaneous "bands," each of which can contain one filter. It's not uncommon for the bands in real-world EQs also to restrict your choice of filter type and to limit the frequency range over which that filter can operate. In such cases, you may find that shelving filters are only available for some of the EQ bands or that other filter types are selected by default. However, it's rare for an equalizer to have no shelving filters at all, so you should be able to dig one up if you look hard enough.

11.2.2 Balancing with Shelves

Once you've got suitable shelving filters on hand, return to that unstable fader and pull it right down. Slowly bring it back up into the mix again until you get to a level where it feels as if some frequency range of that instrument is too prominent in the mix. Take a guess at which half of the frequency spectrum is the culprit ("Is it treble or bass?"), and then load in an appropriate shelving filter to try to get some control over it. (If you're dealing with a stereo signal, then make sure to insert the processor in such a way that both sides of the stereo file are processed identically; otherwise the stereo image may become less solid.) First wiggle the filter's Gain control around a bit so you can hear what frequencies it's operating on, and then tweak its Frequency knob so that you get some useful handle on the region that's out of kilter, while at the same time minimizing changes to the rest of the spectrum.

It's normal for you to lose a certain amount of perspective during the frequently odd-sounding process of finding a good corner frequency for your shelving filter, so once you've set it, return the filter's gain to zero and give your ears 10 seconds or so to reattune themselves to the actual scale of the original balance problem. When you're back in the zone, gradually start lowering the filter's Gain control to pull the overprominent frequency region down to a more suitable level in the balance. Now turn up your fader a bit more; if that same region pops out of the balance unduly again, then pull back the filter's Gain control some more. Keep turning the fader up, and the filter Gain control down, either until you achieve a good balance for that track or until some other balance problem becomes more pressing.

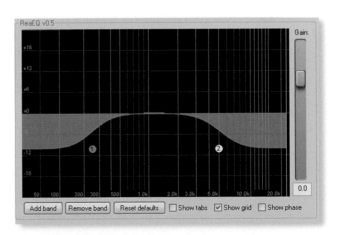

FIGURE 11.2
An EQ curve showing both a low shelving filter (*left-hand side*) and a high shelving filter (*right-hand side*).

If you do achieve a good balance straight away, then you need to make sure you're being as objective as possible about the decision. You should switch off the EQ now that you've finished adjusting it, and once more give your ear a few seconds to acclimatize to what the original balance problem sounded like. Then spend a few seconds trying to imagine what the fixed balance will sound like before switching the EQ back on. What you'll find surprisingly often is that your ear lost some perspective while you were fiddling with the filter gain, so your balance fix turns out to be lamer than a one-legged Vanilla Ice impersonator. It can be a bit dispiriting, I know, but your mix (and indeed your mixing skills) will progress more quickly if you force yourself to swallow the bitter pill of fine-tuning your EQ decisions properly in this way. Checking the validity of the balance on your different monitoring systems will also increase your objectivity.

Should a great balance remain elusive, then decide whether the shelving filter is actually helping at all. If the filter is improving the balance in some way, then you might as well leave it in place, albeit subject to the same objectivity checks mentioned in the previous paragraph. If the filter isn't helping you achieve a more stable fader setting, then don't think twice about abandoning it. Remember that shelving filters go on working all the way to the extremes of the audible spectrum, so listen particularly for any undesirable loss of low-end weight or any snide remarks from the studio bat. Whichever of these outcomes you're faced with, though, you've still got an unstable fader, so you're going to need further processing to deal with that.

> Remember that shelving filters go on working all the way to the extremes of the audible spectrum, so listen particularly for any undesirable loss of low-end weight or any snide remarks from the studio bat.

In the event that restraining one protruding frequency region with EQ reveals another different region poking out too far from the mix, then by all means have a go with additional shelving filters to see whether you can solidify the balance further. As long as you concentrate on making sure you're actually improving the balance, there's no reason not to have several shelving filters active at once on the same track—the gain changes of each new filter will simply accumulate as you'd expect. Normally, though, there's a limit to how much you can do with wideband processing like this, in which case the more surgical action of a peaking filter may be required.

11.2.3 Adding in Peaking Filters

The peaking filter is a more flexible tool for dealing with localized frequency–domain problems. It creates a peak or trough in the frequency response (depending on the gain applied), and you can change the center frequency of this adjustment region to target spectral problems. Well-specified models also let you control the frequency bandwidth that the filter affects, and some designs allow such precision in this regard that you can actually rebalance individual harmonics within a sound if you so wish. For mixing purposes you should choose an EQ that allows you to access all three important variables of its peaking filters—the first two controls will usually be labeled Gain and Frequency, but the last may go under the name of Bandwidth, Q, or Resonance, depending on the manufacturer.

You can follow exactly the same procedure when using peaks for balancing as
when using shelves, the only additional
consideration being how to set the
Bandwidth control:

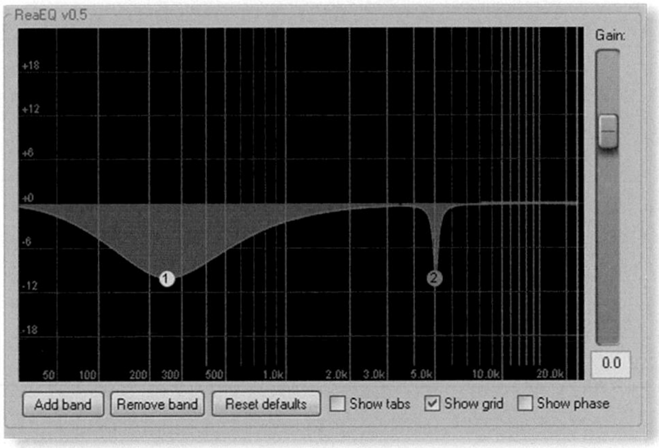

- Turn up your fader until some frequencies start to poke out more than
  others.
- Switch on the peaking filter and wag-
  gle the Gain control around while
  you try to find sensible frequency and
  bandwidth starting points.
- Once the filter is pointing pretty
  much in the right direction, reset its
  gain and give yourself a few seconds to become reaccustomed with the
  original balance problem.
- Turn the peaking filter's gain down (and possibly turn the fader up too) in
  search of a better balance, and make use of all your monitoring systems to
  keep perspective.
- When you think you've homed in on a good filter setting, deactivate it for a
  few seconds and then reengage it to check its validity.
- Refine the filter settings and repeat the previous step if necessary. Remember
  to keep switching monitors and trying different monitoring volumes.

FIGURE 11.3
An EQ curve showing
two peaking-filter cuts.
The left-hand one has a
wide bandwidth (in other
words, low resonance or
low Q value), whereas
the right-hand one has
a narrow bandwidth
(high resonance or high
Q value).

If possible during this last step, try to keep the bandwidth as wide as you can,
while still getting the job done, because undesirable filtering side effects tend
to be worse for narrower settings. "I generally use very wide Qs on everything,"
says Chuck Ainlay. "I don't usually get into narrow-bandwidth boosting or cutting because it takes the music out of most instruments, unless there's some
real problem element to an instrument that I need to get out."[6]

EQUALIZING BASS INSTRUMENTS

Judging by the problems I most commonly hear in homebrew mixes, bass
instruments typically present the biggest EQ challenges for small-studio occupants.
One of the big misconceptions is that all the important EQ adjustments for bass
instruments are at the low end. "I also add plenty of top end, so it fits in the track,"
says Chris Lord-Alge. "You may think that the bass sounds bright when you solo it,
but once you put in the heavy guitars it always seems dull all of a sudden."[7] Rich
Costey stresses this point: "It's sometimes quite shocking to realize how much top
end you need to add to bass to make sure it cuts through a track. If you then hear
the bass sound in isolation it may sound pretty uncomfortable, but in the midst of
the swirling din of a dense track that amount of top end usually works fine."[8] The

(continued)

additional advantage of defining bass instruments using their upper spectrum is that they will come through much better on small speaker systems, which helps the bass line reach the widest audience. Similar principles also apply to kick drums, of course.

One straightforward way of getting great-sounding bass instruments at mixdown is simply clearing enough space for them at the low end. "You can have sounds fighting for their place in almost every other register, and somehow they will find their place," explains Shawn Everett, "but elements that are disagreeing in the low end don't only fuck up the bottom end, but can mess with the entire recording."[9] "[Try] keeping pianos and all the other instruments from taking up that space where the bass starts, its fundamentals," suggests Bob Horn, "keeping that area clean or tighter to where the bass can have its own spot. I like to feel like, when I mute the bass, that there's not a huge hole missing, but a little bit of the weight has gone from the mix, and then you put it back in and it fills up again."[10] Beyond that, a critical EQ task in most mixdowns is finding a way to fit the kick drum and bass together. "Things can become very tricky with the bass," says Marcella Araica. "It's very easy to lose your kick drum when you add the bass. If things are not EQ'd right, you get more of the bass than the actual thump of the kick. You need to treat the bass and the drums completely differently, otherwise you'll get into trouble."[11] "No two atoms can exist in the same space," muses Young Guru. "Sometimes the client will walk in and they'll hear me taking frequencies out of the bass, and they're like 'What are you doing?!' and I'm like 'I'm making room for the kick, so that when these two come together they're going to wrap around each other.'"[12]

Taking a wider view, it's also worth realizing that many bass-processing decisions are inevitably driven by genre-specific expectations. "The major difference between a rock mix and a rap mix," remarks Noah "50" Shebib, for instance, "is that in the rap world you want that kick to crush you, and that's your priority, so of course . . . you're going to have to make room for it."[13] Indeed, the low end of a mix is such a strong stylistic cue that it can help make even retro-themed productions sound modern. "I'll have the midrange and the top end feel more like a sixties record, but put a modern low end on it," explains Eric Valentine, for example. "If you put that modern low end, get that sounding right, you can put almost anything on top of it and it'll be a modern record. It's all about the low end for me."[14]

How much low-midrange energy the main bass part contributes to the mix relative to other instruments is frequently a critical consideration too. Again, this is partly a question of musical genre. For example, mainstream chart mixes will often give the bass a comparatively free rein in this spectral region to clarify the musical harmonies, accentuate the bass part's melodic features, and allow a punchy and uncluttered rhythmic impression—as you can hear on things such as Justin Bieber's "What Do You Mean," Shawn Mendes's "Stitches," or James Bay's "Hold Back The River." Move into heavier rock and metal genres (say Metallica's *Hardwired To Self-destruct* or Icubus's *8*) and you'll find much more emphasis given to the low midrange of the guitars instead. The low midrange of the bass part will often also impact upon the stereo image, on account of the widespread convention of panning bass instruments centrally. If you cut low midrange out of the bass, you can fill that space with other parts that might provide more image interest, perhaps wide stereo keyboards, opposition-panned guitar double-tracks, or panoramic spreads of layered backing vocals.

11.2.4 Some Applications of Notching

Ainlay's final remark leads us to what is probably the peaking filter's most useful application: making narrow-bandwidth cuts, sometimes referred to as notches. "[I'll often apply] hard notch EQ," says Tom Elmhirst. "I try to find things that I hate, and the way I do that is by adding specks of EQ with a high Q setting, find what I don't like and off it goes."[15] Notches are regularly a lifesaver on project-studio drum recordings, where there always seem to be a few dissonant drum resonances that stick out of the mix awkwardly even when everything else appears to be at the right level. Because you're usually just dealing with a single frequency, you can position a super-narrow peaking filter precisely on top of it and pull it down out of harm's way without affecting the overall tone of the drum at all. To do this kind of frequency sharp-shooting, though, you'll need a peaking filter that can be made extremely narrow and some means of working out which exact frequency you need. Miss the right setting by a few Hertz, and you may not get the desired

effect. One method of finding drum resonances is to boost heavily with your narrowest possible peaking filter, and then hunt for the offending frequency by sweeping the filter around. Although this method certainly works, I've given up using it now because I can find these lingering frequency peaks more quickly and reliably using a high-resolution spectrum analyzer.

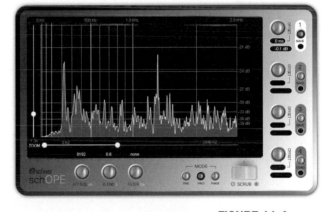

FIGURE 11.4
A high-resolution zoomable spectrum analyzer such as Schwa's Schope is a great help when you're hunting for unwanted narrowband resonances—the snare-drum hit in this screenshot has one at around 1.3kHz, for example.

Tampering with the levels of individual frequencies doesn't usually help you when dealing with pitch-varying parts, because a fixed-frequency notch will inevitably affect different harmonics on different notes. Nevertheless, don't rule out notching if the part in question is simple. For example, I quite often find notches handy for bass parts where the fundamental frequencies of different notes need evening out. Any part with a static drone or repeated note within it is also fair game if you want to rebalance the stationary note against any others, although you might find you need to attack more than one of the note's harmonics to achieve the balance you need. Induced noise from the mains electrical supply is another common candidate for multiple notching too, as this kind of noise often comprises a strong hum component at 50Hz (60Hz in the United States) with a series of harmonics at multiples of that frequency.

Narrow-bandwidth peaking cuts are great for counteracting unwanted recorded resonances, such as the nasty honks that room modes can overlay onto a recording that's been carried out in unfavorable surroundings. Guitar amp cabinets sometimes have unattractive resonant lumps in their frequency responses too, which make anything but the simplest part sound uneven. Sometimes singers overplay certain resonances of their vocal apparatus, so that a few single frequency components pop out from time to time and prevent the voice from sitting properly in the mix. (Tom Elmhirst used four different notches to deal with a situation like this while mixing the lead vocal on Amy Winehouse's "Rehab," for instance.[16]) Overall, you just have to be aware that even a single frequency can stymie your balance if it's far enough out of line, so try to make a point of listening for both wideband and narrowband frequency imbalances while working with peaking filters.

GRAPHIC EQ

The type of EQ that's of most use when mixing is called "parametric" EQ, and it is the main subject of this chapter. However, there is another type called "graphic" EQ, which abandons the idea of Gain, Frequency, and Bandwidth controls in favor of a more visually intuitive setup whereby the audible spectrum is split into lots of little slices, the gain of each being controlled using its own tiny fader. With all the faders arrayed in a line, you get a physical representation of the way the frequency response is being processed, which can be read and understood at a glance. This visual immediacy endears graphic EQ to front-of-house engineers, who already have their work cut out trying to stop Florence scaling the lighting rig, but it comes at the sonic expense of coarser control and increased phase/resonance problems, so I'd recommend steering clear of it at mixdown if you can.

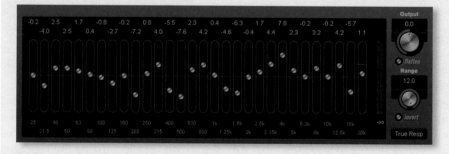

FIGURE 11.5
Steinberg Cubase's built-in 31-band graphic EQ plug-in.

11.2.5 Shift Your Listening Perspective

So far we've looked at how to maximize the audibility of each new instrument you add into the mix by ensuring that all its key frequency components are at the right relative levels. Although that's one big part of what mix EQ is about, the other vital step is to check whether the latest track has masked any indispensable aspects of the more important tracks that preceded it. The best way to assess this is to consciously direct your attention to each of the more important parts, listening for undesirable changes as you mute and unmute the latest addition alongside. With some experience you may be able to speed up this process by listening for changes across the board while muting/unmuting the most recently added track, but if the arrangement is complex, then a piecemeal approach will still justify the time required.

Let's take an example. Maybe you've just added some piano chords into the mix and EQ'd them to stabilize their fader and make them as audible as you want within the balance. However, in doing so you realize that they're masking the upper frequencies of your lead vocal, making it sound dull (or in the case of Morrissey, even duller). What this means is that there's simply not enough space in your mix for your ideal piano tone, and you're going to need to compromise some of its upper frequencies to clear enough room for the lead vocal timbre to shine through. How much you sacrifice the less important track will be a judgment call, of course, depending on which vocal frequencies are really critical and how much they're being masked. You might be willing to jettison a small amount of vocal clarity in order to avoid inflicting grievous bodily harm on the piano timbre, and that's a valid decision—after all, most listeners have highly developed expectations of how a piano should sound. On the other hand, you may be able to utterly purée a synthesizer or electric guitar with EQ under similar circumstances, because a listener's expectations of these instruments are usually much less well defined.

> You may be able to utterly purée a synthesizer or electric guitar with EQ, because a listener's expectations of these instruments are usually much less well-defined.

Once you get the idea of shifting your listening perspective so that you're listening to one track while equalizing another, this kind of masking problem should prove no trickier to address than any of the other EQ'ing we've talked about so far. It is, however, easy to forget to check for masking as you go, and by the time every track is mixed in it can be almost impossible to disentangle complex interwoven masking effects if the mix feels generally cluttered and lead instruments are struggling to sound their best. So try to be as disciplined as you can about this while you build your balance.

11.2.6 Good EQ Habits

You've probably already noticed that my suggested EQ procedure has so far used only EQ cuts, and that's no accident. Although it's actually more natural to think in terms of boosting frequencies you want more of, there are good reasons to retrain yourself to the opposite approach of cutting frequencies you want less of. As I mentioned in Section 4.2.3, all of us have a strong tendency to like the louder of any two sounds we're comparing, so if EQ boost makes your overall sound louder, then that'll add an element of bias to your judgments. Any EQ boost will also always make the processed track more audible

WHAT ORDER SHOULD I PUT EQ AND DYNAMICS PROCESSORS?

The main issue to consider when ordering EQ and dynamics plug-ins within a track's processing chain is that your dynamics processors may respond differently if you alter the frequency balance they are fed with. For this reason it makes sense to put EQ last in the chain if you're already happy with the way the dynamics are operating. However, it's not unusual for a frequency imbalance to prevent successful dynamics processing. A cabinet resonance on an electric guitar recording may well cause a single frequency to dominate the timbre only sporadically, depending on which notes happen to hit the resonance. If you compress a recording like this, you may be able to even out the signal level from a technical standpoint, but the tone will still change from note to note, so the subjective level of the part will remain inconsistent. Dipping the resonance with a narrowband peaking filter precompression would improve this situation.

So the general principle is this: if you're happy with the way your dynamics processors are responding, then EQ after them; if you aren't, then try EQ'ing earlier in the chain to see whether you can improve things. In the latter case, you may need to reassess your dynamics settings in the light of the altered frequency balance. There's no rule against EQ'ing at multiple locations in the chain either, because you may need one EQ setting to achieve a musical compression sound, but another completely different one to slot the compressed track's frequency spectrum into the mix balance as a whole.

in the balance, so it's trickier to be sure whether you've chosen exactly the right settings. The bigger the boost, the more difficult it is to keep your perspective. Sticking to EQ cuts avoids this bias, or rather biases you *against* your own EQ curve so that only settings that really work are likely to pass muster when you switch the plug-in on and off. "People need to know how to use their EQ," says Ed Cherny, "learning how to dip some frequencies and move the fader up instead of boosting all the time. . . . That's the biggest mistake that rookies make—they just reach down and boost the EQ and say, 'That's better!' because it's louder now, but they're not necessarily listening to how the sound is being shaped."[17] To be fair, there are now some EQ plug-ins that address loudness bias by using an automatic make-up gain function similar to the one I mentioned in relation to compressors back in Section 9.1.1. Again, though, I've found such gain compensation routines a little wayward in practice and, besides, I just find it a bit disconcerting when my EQ feels like it's adjusting my channel fader as well as the frequency balance!

There's a second substantial reason to avoid EQ boosts, though: it reduces the damage from a number of troublesome EQ-processing artifacts. Equalizers do a lot of things beyond just adjusting the frequency balance. For a start, most of them adjust the phase relationship between the track's frequency components (often called the track's "phase response"), and it turns out that the biggest phase shifts for many filter types are applied to the frequency region that's being adjusted. This clearly has knock-on effects for the way a track combines with others in a multimiked recording, because the EQ will alter the phase-related comb filtering between the processed mic and others in the setup. If you boost to give a frequency region in such a track extra importance, not only will the comb filtering at that frequency alter unpredictably (and potentially for the worse), but you may also increase the severity of the comb filtering in that region by virtue of increasing its relative level. Cutting, on the other hand, concentrates the phase shifts into frequency regions that are less crucial (that's usually why you're cutting them, after all), so any undesirable changes to the nature of the comb filtering won't matter as much and will at any rate be less severe by dint of the level drop.

Phase response doesn't just matter in multimiked recordings, either. Messing with it even on individual overdubs can make the sound appear less focused and solid in the mix, and transients in particular can start to become undesirably softened. So here too it's sensible to cut rather than boost, and also to sort out any frequency imbalances with shelving filters first if possible, because peaking filters usually cause more destructive phase shifts than shelves by nature of their design.

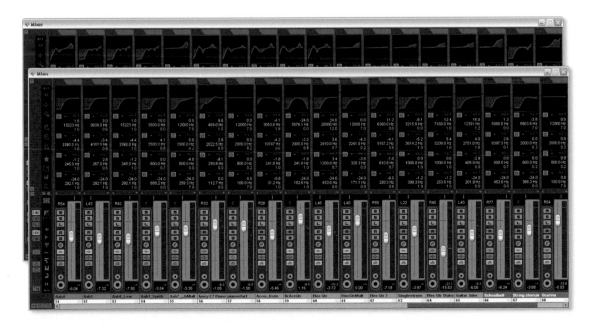

FIGURE 11.6
A before/after comparison of a range of EQ settings from one of my *Sound on Sound* "Mix Rescue" remixes. The addition of lots of EQ boosts on the original mix, especially at the high end, has added harshness, whereas the remix produced a smoother sound by concentrating on EQ cuts instead.

But there's more. Filters are also resonant devices, which means that any adjusted frequencies ring on slightly, extending over time—especially if you use narrow bandwidths. Again, this time smearing isn't actually desirable in some cases, so it makes sense to position it in a region of the frequency response where it doesn't impact as much—in other words, by using cuts instead of boosts. In addition, filters can often add distortion to the frequency regions they process, and while this is actually part of the appeal of celebrated

LINEAR-PHASE EQ: PROS AND CONS

Most equalizers will inevitably change a sound's phase response as soon as you try to tweak the frequency balance, but that doesn't mean all of them do. An increasing number of digital "linear-phase" equalizers, for example, have been designed not to alter the phase of a processed signal at all, and these can provide more predictable control in situations where there is the potential for phase cancellation between mixer channels, such as when processing individual tracks in a multimiked recording. Linear-phase processing can also help when it proves necessary to apply heavy corrective EQ to a lead part, because normal "minimum-phase" EQ designs can make the processed sound less subjectively solid in this scenario, especially if a number of narrow EQ notches are active at once. However, it's important to realize that linear-phase EQ isn't some kind of cure-all—it's just as likely to introduce distortion and time-smearing side effects, so it needs to be used with the same degree of care. In particular, you should be on the lookout for any resonant filter ringing, because in linear-phase designs this usually precedes rather than follows each sonic event in the track—an effect that sounds particularly weird on drums and percussion. Linear-phase processing also tends to be much harder on your computer's CPU than normal EQ, so you may have to choose your moments carefully unless you want an ear bashing from Scotty in the engine room.

FIGURE 11.7
Linear phase equalizers, such as the IK Multimedia T-RackS device shown here, can be very useful when working with multimiked recordings, as long as you can handle the higher CPU munch.

analog EQ designs, cheaper analog equalizers and low-CPU digital plug-ins (which small-studio denizens usually have to rely on when dealing with high track counts) may therefore sound harsh and unmusical if too much boosting goes on. Much better to use cuts instead so that any nasty sounding distortion components are kept at a low level.

Now I'm not saying "Thou Shalt Never Use EQ Boosts!" because there are some situations where it makes more sense to use boosts than cuts for mix-balancing purposes. For example, it's a lot easier to handle a peaking boost than trying to implement a similar spectral change using a pair of shelving cuts. With some classic designs, the sonic side effects of boosting may actually be desirable and improve the subjective tone of the processed track. However, I think these situations are best treated as the exception rather than the rule. You need to exercise caution wherever you use a boost, in order to minimize any degradation to the sound and to reduce the likelihood that you'll play mind games with yourself. In particular, try to keep peaking bandwidths as wide as you can, and be wary of judging the effectiveness of any filter boost simply by switching it on and off. In addition I'd also suggest a general rule of thumb: if you're boosting any track with more than one filter or with more than 6dB of gain, then you're probably overdoing it. "If I'm having trouble with a mix," says Gus Dudgeon, "probably I'm using too much EQ. That's a pretty golden rule . . . Basically, if you use too much EQ, you're trying to force something through a slot that's too small. . . . What [you] should be doing instead is backing off the EQ and turning the volume up."[18]

11.3 EQUALIZING MULTIMIKED RECORDINGS

Although the EQ techniques I've looked at thus far are fine for most simple recordings, there is an additional complication to contend with if you have multimiked (or DI-plus-mic) recordings among your multitracks: namely, if your EQ processing changes the phase relationships between the different mics and DIs in the recording, the outcome of the equalization may end up being as intuitive and predictable as a rollerblading platypus. "If you crank in a lot of EQ on the individual instruments, you really hear it," says Shawn Murphy, a specialist in mixing large-ensemble film scores, "and it alters the characteristics of the leakage, so you're fighting the fact that in order to make the direct sound right on a certain mic, you've got badly EQ'd leakage to deal with."[19] So let's investigate this subject properly.

11.3.1 Multimiked Instruments

The simplest situation is where all the mics and DIs in the recording are capturing the same instrument without appreciable spill from any other instrument. With any luck you'll already have taken the opportunity at the balancing stage to buss all the component mic/DI tracks through a communal

TONAL EQ

Although this chapter primarily centers around the balancing role of EQ, there is also a legitimate place at mixdown for purely subjective tonal processing. In other words, equalizing a sound to make it more attractive, rather than just to fit it into the mix balance. However, there's not a tremendous amount of useful advice anyone can give for this kind of EQ work. Just crank knobs until it sounds better!

It's in this context that classic analog equalizer designs really come into their own, because the processing by-products of these are often as much a part of their appeal as their ability to sculpt frequency response. Which equalizers suit which tracks is a subject that frequently seems to lead to pistols at dawn among different practitioners at all levels, so only your own experience can really guide you. Experimentation is the key to success here, and there's little excuse not to have a go now that digital emulations of classic analog devices are so numerous. My advice is to make a habit of trying out two different "character EQ" plug-ins every time you're looking for extra tonal magic. Do your best with each one, and then switch between them to assess their relative merits before choosing your favorite. This doesn't take too much extra time to do, but it quickly builds up your own knowledge of how different classic EQ designs respond to different sound sources.

Oh, and feel free to completely disregard my tutting about boosts when you're using EQ solely for tonal coloring (rather than for balancing), because they'll alter the attitude of a sound more than cuts will, especially when you're using emulations of classic EQ units. You might also want to chain several contrasting EQ plug-ins together on a single track, and that's all right too as long as you're convinced that the sound (and not just the loudness) is improving. "I think you can get [better results] by not trying to get everything out of one EQ," says Dave Pensado. "Every EQ is good for certain things. Even the cheapest EQ has a use. Don't judge gear by its cost or inherent quality; judge it by its uniqueness."[20] However, when chaining EQs do try to be clear in your mind whether each band of processing you use is for tone or balance, because you're likely to come a cropper if you try to balance your mix with massive character-EQ boosts or attempt to overhaul a sound's subjective timbre using bland balancing-EQ cuts.

Finally, remember that we humans tend to judge the tone-quality of a sound according to its context, so you can change how any instrument sounds in two ways: either you can process the instrument, or you can process its context. "If you want something to sound bright," says Dave Bianco, "put something dark behind it."[21] So, for example, you might make a lead vocal seem brighter not only by boosting its high frequencies with EQ, but also by cutting high frequencies from the instruments accompanying it to enhance their contrast with the singer's timbre—a technique powerfully demonstrated during the opening of Adele's song "Hello," where her high frequencies are effectively enhanced, subjectively speaking, by the much duller-sounding piano timbre.

group channel to allow global control over them all, and you'll have set each track's relative level to achieve the best combination of timbre and balance you can. However, do double-check that this internal level relationship still holds water in the light of any processing you've done since constructing your first balance.

If the group channel fader of the instrument still feels unstable in a way that appears to demand EQ, then you've got three options available:

■ You could tamper creatively with the timing, phase, and polarity of the individual mics to achieve a more suitable frequency balance in the combined sound. The upside to this option is that you can achieve dramatic changes if necessary without incurring any processing artifacts, because no actual processing is involved. The downside, though, is that this process is very hit and miss—you can experiment all you like in the hope of finding a better sound, but there's no guarantee you'll find it.

■ You could EQ the individual mics/DIs in the setup separately to achieve the best balance of desirable frequency components from each. If you have a target balance in mind for the instrument, it may help to listen to each of its component tracks separately to ascertain which track is most appealing for each frequency region, so that you can use EQ to favor the best bits on each track. However, this approach may not really bear fruit unless you've minimized any comb filtering between the tracks (so that they seem to add together fairly predictably) and the EQ processing doesn't unduly alter the phase response. Although you sometimes get lucky, in practice I've found that this kind of equalization inevitably ends up being a bit of a lottery most of the time and often necessitates remedial phase/polarity adjustments.

■ You could EQ the group channel that carries the combined instrument signal. This will be the most predictable of the three options, because you won't skew the phase relationships between the individual mics/DIs— you can process it just like a single mono/stereo recording, as described in Section 11.2. On the other hand, though, this method also gives you much less scope to alter the instrument's balance or subjective timbre.

Clearly, there are many ways to skin this particular cat, and it's fair to say that each individual engineer will have preferred working methods here, which will also change depending on the exact nature of the recording. The only real guidance I can give therefore is to say that a blend of all three tactics tends to bring about decent results most efficiently. So I might start off tinkering with timing, phase, and polarity adjustments to get the overall character and tone of the instrument into the right kind of ballpark, but without spending all year on it; then I might apply EQ to massage the balance of each mic/DI within the most important frequency ranges, favoring settings that don't impact too heavily on the phase response; and finally I might EQ the instrument's global group channel to deal with masking of, or by, its frequency response within the mix as a whole.

EQ FOR LENGTH

Broadly speaking, the high frequencies of musical sounds tend to decay more quickly than their low frequencies, which means that changing the frequency balance of an instrument can have an appreciable effect on its subjective note durations. As Dave Way explains, "When I think of bass, I think of length; like how long a note should be, because you can make a note longer by boosting 20 or 30Hz. Those frequencies are so big, they ring out longer and make the note actually longer. Sometimes . . . you might have to clean up the bass and take out some of that 20–30Hz to make it tighter."[22] Eddie Kramer gives a common example where the subjective length of a note may need to be changed in this kind of way: "If the bass drum pattern is busy, you don't want too much bottom end. If the pattern is such that there's a lot of air and space between each note, you can afford to put a little more woof into it."[23]

11.3.2 Multimiked Ensembles

Although there are no additional EQ tactics for multimiked ensembles beyond what we've just looked at in relation to multimiked instruments, there are further decisions to be made about which tracks are to be EQ'd separately and which are better processed in combination. In a drum kit recording, for example, you might wish to group three snare drum mics together to gain global EQ control over their combined frequency balance, while using separate phase/polarity adjustments and equalization on each individual mic to determine the exact nature of that timbre; but you might then further EQ the snare sound from the main drums group channel where your adjustments won't affect the snare microphones' phase relationship against the overheads. At the end of the day, your ears must decide on the most successful outcome, and even the professionals disagree as to the best approach. For example, many rock engineers think nothing of processing individual instrument mics in an ensemble, but veteran producer Renaud Letang expresses concerns with this approach: "I find that if you apply EQ or compression to individual [instrument] parts you lose the impression of . . . a unified section, and it becomes really easy to lose the overall balance. But when you pre-mix and then EQ or compress the whole section, the sound remains much more natural."[24] My advice is just to keep a cool head, add in the mics/DIs as you did during the first balance, but listen carefully whenever you mute or unmute each track/group in case of unwanted knock-on effects.

One other point to make is that you may wish to reassess the high-pass filter settings you initially decided on while balancing, in the light of your increased knowledge of the mix. If you remember, we set them fairly conservatively to start with in order to defer dealing with additional phase problems. Now that we're better equipped to handle those issues, you should feel free to clear out unwanted low end more astringently should you so desire.

11.4 ADVANCED EQUALIZATION TOOLS

Although traditional parametric equalizers continue to be the mainstay of most mixing engineers, new EQ technologies continue to be developed. Some of this innovation is driven by a desire to improve the usability of the familiar facilities, but there are also new tools being designed to cater for more specialist processing tasks.

11.4.1 Specialized Filter Shapes

One of the advantages of digital equalizers is that they can include filter curves that would be all but impossible to implement in the analog domain, and some of these more exotic filters can be very useful at mixdown. I already touched on the concept EQ "notches" back in Section 11.2.4, but there I used the term colloquially (as most engineers do) to refer to narrow-bandwidth EQ cuts in general. To be more accurate, the term "notch filter' applies to a filter that completely removes energy at its center frequency, rather than just cutting it, and some EQ plug-ins now offer that option too. I find it most useful for getting rid of discordant pitched resonances in percussion sounds, but it's also occasionally very handy if I need to remove one or two notes from a guitar or keyboard

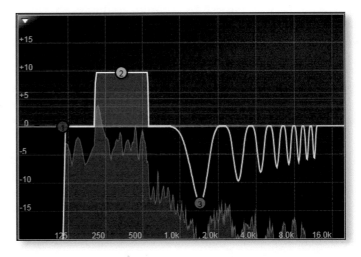

FIGURE 11.8
Here you can see a selection of less common digital filter types operating in Tonebooster's TB-Flx plug-in: band one is a brickwall high-pass filter; band two is a rectangle filter boost; and band three is a harmonically related multi-peaking filter cut.

chord because they clash with the music's underlying harmonic progression. A problem in the latter scenario, though, is that each musical note will generate its own entire harmonic series, not just a single frequency, which means you need a whole chain of harmonically related notch filters to remove each note. In response to this, some plug-in developers have started introducing a special filter type that combines a series of notch filters at multiples of a user-specified frequency. If you position the lowest notch of this multi-notch filter such that it removes a note's fundamental frequency, it'll also remove all the harmonics of that note as well, silencing it extremely effectively. A few manufacturers have extended the idea further by chaining fully featured peaking filters in a similar way, so you can rebalance notes in a more nuanced way, rather than just eliminating them.

If you hunt around, you'll find plenty of other weird and wonderful EQ shapes besides: brickwall and rectangle filters with super-steep cutoff slopes; logarithmically distributed chains of notches/peaks with the filters spaced in octaves; and "tilt" filters that skew the whole frequency response one way or the other. To be honest, though, I rarely find those of much practical use for mixing, or else they provide little advantage over more familiar filter types when dealing with musical signals.

11.4.2 Automated and Matching EQ

Wouldn't it be great if your DAW could handle all the EQ settings on its own? This question has inspired several recent EQ plug-ins that analyze the audio signal they're given and suggest an appropriate EQ curve automatically. Needless to say, their introduction has triggered a great deal of interest amongst small-studio operators, so I've sought out several such products for testing, but I have to say that none of them substantially eased the mixing process, in my opinion. The problem isn't so much with the analysis and machine-learning technologies employed, which are in some cases fantastically sophisticated, it's that the plug-in takes little (if any) account of the audio signal's overall mix context—in other words, how the sound reacts to other processors in its plug-in chain; how the sound interacts with other sounds in terms of masking and buss-processing; how the sound's frequency content reflects its musical role; and how the sound relates to your mix references and wider genre expectations. Honestly, I think such EQ algorithms offer little more value than the simple presets I discussed in Section 11.1, despite plenty of seductive marketing hype full of words such as "smart" and "intelligent."

There is another type of "automatic" EQ that has clearer practical applications, though, and that's the family of products that can compare two signals and then suggest an EQ curve that remodels the spectral balance of one to match the other. For example, let's say you have an ensemble recording that includes

---

EQ FOR DISTANCE

Because high frequencies are so easily absorbed in nature, this affects the frequency response of sounds arriving from a distance. Tony Maserati explains how he learned to use this psychological effect to advantage at the mix: "I learned from listening to Roger Nichols and Steve Hodge . . . They would use EQ to move things front and back. Obviously, the brighter something is, the closer it is; the duller it is, the farther away."[25] Jaycen Joshua provides a common application of this principle: "I rarely get a sound on the lead vocal and then duplicate it and put it on the backgrounds. You kind of want the backgrounds to have a little bit less of the top end . . . It gives [them] less presence so [they sit] behind the track a little bit."[26]

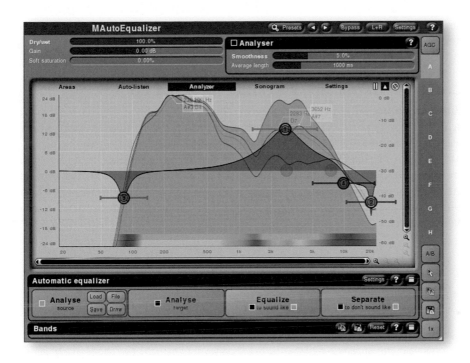

FIGURE 11.9
Matching EQ plug-ins such as Melda's MAutoEqualizer can provide useful inspiration when dealing with tricky tonal issues, as long as you're careful not to trust their processing suggestions over the evidence of your own ears!

both microphone and DI signals for an acoustic guitar, but the microphone has nasty-sounding spill on it from other instruments. In this case you could analyze the frequency content of the microphone signal (perhaps from a more sparsely textured section of the arrangement, in order to minimize the spill's effect on the analysis routine) and use this EQ curve to match the DI signal's inevitably less natural frequency balance to that of the microphone signal's truer-sounding timbre. Another situation where matching EQ can be a godsend is where any voice or instrument track has been stitched together from overdubs captured in a different location or with different recording chains. Some matching EQ plug-ins can also suggest an EQ curve that further differentiates the frequency responses of two sounds, rather than matching them, which can be helpful when you're troubleshooting complicated masking effects between instruments at mixdown.

My main advice for getting the best out of matching EQ plug-ins, though, is to take their algorithmically generated curves with a hefty pinch of salt! Although I find that matching EQ often helps me more swiftly find specific frequency ranges that need attention, it will also happily create curves that sound utterly loony from time to time! So keep your ears open and be careful not to trust the analysis over the evidence of your own hearing. One tactic that can work well here is to use the matching EQ for its analysis algorithm, but then to experiment with its frequency-curve suggestions using a different EQ plug-in, because you're more likely to evaluate each EQ band on its own sonic merits that way.

11.4.3 Pitch-Tracking and MIDI-Triggered EQ

With their pioneering SurferEQ plug-in, the company SoundRadix have popularized a new type of EQ whose bands can dynamically follow a monophonic instrument's fundamental frequency, or indeed any proportion or multiple of that. At a stroke this facility gives EQ the power to adjust timbre in a totally different way, because it can consistently target certain elements of an instrument's harmonic series irrespective of what note it's playing. For example, I most commonly use this with bass lines, where it gives me independent control over the levels of the instrument's lower harmonics. It's also great when you're

MASKING METERING

The software manufacturer Izotope recently launched a new kind of meter that can analyze two different signals and show a real-time graph of how much they're masking each other at different frequencies. This can speed up finding good frequencies for anti-masking EQ cuts, but it's crucial to be selective about which areas of masking you choose to tackle, because much of the masking that happens in a mix actually has a positive effect—it's an important part of what creates the production's sense of blend, coherence, and fullness. If you start trying to eliminate all the masking meter's hotspots at mixdown, you'll quickly end up with woefully unblended and hollowed-out results. As with all metering software, use it as an aid in solving problems you've already sensed by ear, and beware of it luring you into subjectively counterintuitive mix moves.

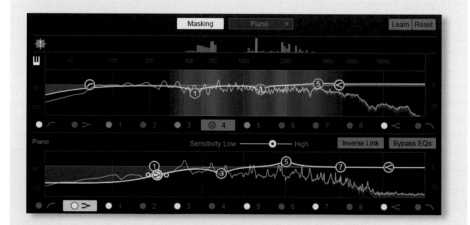

FIGURE 11.10
The Masking Meter available in Izotope's Neutron plug-in.

FIGURE 11.11
SoundRadix's SurferEQ plug-in provides both pitch-tracking and MIDI-triggered EQ functionality.

layering two parts in unison and want to eliminate the fundamental frequency of one layer so as to reduce the potential for tonal inconsistencies on account of phase-cancellation. The Sound Editor within newer versions of Celemony's Melodyne allows you to achieve similar results by allowing you to rebalance individual harmonics directly, and its offline processing methodology has the advantage of working with polyphonic audio too.

More recent versions of SurferEQ have expanded on the pitch-tracking idea by allowing EQ bands to follow incoming MIDI notes instead. From a mix-down perspective, I find this most helpful when I want to rebalance individual notes within recordings of polyphonic instruments such as acoustic guitar or piano. Yes, it's perfectly possible to carry out this job using a traditional EQ, perhaps in conjunction with a high-resolution spectrum analyzer, but the MIDI-tracking function makes the task a whole lot quicker and easier.

11.4.4 Independent EQ for Periodic and Nonperiodic Components

Another recent development in equalization technology is the option to apply independent filter curves to the periodic (i.e., pitched) and nonperiodic (i.e., transient/noisy) components of an audio signal. Although it can take a

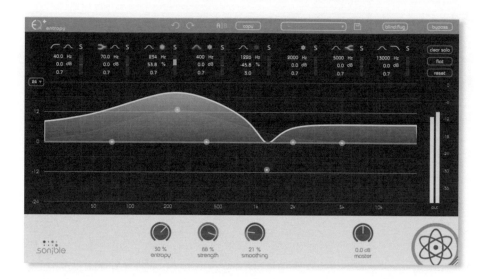

FIGURE 11.12
Sonible's Entropy:EQ+ plug-in lets you equalize periodic and nonperiodic elements of a sound independently.

while to get used to thinking about EQ in this way, some of the plug-in implementations of this approach (such as Sonible's Entropy:EQ+, Boz Digital's Transgressor, and Eventide's Fission) will let you pull off a selection of extremely nifty mixing moves once you get the hang of them. A flabby, rolling kick drum can often be tightened up dramatically by simply removing its low-frequency periodic components, for instance, while lead vocals can be drawn subtly forwards in a high-energy rock mix by pulling out some of the midrange noise components from layered electric-guitar textures. In many cases, though, such plug-ins begin to blur the line between EQ and dynamics processing, a subject I'll return to in Chapter 13, where I explore the subject of frequency-selective dynamics processing in detail.

11.5 THE LIMITS OF EQ

For mixing purposes, that's pretty much all you need to know about EQ. However, there's inevitably a limit to what it can do against some tonal and balance problems, for two reasons. First, it doesn't actually add anything to the sound—it just adjusts the levels of the frequencies already present. For example, a dull-sounding electric piano feeding a beat-up keyboard amp is never going to sound very bright, no matter how much you boost the high end, because there's nothing much happening above 4kHz other than noise. The second problem with traditional EQ is that it's static, so it can't deal with frequency imbalances that alter over time.

Again, don't spend weeks trying to solve every mix problem with EQ. Be happy to improve things as far as you can, and leave any trickier troubleshooting to more advanced processing. My own motto is this: if I've spent more than five minutes trying to equalize a track, then EQ is probably not the whole solution to its problems.

CUT TO THE CHASE

- The main purpose of equalization at mixdown is to address frequency-domain problems with the mix balance. Although there's a place for more subjective tonal EQ'ing at mixdown, this should be a secondary consideration.
- Add your instruments into the balance in order of importance, so that it's more straightforward to deal with frequency masking between different tracks. Always confirm the validity of your EQ judgments within the context of the mix, rather than in solo mode. Avoid graphic EQ designs for mixing, and reach for shelving EQ before peaking EQ. Favor cuts over boosts wherever possible. If boosting is unavoidable, then do your best to limit it to one EQ band, keep the gain below 6dB, and use as wide a bandwidth as you can get away with. Try to remain sensitive to both wide-bandwidth and narrow-bandwidth frequency-balance issues. To maintain your objectivity, bypass each filter frequently as you set it up, and make use of different monitoring systems and listening levels. Once you have an ideal balance for a processed track in its own right, switch it in and out of the mix while listening for unacceptable masking of other more important sounds, applying further EQ cuts at conflicting frequencies if necessary.
- When equalizing multimiked recordings, EQ may alter the tone and balance of the sound unpredictably because of changes in the phase relationship between the individual mic/DI tracks. Applying EQ to individual tracks will give a different outcome than processing group channels. Linear-phase EQ may help increase your options here, but it is not without its own processing side effects.
- If a track's dynamics plug-ins aren't doing the job properly, then equalization earlier in the processing chain may improve the musicality of the results. Postdynamics EQ, on the other hand, is a better bet if you're already happy with the dynamics action. It's not uncommon to need both pre- and post-dynamics EQ.
- When the main purpose of an equalizer is to make a sound subjectively "better," rather than just to balance its frequencies in the mix, the rule book pretty much goes out the window—no setting is wrong if it achieves a sound you like. Make a point, though, of trying out two different equalizers every time you're equalizing with this intent, so you begin to build experience concerning which models suit which types of sound. Also be sure you're not fooling yourself into thinking something sounds better just because boosts are making it louder.

■ New developments in EQ technology are appearing all the time, and many offer trouble-shooting facilities that reach well beyond the power of traditional equalizers. Multi-notch/peak filter types can be particularly useful for internally rebalancing polyphonic parts; matching EQ can help diagnose complex tonal problems; pitch-tracking EQ can resculpt the harmonic balance of melodic instruments; and some specialist processors can independently process the periodic and nonperiodic components of a sound.

Assignment

■ Track down all the equalizers on your DAW system so that you know what options are available. If you have little choice, then consider supplementing your selection with third-party plug-ins. Try to get hold of a linear-phase equalizer too if you work with a lot of multimiked recordings.
■ Mute all the tracks in your mix and rebuild the balance again, this time applying equalization in search of a more stable balance.
■ Make a note of any faders that still feel unstable.

Web Resources

On this book's companion website you'll find a selection of resources to support this chapter, including:

■ a special tutorial video to accompany this chapter's assignment, in which I demonstrate how to use EQ for balancing a simple multitrack session;
■ a selection of audio examples illustrating EQ techniques such as unmasking, notch-filtering, and pitch-related processing. There's also a demonstration of the pre-ringing side effects of linear-phase equalizers;
■ links to a wide range of affordable EQ and spectrum analysis plug-ins, as well as to some of the more advanced EQ processors mentioned in Section 11.4;
■ a number of mix case-studies from the *Sound on Sound* "Mix Rescue" column that focus specifically on the distinction between creative "tonal" EQ and more technical "balance" EQ in practice.

 www.cambridge-mt.com/ms-ch11.htm

CHAPTER 12
Beyond EQ

As I said in Chapter 11, equalization is extremely useful for mixing, but it can only adjust the balance of frequencies that are already in the recording. Much of the overprocessing in typical small-studio mixdowns arises from people maxing out their EQ settings in a misguided attempt to add something new to the sound. So if EQ isn't giving you a sound you like or is unable to deliver a satisfactory mix balance, then it's time to stop footling about with what's already there and actually add information to the mix instead.

12.1 DISTORTION AS A MIX TOOL

The textbook situation in which EQ has nothing to offer is where you want to brighten a bass guitar/synth line with no appreciable high frequencies. This might be because the synth's low-pass filter has cut away all the high end or simply because a limited-bandwidth oscillator waveform was used to start with. In such a scenario, there may well be no energy at the upper end of the

FIGURE 12.1
The upper graph has less harmonic density than the lower graph. If the harmonics of an instrument aren't dense enough, then distortion may help thicken it up.

frequency spectrum, so equalization is powerless to alter the subjective tone. The only alternative is to create new high-frequency components that weren't originally there.

That's just the most obvious case where EQ doesn't deliver, though, because even where an instrument does generate frequency components in a required spectral region, they may not be dense enough to give that spectral region a sense of fullness within the balance. In this kind of case, you'll be unable to set a satisfactory EQ boost. It'll either feel too small, because the frequency region isn't dense enough to compete with other sounds in the mix, or it'll feel too big, because the few frequencies that are in that spectral range are poking too far out of the balance.

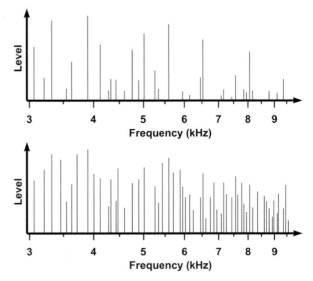

Increase the harmonic density, however, and that spectral region can remain solid in the mix without individual frequency components sticking their heads over the parapet.

The tool that can help in both these scenarios is distortion, an effect widely used in professional mixing circles but pretty much ignored in a lot of small studios. Although people take great pains to minimize distortion a lot of the time in record production, that's no reason to neglect its creative possibilities at the mix. The bottom line with distortion devices is that they add new harmonics to the processed signal, and because the frequencies of these are usually related in some way to the frequencies of the unprocessed signal, they retain a musical connection. In short, distortion allows you to change the timbre of a track without substantially altering its musical function in the arrangement. As such, its use is almost as widespread as EQ. "Distortion is our friend," says Jaquire King. "I love distortion on pretty much everything in some degree. It's our friend because we're able to record so clean now with going to digital."[1]

Bass instruments are a common target, often to give them more attitude. Morten "Pilo" Pilegaard distorted the bass on Lukas Graham's "7 Years," for instance: "If you were to solo it, you'd hear only grit and fret noise and distortion. It is there for texture."[2] Dave Ogilvie[3] did the same on Carly Rae Jepsen's "Call Me Maybe" "for some grit," while Kevin Kadish[4] added distortion to the upright bass in Meghan Trainor's "All About That Bass" because he'd "rather have things sound too dirty than too nice." Many urban-music producers also make a habit of distorting the ubiquitous Roland TR808 kick drum sound. Young Guru explains the logic here: "The biggest misconception about 808 is that they're all bass. That's the easiest part of the 808, but you're not always going to hear that low, low end on a radio . . . You have to give it a little bit of distortion so that that thing can creep into the 120Hz and 140Hz, and then go even further into your midrange."[5] Rich Costey is among many engineers who regularly distort even lead vocal tracks: "[It's] so commonplace nowadays that to me it's the same as distorting a guitar or a bass. And much of the time when I'm mixing records for other people I'm distorting the vocal, whether they know it or not!"[6] The big challenge when using distortion in a mix setting, though, is restraining it so that it adds harmonics only at those frequencies you want it to. Here are some tips for keeping it under control.

First, it's nice to have different distortion devices to choose from, so that you can mix and match. There are lots of electrical devices that can create musically useful distortion, and many of these have now been digitally modeled for use in plug-in form: tape machines, mixing consoles, hardware outboard units, guitar amps, effects pedals, soft-clippers, vacuum tubes, transistors, and transformers, to give but the briefest of lists. Many of these plug-ins also have selections of presets that offer widely varying sounds, and you don't need to understand anything about how any of them work just to browse through and find one that seems to add something useful to the track you're processing.

FIGURE 12.2
A selection of different distortion flavors (*clockwise from left*): tape saturation (Toneboosters's TB_RcclBus), valve distortion (Voxengo's Tube Amp), variable-knee clipping (GVST's GClip), emulated mixing-console saturation (Sonimus's Britson Channel), and digital aliasing distortion (Sonalksis's Digital Grimebox).

Even a lo-fi digital "bit-crusher" distortion can occasionally be useful. "What that does is create some contrast between that track and the rest of the song," says Serban Ghenea. "You don't need much of it, but it can make a world of difference."[7] Be aware, though, that the digital distortion products from these devices tend to be less closely related to the input signal on a musical level, so be on the lookout for uncomfortably discordant added harmonics, especially on pitched instruments.

Second, some distortion processors have loads of controls, but to be honest life's too short to spend ages learning what most of them do, so you might as well just laugh maniacally and twiddle with abandon. What actually matters more than anything with distortion devices is how hard you drive them. If you hit a distortion process gently, you may just get a subtle thickening of the sound, whereas if you slam your track in at high level, you may barely be able to recognize the original sound among all the added frequency-domain hash. If there's any control on your distortion unit labeled Drive or Gain, then

> ## MORE ADDED HARMONICS: HIGH-SPEED MODULATION EFFECTS
>
> Another way you can add harmonic density to a signal is to treat it with audio-frequency modulation effects. There are a number of different possibilities to choose from, including frequency modulation, amplitude modulation, ring modulation, and any vibrato/tremolo/auto-wah plug-in that can set its modulation frequency above 20Hz. If you fancy getting to know the mechanics of how these effects work, go ahead, but there's no obligation—you can use them just fine in a mix situation by surfing the presets. The only really important control you need to know about will be called something like Modulation Depth, and it basically determines the amount of frequency information added.
>
> What you'll notice pretty quickly, though, is that frequency additions from these effects aren't musically related to the input signal, so you'll usually introduce dissonant tones into pitched parts. Although simple melodic parts can nonetheless still be coaxed into responding musically if you experiment with the effect's modulation frequency, in practice you're more likely to get useful results with unpitched sounds. That said, the very unmusicality of some of these effects can be a boon on occasion. For example, I've used ring modulation to add growl to aggressive lead vocals on more than one occasion.

that's the one you should pay most attention to. Occasionally no such control is provided, in which case you can vary the degree of distortion by adjusting the signal level before the distortion device's input.

I'm frequently amazed by how effectively you can remodel or enhance a lackluster track simply by browsing a few different distortion plug-ins, choosing your favorite, and then adjusting the drive to taste. It's so simple that it almost feels like cheating! However, there are significant limitations with this basic insert-effect configuration. First, increasing the distortion's drive will simultaneously increase both the level and the density of the added harmonics, whether you want that or not. Second, you can't directly specify how much the distortion products in each spectral range balance with the track's original frequencies. For these reasons, a parallel processing approach (in other words, using the distortion as a send effect) is usually my preferred choice, because this provides much more scope for deciding exactly what it is the distortion adds. You might, for example, decide to drive the distortion hard to create denser harmonics, but mix those in with the unprocessed track at only a moderate level. You are also free to EQ the distortion channel to funnel the added spectral energy into those areas where it's really needed. You might also consider compressing the signal before it reaches the distortion to maintain a more consistent density of added harmonics. As with all parallel processing, though, you do have to be careful of phase mismatches between the processed and unprocessed channels (particularly when using guitar amp simulators). I often end up using phase rotation in such cases to get the best combined sound, especially if I've resorted to EQ to carve away unwanted regions of the distortion sound.

One type of parallel distortion processing warrants special attention, and that's the high-frequency enhancement technique pioneered by Aphex in its Aural Exciter. In simple terms, with this type of parallel distortion technique the signal feeding the distortion is aggressively high-pass filtered and then compressed before it reaches the distortion. Mixing this processed signal in with the unprocessed track gives an unnatural level of high-end density, which can help sounds seem detailed, airy, and more upfront in the mix.

> High-frequency enhancement can be tremendously fatiguing on the ear, and your hearing will quickly become desensitized to it.

Needless to say, the Aphex process has long been a common choice for lead vocals. If you're going to use this effect, however, there are a few issues to beware of. Most of all, you need to realize that this effect can be tremendously fatiguing on the ear, and your hearing will quickly become desensitized to it, so it's more important than ever to keep checking your objectivity. Switch the effect out of circuit, allow your ears to adjust to the preprocessing balance, imagine the sound of the processing you're after, and then pop the enhancement back into the mix to check that it's actually doing what you want it to. If you don't take this precaution, you'll fall into one of the oldest traps in the book: turning up the enhancement too far because your fatigued ears have become less sensitive to it and then discovering after a night's sleep that your singer actually sounds like a wasp shaving!

Another point to bear in mind with the majority of distortions is that they can reduce dynamic range, so you may need to apply additional balance processing to compensate. In many situations, though, this side effect can actually be beneficial, particularly when processing transient signals. For example, I frequently find myself using clipping to round off over-spiky percussion tracks.

REAMPING

The practice of capturing DI signals from electric guitars during performance and then using those to create and record amplified timbres later on (a process known as reamping) has gained popularity in recent years, largely because you can control a guitar amp's sound much more easily while you're recording it than you can by throwing mixdown processing at it. While a detailed explanation of guitar-recording techniques is outside the scope of this book (check out *Recording Secrets For The Small Studio* if you want more info about that), it's worth pointing out that plenty of high-profile mix engineers don't just reamp guitars, but will happily take advantage of reamping's tonal coloration and distortion components to sculpt any other sound in their productions too. Moreover, room reflections captured during the reamping process (perhaps with dedicated room mics) may also obviate some of the need for reverb processing, as we'll see in Chapter 16.

12.2 LOW-END ENHANCEMENTS

By its very nature, distortion processing tends to add most, if not all, of its added energy above the lowest frequency of the unprocessed sound, and when you're looking for tonal changes, this is usually perfectly appropriate. However, there are some situations where you might want to add "oomph" at the low end of an instrument's frequency range, or even below that. This is where a type of processor called a subharmonic synthesizer can help by artificially generating additional low-frequency energy that is still musically related to the unprocessed sound.

Although high-profile engineers such as Jack Douglas,[8] Jon Gass,[9] Josh Gudwin,[10] Alan Moulder,[11] and Mark Lawson,[12] have all professed themselves users of such processors, I have to be honest and say that it's rare to hear anyone applying them effectively in the small studio. This is partly because you need good low-end monitoring to gauge their true impact, but it also has to do with the fact that every manufacturer of a subharmonic synthesizer appears to use a slightly different algorithm, and you can never be quite sure how well each will respond to a given bass-instrument recording. One plug-in might work great on kick drums but churn out a load of woofy rubbish when presented with a bass line; another might handle only midregister bass parts; another might handle a good range of bass notes but throw a tantrum if there's any drum spill on the bass mic.

FIGURE 12.3
A couple of subharmonic enhancer plug-ins: Waves LoAir (*top*) and Voxengo's LF Max Punch (*bottom*).

Therefore, I personally rely on alternative strategies most of the time. For kick drums I insert an additional sample, whereas I supplement melodic bass instruments with a MIDI "subsynth." Not only do both these techniques allow complete pitching and timing reliability, but they also offer much wider scope for tailoring the timbre of the enhancement to suit the production as a whole. Let's deal with each in turn.

12.2.1 Drum Triggering

There are now lots of software plug-ins that will automatically detect drum hits in a track and play back a sample in time with them, but with kick drums in particular you'll usually get the best results if you refine the placement of each triggered sample by hand. "Software is very time-saving, but the drawback remains that it's not as good as doing it manually," says Cenzo Townshend. "You still have to go through each hit and move it in time to make sure it's 100 percent correct and the phase is correct. You look at the waveforms, but you also need to use your ears."[13] Of course, as I already discussed in Chapter 8, the same principle applies in any situation where more than one kick-drum sound is layered. "It's super-monotonous," says DJ Swivel "but it's the best way I can do it . . . I mean, I am literally shifting things in terms of a few milliseconds to make absolutely sure that the kicks line up . . . and that there's no phase inversion. This ensures that each drum punches through as strongly as the last."[14] Even once you've made sure that the phase relationship between the different kick sounds remains fairly constant, though, it's as well to investigate whether additional overall timing/ phase adjustments will further improve the combined sound. "It's not really an EQ thing to get the strongest hit," comments Young Guru. "You just put everything in phase."[15]

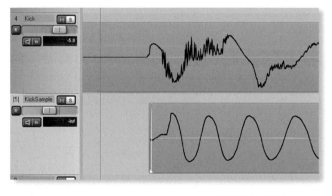

Of course, sample triggering can also be used for much more than just filling out the low-end of kick drums, because it allows you to add new characteristics to any drum sound. It's particularly useful with live drum recordings which, for whatever reason, need to rely heavily on the close-mic signals—for example, if the overhead mics are picking up mostly cymbals. Close mics positioned just a few inches from each drum will spotlight only a very small portion of the instrument's complete sound, so the recordings from these mics will usually be rather unnatural and uninspiring to listen to. (Certainly, nine out of ten snare close-mic recordings I hear from small studios just go "donk"!) Triggering a sample alongside, or instead of, the close mic can help to provide a more pleasing overall tone in such cases. Andy Wallace uses an interesting variant on this technique to supplement his snare ambience: "I use the samples to drive reverbs . . . The thing I like is that I can

FIGURE 12.4
If you're layering a sample against a live kick drum, you need to make sure that the polarity of the sample matches the live track for every hit. In this screen you can see an instance where it's not matched, and this results in a noticeable loss of low-end weight.

EQ them so that I can really tune the ambience and where it sits in the whole frequency response . . . More so than I can with the overheads, because I normally EQ those so that the cymbals sound the way I want them to sound . . . [I can] shade with a little more control using my ambient sample."[16] He's used similar techniques for kick drum too. "I may also throw in a kick sample to generate reverb," he adds. "Occasionally the kick needs to be bone-dry . . . but if it's a rock track I like to add a room sound."[17]

But there's another good reason for triggering samples within the context of a live drum kit recording: they have no inherent phase relationship with the other microphones and no spill. This means that you can freely balance and process samples without fear of destroying the rest of the kit sound—great if you only need a sample to supplement the sound in a specific way. For example you might apply slow-attack gating and assertive high-pass filtering to isolate just a snare sample's high-frequency sustain components, or add electronic drum samples to give a hip-hop oriented low end (as Steve Fitzmaurice did on most of the songs on Sam Smith's *In The Lonely Hour*).[18]

The Internet being what it is, it's child's play to find a selection of decent samples for most drums. However, on the offchance that typing "free drum samples" into Google doesn't provide more choice than you could ever want,

FIGURE 12.5
If you don't have suitable drum samples for mix triggering, then it's easy to lay hands on commercial sample libraries from outlets such as Best Service.

> ## PITCH-SHIFTING FOR DENSITY
>
> Another way you can add harmonics to an instrument is to use a pitch shifter to duplicate it into a different pitch register. With melodic and harmonic parts, the most foolproof interval will usually be an octave, but major/minor thirds (shifts of three/ four semitones) and perfect fourths and fifths (shifts of five and seven semitones) can also work well in some instances. Upward shifts tend to be more useful than downward ones, and you'll typically want to mix the shifted channel at a low enough level that it's perceived as part of the unprocessed track, rather than as a new set of notes in its own right. This technique is usually less successful with drum and percussive sounds, because signals full of transients and noise tend to stress pitch shifters beyond the point where they sound any good. That said, if you're programming your own electronic music, then the facility to adjust the tuning of samplers and drum-machines can fulfil a very similar function without the same pitch-shifting artifacts. "Sometimes if a kick-drum isn't right," says DJ Swivel, for instance, "and it doesn't have the low end—you throw an EQ on it and try and boost it and you're still not getting what you want—well, if you just drop it down an octave, now all of a sudden you have all these new low harmonics that didn't exist before and it helps."[19]

there are lots of commercial outlets that offer tons of material. Check out www.bestservice.com or www.soundsonline.com, to mention but two large-scale international distributors.

12.2.2 Incorporating a MIDI Subsynth

Propping up a lightweight bass line with a low-frequency synth part is a stalwart of many urban styles in particular. "The bass is very important," explains DJ Premier. "If the low end of a [loop] sample isn't really heavy, I'll always follow the exact bass line of the song and put that underneath. A lot of people ask me what EQ I use to get the bottom end of my samples to come through so strongly, but I'm like 'Man, it's not EQ! I'm playing the same notes verbatim.'"[20] This is often pretty straightforward to do if the bass is repetitive and tightly locked with a regular song tempo, but if not then there may be some initial donkeywork involved in sorting this out. Some audio editors (Celemony's Melodyne, for instance) provide an option to generate MIDI directly from an audio track, so this is certainly worth investigating if your bass part's more complicated, but even where you're forced to painstakingly program the MIDI by hand for whatever reason, that time investment is usually worth the peace of mind that comes from knowing that the low-end frequency balance will be totally dependable. For my part, only the freakiest of extended jazz-fusion free-for-alls would have me waving the white flag and scurrying back to traditional subharmonic processing these days.

Once the MIDI part is in place, you need to consider what kind of sound the subsynth should use. One common option is to double the bass line at the

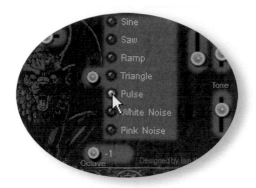

FIGURE 12.6
When designing subsynth parts, your choice of synth waveform is an important decision.

octave below using a simple sine-wave patch. In this case, there's nothing much to do other than set the sine wave's level and listen to it in solo just to check that the synth's envelope settings aren't so fast that they create unwanted clicks. Things get more complicated if you're using a subsynth to try to beef up the existing fundamental frequency of a bass part, because the subsynth will overlap the main bass part's frequency range. If the subsynth's waveform isn't in phase, then phase cancellation might actually cause the combined sound to become even skinnier! Moreover, because this phase relationship will usually change from note to note, you could end up with an uneven low end that's all but impossible to balance with the rest of your mix. If the existing bass sound's fundamental frequency is strong enough to cause difficulties such as this, then try to get rid of it to make space for the subsynth at the low end. A steep high-pass filter on the main bass part is one solution here, but at times you may need to use the more surgical approach of notching out individual frequencies (with the assistance of a high-resolution spectrum analyzer) or investing in specialized EQ software that can track the instrument's pitch (as discussed in Section 11.4.3).

In the event that you want to do more than just add a suboctave or emphasize the existing bass fundamental, a sine-wave synth oscillator may not be the best bet; other waveforms will fill out the overall tone better. I like using a triangle wave most of the time, because it doesn't have such dense harmonic spacing as a sawtooth and it is duller-sounding and less characterful than a square wave—both of these properties seem to make it better at blending with (rather than overwhelming) the sound it's layered with. Whatever waveform you use, though, you should also take the same precautions against phase cancellation of the bass part's fundamental frequency. I'd steer clear of detuned multioscillator patches too, because the "beating" between the detuned layers may cause the subsynth's fundamental frequency to fluctuate unacceptably in level. It probably makes sense to stick with mono patches as well so you don't get mono-compatibility problems. These restrictions mean that you only really need a simple instrument to generate subsynth parts. Whether you engage the subsynth's own filter or sculpt its output with EQ is purely a question of what kind of low-end tonal enhancement you're looking for. With a triangle wave in particular, you might not feel the need to filter it at all, although I do personally employ some kind of gentle low-pass filter most of the time to restrict its contribution to the lower octaves.

One final point to make is that subsynth parts usually need to be controlled tightly in terms of dynamic range or else they can really eat into your track's overall headroom. This is where the MIDI-triggered method really pays dividends, because it's simple both to restrict the dynamics of the MIDI part and to compress the subsynth's output. So even if your low-frequency monitoring facilities aren't up to much, you can still rely on a subsynth to give consistent low end if you pay attention to your DAW's meters.

COMB FILTERING AND RESONANCE AS TONAL TOOLS

Another method of adjusting the subjective tone of a track is to deliberately introduce the complex frequency-response undulations of comb filtering by mixing in a delayed version of the signal. All you do is set up a delay plug-in as a send effect and feed it from the track that needs timbral adjustment. Choose any old preset you like, and then set the effect's Delay Time parameter to anything less than about 20ms. This should result in obvious comb filtering when the delayed signal is mixed in with the unprocessed track. The way the filtering sounds will depend on the exact preset and delay time you've used, so tweak both to taste while keeping a hand on the delay return channel's fader to control the severity of the filtering. Some of the presets may introduce pitched sympathetic resonances, and if these are sensitively tuned using subtle adjustments of the Delay Time control and subtly applied, they're capable of adding extra musical sustain. Indeed, a few manufacturers actually offer specialized resonator plug-ins built on this concept that allow you to control several pitched resonances at once, in a more convenient way. Whatever tactic you use, though, do keep in mind that you can EQ the delay return too, should you wish to curtail the comb filtering and sustain artifacts in some areas of the spectrum.

Related to this kind of short delay effect are the three common modulation treatments: chorusing, flanging, and phasing. Fun as they are for special effects, from a mix perspective they don't really serve much of a technical purpose, so you might as well just surf the presets and tweak things erratically in search of something you like. Some people say that chorusing can create the illusion of a larger ensemble, but it never seems to work like that for me! (Modulation treatments can be useful for adding stereo width, too, as I'll discuss in Section 18.4.)

Artificial reverb send effects can also be coaxed into producing a wide range of changes to tone and sustain. Again, feel free to process the reverb return channel, and consider in particular narrowing its stereo image if you're processing mono sounds, so that the added width doesn't reveal the effect's synthetic nature. No need to fuss with understanding the controls at this point, just look for a preset that sounds good—each will have a unique tonality, echo pattern, and overall resonant character, so you may be surprised how different they sound in this context. Don't ignore any rubbishy reverb plug-ins either, because they often yield the most dramatic timbral changes. As with many mix effects, reverb doesn't have to sound any good on its own as long as it combines musically with the track you're processing. (If you want to get deeper into using tonal delay/reverb effects, skip ahead and browse Chapters 16 and 17.)

12.3 SYNTH PADS

Use of MIDI sound sources for mixing purposes doesn't stop at sub-synth parts, by any means, because chordal synth pads in particular can be immensely useful across a surprisingly wide range of different music styles. Despite their inglorious history, pads don't have to turn your mix into some kind of 1980s bubble-perm nightmare as long as you design and apply them effectively. Indeed, pads can easily add warmth and sustain to thin-sounding acoustic guitar parts, for example, or improve the pitch definition of mushily overdriven electric guitars, without anyone being the wiser. There are very few parts that you can't supplement with a pad synth if needs be, just as long as you pare each pad down to its bare essentials—that's the key to keeping pads from becoming too audible in their own right or trespassing unnecessarily on other aspects of the production sonics.

> The first mistake most newcomers make is neglecting to equalize a synth pad's frequency response, because it's vital that pads only operate precisely where they're required.

The first mistake most newcomers make is neglecting to equalize the synth's frequency response, because it's vital that pads only operate precisely where they're required. A neat trick here is to turn the pad up a couple of decibels too loud at first, and then return it to a more appropriate subjective level using EQ cuts, as that'll concentrate its energy in the areas of the mix where it's most useful. The exact note pitches and synth timbres also warrant careful thought, as these can dramatically affect the density of harmonics within the target frequency ranges. (For more detailed tips here, check out this chapter's web resources.) Pads don't have to be long held chords, either, because in many cases a simple rhythmic repetition or arpeggiation of synth chords is better at reinforcing the track's groove. When it comes to deciding on a fader level for the pad, you don't often want the general public to notice you're using a pad at all—bands who specialize in natural-sounding acoustic styles or hard-hitting rock music don't really like to brag about their synth chops, for instance, despite the improvements that pads can bring to guitar textures in both those genres. So my normal rule of thumb is this: if you can hear the pad, it's too loud! To put it another way, you should only notice that the production has a pad in it when you toggle that channel's mute button.

WHEN ALL ELSE FAILS: RERECORD!

The range of tone-manipulation technology available to small-studio owners is truly mind-boggling compared with what was available 20 years ago, but there are times in every mix engineer's career when you're presented with a recording that no amount of clever DSP can salvage, at least not without wasting huge amounts of your time. In those situations, don't forget the option of replacing the troublesome part with an online session musician, as mentioned back in Chapter 7. While no-one likes spending money to replace a recording that's already ostensibly in the bag, the cost of a bespoke overdub gets easier to swallow if you set it against the time (and soul) you'd otherwise lose to salvage drudgery. Believe me, as someone who's done more than 50 *Sound on Sound* "Mix Rescue" remixes, I speak from bitter experience on this one . . .

CUT TO THE CHASE

- Equalization can only adjust frequencies that are already there; if you need to add something new to the sound instead, then you'll need different processing.
- Distortion is something of a secret weapon for filling out small-studio productions. You can achieve quick mix improvements using distortion effects as insert processors, but it's usually preferable to use a parallel processing configuration instead. Just be wary of unwanted comb filtering should your distortion plug-in tamper with the processed signal's phase response.
- Pitch shifting can add complexity to a track's frequency response if you take care with the shifting interval used. This tactic usually sounds best for simple pitched instruments and upward shifts. High-speed modulation effects are another option, although care must be taken with pitched instruments if you wish to avoid unmusically dissonant frequencies being added. Short delays and reverbs can deliver tonal changes through complex comb filtering and have the potential to add subtly appealing sympathetic resonances in addition.
- Subharmonic synthesizer plug-ins are useful on occasion, but adding supplementary sample/synth parts is usually a more successful way to add low-end weight to bass instruments.
- Added drum samples not only provide the opportunity for timbral overhaul, but they also allow independent manipulation of one instrument within a multimic recording without compromising spill from other sources. Although automatic drum-triggering software is available, you should refine the timing of each sample by hand to avoid phase cancellation against the trigger track.
- MIDI-driven synthesizers are excellent tools for adding spectral complexity. Subsynth patches can help provide small-studio productions with powerful yet controllable low end, whereas pads can usefully fill out the sound of almost any pitched instrument, even in musical styles where you'd least expect it. It's rarely desirable to hear a pad in its own right, so you should use your MIDI programming, sound-design, and equalization skills to tailor its frequency additions to the specific task at hand.

Assignment

- Do a quick survey of the processing available on your mixing system to get an idea of your options with respect to the various different effects mentioned in this chapter: distortion, high-speed modulation, pitch-shifting, subharmonic synthesis, delay, and reverb.
- Investigate your DAW's facilities for triggering drum samples in response to recorded audio, and for generating MIDI parts to shadow specific pitched instruments.

- Start building a library of good drum samples for mixing purposes so that you don't have to hunt around for ages every time you need one. Also check out what synths you have available to you, and make a note of any presets that turn out to be more useful than others in practice.
- Mute all the tracks in your mix and rebuild the balance, addressing any frequency-balance concerns that careful equalization was unable to resolve.
- Make a note of any faders that still feel unstable.

Web Resources

On this book's companion website you'll find a selection of resources to support this chapter, including:

- a special tutorial video to accompany this chapter's assignment, in which I showcase how some of the timbral techniques mentioned here can improve the balance of a simple multitrack session;
- dozens of audio examples demonstrating how distortion, pitch-shifting, and additional synths/samples/overdubs can be used to overhaul the tonality of different sounds at mixdown.
- links to a wide range of affordable plug-ins: distortions, exciters, high-frequency modulators, pitch-shifters, resonators, and subharmonic synthesizers. There are also some suggestions for drum triggering software, as well as some of my own sample-library recommendations.
- further reading about designing and using synth pads.
- a number of mix case-studies from the *Sound on Sound* "Mix Rescue" column that focus specifically on the distinction between creative "tonal" EQ and more technical "balance" EQ in practice.

 www.cambridge-mt.com/ms-ch12.htm

CHAPTER 13
Frequency-Selective Dynamics

Chapters 9 and 10 demonstrated various ways to handle time-domain balance problems, and Chapters 11 and 12 focused on frequency-domain issues. Although those techniques alone may be enough to deliver you a workable balance, usually at least one instrument will remain that can't be persuaded to maintain its proper place in the mix by these means. This is because some balance problems cannot be overcome unless you process both the time and frequency domains at once. To put it another way, not one of the processing methods we've discussed so far is much use against a dynamic-range problem that exists only at high frequencies—a normal dynamics processor's gain reduction won't target that frequency region separately, and the static frequency-balancing facilities of equalization won't tackle the level inconsistencies.

To pull off this kind of stunt, you need frequency-selective dynamics processing. Plug-ins that provide this feature do tend to look a bit intimidating, so it's hardly surprising that many small-studio operators leave them well alone, whereas others just pick a preset and hope for the best. However, these devices are too powerful to ignore if you want commercial mix results, but they are also powerful enough to completely wreck your tracks if applied in a scattergun fashion. This chapter shows how you can combine time-domain and frequency-domain treatments for yourself to tackle some of the toughest balancing tasks that small-studio productions face.

> Frequency-selective processors are too powerful to ignore if you want commercial mix results, but they are also powerful enough to completely wreck your tracks if applied in a scattergun fashion.

13.1 FREQUENCY-DOMAIN TWEAKS FOR FULL-BAND DYNAMICS PROCESSORS

Before I introduce any new types of processor, let's have a look at how the dynamics devices we've already covered (often called "full-band" processors, because they operate on the entire audio spectrum) can be coupled with EQ to increase your balancing options.

13.1.1 Equalizing Parallel Processors

One simple way to begin exploring how dynamics can be aimed at specific frequency regions is by elaborating on an idea we've already dealt with quite a bit: parallel processing. If you set up any dynamics processor as a send effect, then you can EQ its return channel to tailor the processed tone. So where a fast-attack/fast-release compressor working in a parallel configuration would normally just emphasize overall sustain, a low shelving cut to the return signal would direct that sustain more toward the upper frequency regions—a good way of compensating for the shorter sustain of higher piano notes, to give one example. Dave Pensado supplies another application, this time for the kick drum. "You can have the main kick drum fader, and another fader where you compress the dog snot out of the kick, and add a lot of top end to it. Then, maybe in the chorus you add in a little bit of that second chain. It gives a tiny bit more attack . . . as if the drummer was hitting it harder."[1]

Dylan Dresdow combined compression, EQ, and gating to achieve frequency-selective transient enhancement for his work with Black Eyed Peas: "I'd mult the kick drum and mult the snare, compress them really hard, and roll off a lot of high end and low end, so they become very midrangey, and I'd put a tight gate on them. I'd sculpt the attack in this way and would add this into the track."[2] Of course, you might achieve a similar kind of effect by EQ'ing a parallel transient enhancer. In all of these situations, a little further EQ may be desirable on the combined sound to compensate for any overall level increase in the enhanced frequency region—but that shouldn't be a big deal, because you already know exactly where that region lies.

As usual in parallel setups, you do need to be careful that EQ phase shifts don't cause any unpleasant comb-filtering side effects when the processed and unprocessed signals are combined—this happens often enough for me that I now use linear-phase EQ for parallel dynamics return channels as a matter of course. There's also the question of whether to use the equalizer before or after the dynamics, although this should hold no new mysteries if you've already absorbed the advice in Chapter 12: if the gain changes seem musical, then EQ after the dynamics; if not, then try using EQ earlier in the processing chain. Returning to my earlier piano example, if you want to add extra high-frequency sustain with a filtered parallel compressor, it makes sense to put the filter before the dynamics. That way the compressor stamps on the high-frequency signal peaks from high-pitched notes and smoothly resets its gain reduction between them to allow the interpeak sustain tails to shine through. Had the high-pass filtering been placed after the dynamics, the lower notes would also trigger gain reduction, causing unnatural lumps and bumps in the high-note sustains.

13.1.2 Refining Your Gating Action

Another common hybrid technique is where equalization is applied to the dynamics processor's level-detection circuit (or "side chain"), but not its

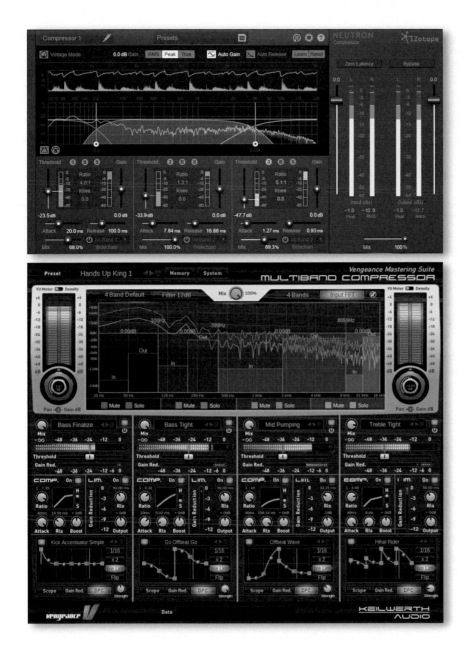

FIGURE 13.1
Complex frequency-selective dynamics processors such as Izotope Neutron Compressor or Keilwerth Audio Vengeance Multiband Compressor may appear intimidating, but there's actually no great mystery to using them if you remain focused on the central issue of mix balance.

main signal path. This allows you to adjust the sensitivity of the processing to different frequencies without directly changing the tonal balance of the processed audio. You're most likely to find side-chain EQ facilities on gates

FIGURE 13.2
Like a lot of well-specified gates, Steinberg Cubase's bundled Gate plug-in features built-in side-chain equalization to refine its triggering action.

(usually in the form of simple high-pass and low-pass filters) where they can help a great deal in getting the gate to open and close the way you want it to. Without such facilities it can be well-nigh impossible to reduce unwanted spill signals in multimiked recordings of closely spaced instrument ensembles such as drum kits.

Take the example of a microphone positioned underneath the snare in a drum kit. This will normally pick up the kick drum along with the snare, so if you only like its snare contribution, then it makes sense to gate out everything else. The problem is that the level of spill from the loudest kick drum hits may exceed the level of direct sound from the softer snare hits, so no setting of the gate's level threshold will get it working how it should: either some of the kick spill will make it past the gate, of you'll lose a few of the softer snare hits. Side-chain EQ can improve this situation by filtering the low end out of the signal feeding the gate's level detector, thereby desensitizing it to the powerful low frequencies of the kick drum spill. The great thing about the EQ being only in the side chain is that you can slap it on with a trowel without affecting the tone or phase response of the processed snare sound at all.

Other applications of side-chain EQ for drum kit gating abound: reducing hi-hat spill on the main snare signal, desensitizing the gate on a floor tom to its own rumbling sympathetic resonances and to accents on nearby cymbals, and drying up the room ambience on an out-the-front kick mic. However, there is one thing to be careful of: if a drum's attack transient is mostly made up of high frequencies, then filtering those out of the detector side chain may cause the gate to open slightly late, making your kick all icky. In such cases, a gate with lookahead functionality (as discussed in Chapter 10) may provide a workaround.

13.1.3 Sibilance Reduction

Side-chain equalization of compressors is slightly less common but can nonetheless be a godsend when you're trying to minimize the unwanted side effects of heavy compression. This is nowhere more relevant than with vocals, which have a wide inherent dynamic range that's frequently pulverized in many mainstream music styles for the sake of maximum audibility or attitude. Because most compressors are, by nature, less sensitive to high frequencies, the most common unwanted side effect of heavy vocal compression is that sonic events that contain mainly high frequencies tend to cause a lot less gain reduction and therefore end up being overemphasized by the processing. With vocals, that means that breath noises and consonants such as "ch," "f," "s," "sh," and "t" will all tend to rise up too far in the balance—particularly the "s" sounds (or sibilants, as they're called by studio nerds like me). Moreover, the general preference for bright vocal sounds in modern productions only exacerbates the situation, because any static EQ will be a nonstarter. "If you use straight EQ'ing to take away your esses," says Manny Marroquin, "you take all the life and presence out of the vocal."[3]

FAKING SIDE-CHAIN EQ

Nowadays most compressor plug-ins have some sort of side-chain EQ, or else an external side-chain input so you can roll your own. No need to panic, though, even if you're using a compressor with neither of these facilities, because you can easily fake side-chain EQ functionality nonetheless. Here's the trick. First, insert a simple digital EQ plug-in before the compressor, and use its shelving and peaking filters to create the side-chain EQ curve you want to implement. Then duplicate that EQ plug-in (along with its settings), placing the copied instance after the compressor in the plug-in chain. Finally, invert that second EQ plug-in's gain settings—so if you've used a low shelving filter in the pre-compressor plug-in to cut 12dB at 100Hz, then convert that cut into a 12dB boost in the post-compression EQ plug-in. If you do it right, the second EQ should perfectly cancel out the tonal effects of the first as far as the mixed sound is concerned, but the compressor will still be acting on the equalized version, exactly as if your pre-compressor EQ setting were applied to its side-chain signal.

Boosting the high frequencies of the main vocal compressor's side chain can help in this instance by encouraging the compressor to reduce the gain of high frequencies more stiffly. However, if sibilance is a big concern, then it's usually more effective to use a second compressor specially configured to combat it—the attack and release times that best suit the dynamic-range properties of your vocal performance may not be nimble enough to contain fast-moving high-frequency sibilants. Whichever setup you choose, it's important to work out which side-chain frequencies you need to boost. The easiest way to do this is first to use EQ boost in the main signal path with the aim of making the "s" sounds as overblown as possible. Once you know at which frequencies the sibilance resides, you can transfer that EQ setting to the compressor's side chain instead and boost until the sibilants come down to an appropriate level in the mix. However, there's a danger here: if you soften the "s" sounds too much, the vocalists will sound like they're lisping, so be careful not to go overboard with your de-essing, otherwise people might think you're taking the pith. Sometimes, though, a static de-esser setting simply won't deliver sufficiently natural-sounding results for a whole track, which is why Jon Gass,[4] Manny Marroquin,[5] and Bob Clearmountain[6] have all recommended automating the amount of de-essing to tweak its severity on a case-by-case basis.

> If you soften the "s" sounds too much, the vocalists will sound like they're lisping, so be careful not to go overboard with your de-essing, otherwise people might think you're taking the pith.

13.1.4 Pumping and Breathing

The flip side of compressors being generally less sensitive to high frequencies is that they're more sensitive to low frequencies, and that results in another frequently encountered compression side effect, which is usually most obvious on full-range, transient-rich signals such as drums and piano. What happens is that a low-frequency transient hits the compressor, triggering heavy gain reduction of the entire frequency range and dipping the sustain tails of any higher-frequency events that preceded it. Once the transient has passed, the gain reduction resets, bringing the higher-frequency sustains back into the mix. This lurching of the high-end balance in response to compression triggered from low-frequency events is commonly referred to as "pumping" or "breathing." If you want to hear a commercial example of this effect, check out Stardust's "Music Sounds Better with You," where the whole track is being pumped by the kick drum hits—a standard dance-music effect, although taken to a bit of an extreme in this case!

In some situations pumping can be desirable, because it lends sounds an artificial sense of aggression and power. This is partly because it increases the average level of the processed track, but also because it simulates the way the compression-like safety mechanisms within our own hearing system affect our perception when they're trying to protect the delicate apparatus of the inner ear from damagingly high sound levels. Drum sounds in rock music regularly use compression pumping for these reasons, and many dance tracks create an extra illusion of power for the main kick drum in this way, as in our Stardust example. Naturally, pumping won't be at all appropriate in other situations, and a dose of low-end shelving cut in your compressor's side chain can work wonders in reducing it. Indeed, some celebrated analog compressor designs, particularly those renowned for processing full mixes, incorporate low-end contouring of this kind in their designs.

FIGURE 13.3
DDMF's NYCompressor is one of only a few compressors that include variable onboard side-chain equalization, a feature that allows it to avoid unwanted pumping artifacts that other designs struggle with.

13.2 MULTIBAND DYNAMICS

Side-chain equalization can help a good deal with refining the action of dynamics processors, but it can't overcome the inherent limitation of full-band processing—namely, that it always applies the same gain reduction across the entire frequency range. Consider a drum loop where the kick drum's low frequencies are far too prominent in the balance, yet when you try to compress them back into line you hear unacceptable pumping artifacts on the other instruments. Side-chain EQ may reduce the pumping, but it'll also bring back the balance problem, whereas normal EQ in the main signal path will thin out all the other sounds that happen between the kick drum hits. One way to square this circle is by splitting the frequency spectrum of the track into a number of different ranges (or "bands") so that you can give each range its own independent dynamics processing—a setup generally referred to as "multiband dynamics." In the drum-loop example, a multiband approach allows us to compress just the low end of the loop, reducing the overprominent kick frequencies without affecting the low end of intervening sounds such as toms and snares, or indeed introducing unnatural undulations in the balance of the hi-hats, cymbals, and percussion higher up the spectrum.

The logic of applying different dynamics treatments to different frequency bands is most obvious in the event that different instruments within a single recording occupy contrasting frequency ranges. Slightly more challenging conceptually is the idea of applying multiband processing to a single instrument, so quite a good trick for making sense of this is to think of the different frequency ranges of the processed track as if they were actually separate mini-instruments in their own right, each responsible for a different aspect of the composite sound. So, for example, you might visualize three frequency ranges of an acoustic guitar track as if they were an ensemble of four separate mini-instruments: the sound-hole boom, the soundboard resonances, the string detail, and the pick noise. All three of these mini-instruments need to have their own level in the mix, yet each may need very different dynamics processing to stabilize its own notional fader. So low-frequency limiting might help to make the resonant contributions of the sound hole "instrument" more consistent; low-ratio compression in the midrange might enhance the warmth and sustain of the strings; whereas threshold-independent transient reduction at the high end might help soften unforgivingly spiky picking. Building a mix balance is about gaining control over all the different sonic components of every track, and the power of multiband processing is that it increases the precision with which you can target each of those individual components.

Drum loops and acoustic guitars are good examples of sounds that frequently require the attention of multiband processing in the small studio, but there are lots of other occasions where it can also be warranted:

■ It's common for bass guitar recordings made in small studios to be uneven at the low end, because of performance difficulties, a poor-quality instrument, or resonance problems in the cabinet or recording room. Upright bass can suffer too, although body resonances take the place of cabinet resonances in this case. Either way, low-frequency compression can ride to the rescue.

> ### DO-IT-YOURSELF MULTIBAND PROCESSING
>
> Most DAWs now have a selection of multiband dynamics processors to choose from, but you may occasionally come across something that those plug-ins can't manage. Maybe you need more bands, or you want a type of dynamic processing that isn't provided, or you fancy experimenting with different crossover slopes. One way to get the facilities you feel you're missing is to build your own multiband setup from scratch, and although this might seem a bit hardcore, it's fairly straightforward to do by combining filters and full-band dynamics processors. The main hazard to avoid is that filters and dynamics plug-ins can both mess with the phase response of audio passing through them, so be on the lookout for any undesirable hollowing out of the sound from phase cancellation. "All the while I'm checking phase," says Tony Maserati, who has often used his own do-it-yourself band-splitting setups. "That's the most important thing whenever you're combining two of the same signal that's been EQ'd and compressed differently."[7]

- Kick drums that rumble on for too long at the low end can cause big problems in an up-tempo mix or in arrangements that demand good bass-line definition. Expansion of the drum's low-frequency range can tighten up the low end of your production considerably, improving clarity and punch.
- Synth lines with high-resonance filter sweeps can be a nightmare to balance if the filter's frequency peak is too prominent. You can't use normal compression, because the peak is there all the time, and EQ won't work either because the peak frequency doesn't stay still. If you compress the frequency range over several frequency bands, gain reduction will only occur in the band that the peak inhabits.
- Background synth textures (such as pads) and other atmospheric noises can encroach too much on the foreground of a mix if they include any transients. However, applying enough transient reduction, even with a decent threshold-independent processor, can involve unacceptably jerky gain-change artifacts in the otherwise fairly continuous backdrop. Splitting the transient processing across several bands gives you more transient reduction before side effects become distracting. I find this especially helpful when layering vinyl noise into a production, as many vinyl-noise samples include pops and crackles alongside the more steady-state hiss signal.
- Vocal sibilance problems can be targeted with multiband dynamics too, by compressing the high end. The trick is first to set a threshold that results in gain reduction only for the overloud sibilants and then to finesse the compression ratio so that you get the required gain reduction. Mouth clicks and lip smacking can also be targeted in a similar way, although you may find that dedicated transient-reduction of the upper spectrum targets those more

precisely. (If that draws a blank, though, then there's still detailed audio editing and/or specialist "declicker" restoration software.

- All directional microphones inherently boost the recorded bass levels of sounds that are close to them (a phenomenon known as the "proximity effect"), so if you record any performer who moves backward and forward with relation to the mic, you'll get an uneven low end. "[The bass player] was moving around quite a bit while playing his upright," recalls Mike Shipley, for example, when he was working with Union Station, "changing the distance from the U47 with which I recorded him, and the C4 [multiband compressor plug-in] helped to even things out."[8] Compressing the low frequencies can help salvage a more dependable level, but be prepared to spend time finding the best attack/release times and experimenting with the exact crossover frequency/slope to get the smoothest results.

- Producers seem increasingly to be using multiband compression in a more generic manner on lead vocals to even out and broadly rebalance the overall spectral content. Josh Gudwin[9] used this kind of processing on many of Justin Bieber's vocals, for instance, Delbert Bowers[10] used it on Lukas Graham's hit "7 Years," and Tony Maserati[11] applied it to Robin Thicke's voice on "Blurred Lines."

The biggest single pitfall to avoid with multiband processing is the temptation to process across loads of different bands just because you can. In practice, the majority of mixdown scenarios that benefit from multiband treatment only require dynamics processing on one section of the frequency response. The previous chapters in this book should already have armed you with the understanding to avoid this trap, but it bears repeating that blanket multiband processing seldom makes any sense at all from the perspective of balancing a mix. In fact, one of the big reasons why the presets on all-in-one multiband processors are largely useless for mixdown purposes is that their approach is too indiscriminate—if you finally locate a preset that deals with your specific balance problem, you'll usually discover inappropriate settings wreaking havoc elsewhere in the frequency spectrum.

I also want to stress that the process of setting up and evaluating the action of each dynamic processor within a multiband plug-in should involve exactly the same techniques and thought processes that pertain when using a single fullband processor. Again, this is where it helps to think of each frequency band as if it were a separate track, because that way you're more likely to give it the amount of attention it deserves. If you expect to get results equally quickly from fullband and multiband dynamics plug-ins, then you'll cut too many corners. Attack and release times in particular need to be reassessed for every frequency band, because lots of multiband devices seem to give you identical attack/release values for all bands by default, regardless of the fact that signals in different frequency ranges move at very different speeds. With multiband compressors, you should

DYNAMIC NOISE REDUCTION

Unwanted background noise has vexed small-studio users since time immemorial, but digital technology now gives us many more options for fixing it in the mix. The simplest remedy is gating, but the disadvantage of this is that it only reduces noise levels in the gaps between instrument/vocal sounds. A slightly more sophisticated approach involves a kind of dynamic low-pass filter, the cut-off frequency of which tracks the signal level. Although this isn't that much better than a straight gate, it does work better at reducing noise during note sustains, which are typically less bright and therefore mask hiss less. Multiband expansion can give you another slight increase in the efficiency of hiss reduction, but it's not going to help a great deal with separating desirable high frequencies from the noise floor.

Unless, that is, you can increase the number of expansion bands well beyond the four or five offered by most all-in-one multiband processing plug-ins, in which case you can actually gain a surprising degree of noise reduction with little apparent loss of desired signal. However, controlling hundreds of bands of expansion would be like juggling a family of slimy amphibians (the Kardashians, say), so dedicated noise-reduction processors use a different control method, which semi-automatically configures the controls in response to an isolated sample of the unwanted noise. As such, there's not a tremendous amount to say about using dedicated noise reduction, beyond reiterating that it's usually best tackled from the first slot in any track's processing chain. What can be informative, though, when you first use one of these beasts is to turn the noise reduction up way too far so that you can hear what the processing artifacts sound like. Then when you're trying to gauge the depth of the proper processing, you'll be more keenly aware of what to listen out for.

FIGURE 13.4
One well-specified dynamic multiband noise-reduction plug-in is Izotope's RX Denoiser.

also be wary of automatic gain compensation, because some models don't allow you to switch it off. If you can counteract any automatic gain boosts with manual makeup gain cuts, you'll be at less risk of playing psychological tricks with yourself due to loudness changes.

When it comes to deciding on the crossover frequencies between processing bands, my first recommendation is to use only as many bands as you really need, because this keeps any side effects from the crossover filters to a minimum. If you only need to process the high end of a track, then two bands will be fine, regardless of how many your specific plug-in offers by default or can provide in theory. The exact crossover frequencies will, of course, be critical to gaining the best control over the spectral regions you're concerned with, so sticking to a plug-in's default values makes no more sense that using EQ presets (which I warned against back in Section 11.1). What can assist here is to solo each band in turn to refine its crossover settings, although a decent spectrum analyzer may help too. Some plug-ins offer a choice of slopes for their crossover filters, which is actually a useful option, because sharper slopes tend to work better for narrow-band troubleshooting (such as controlling wayward pitched resonances), whereas gentler slopes normally sound more natural for broadband enhancements (such as increasing the high-frequency sustain of a piano).

> Multiband dynamics can be used for parallel processing, just as full-band models can, and this opens the doors to a wealth of natural-sounding frequency enhancement effects, especially when using compression.

One final tip: multiband dynamics units can be used for parallel processing, just as full-band models can, and this opens the doors to a wealth of natural-sounding frequency enhancement effects, especially when using compression. Just try to choose a multiband plug-in that doesn't spoil the party by messing with the processed signal's phase response and introducing comb-filtering nasties.

13.3 DYNAMIC EQUALIZERS

An alternative way to achieve frequency-selective dynamics is to swap the static gain controls of a regular equalizer for gain-control elements that can respond to the input signal in real time, thus creating what's known as a dynamic equalizer. Dynamic EQ is functionally very similar to multiband dynamics, inasmuch as both adjust the time-domain characteristics of localized areas of the frequency spectrum, so you'll find that many jobs can be accomplished just as readily with either type of treatment. However, dynamic EQ has a few tricks up its sleeve that go beyond the scope of most multiband dynamics processors, so it warrants further investigation.

A dynamic equalizer works by making each filter's gain control act in the same way as the gain-change element in a dynamics processor. Depending on the model, there may be the option for it to mimic a compressor, an

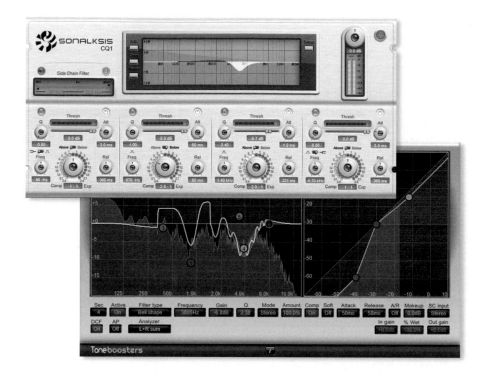

FIGURE 13.5
Dynamic equalizers, such as Sonalksis's CQ1 (*top*) or Toneboosters's TB_Flx (*bottom*), can defeat some of the most dastardly mix gremlins, but they do take some work to set up properly.

expander, or even a threshold-independent transient enhancer. For each band you therefore need to decide on the filter type, frequency, and band-width and then set up the band's dynamics processing type and controls (Threshold, Makeup Gain, Ratio, Attack, Release, and whatever else) to govern how gain is applied. If that sounds like a lot of work, you're not wrong, but the upside is that once you've got the hang of dynamic EQ, it can bail you out of situations that practically nothing else can, by virtue of the fact that it can operate on precise frequency regions—or even single frequencies in some cases.

Lead vocal parts are where dynamic EQ probably sees most use. Sibilance reduction is one of the options on the menu, but although a dynamic EQ can be exceptionally good at that function, the real star turn of this processor is its ability to contain sporadic narrow-bandwidth frequency peaks. The reason this is so useful with vocals is that a singer creates different timbres and vowel sounds by physically altering the resonant qualities of the throat, nose, and mouth. With some singers, you'll find that certain combinations of note pitch, delivery force, and vowel sound can briefly create an enormous resonant peak somewhere in the frequency response (normally somewhere above 1kHz), and this makes it almost impossible to balance that vocal section satisfactorily. When the rest of the vocal sound feels in balance, the resonance pokes you in the ear with a pointy stick whenever it surfaces; if the fader is set to balance the resonance, then the rest of the vocal disappears into the mix. Normal dynamics don't help, because the resonance doesn't add substantially to the vocal's overall level, and normal EQ doesn't help either, because any equalization savage enough to zap the resonance will also administer the last rites to the overall vocal timbre. Mick Glossop is one of several producer/engineers who have touched on this problem in interview: "It's very common for a singer's voice to develop a hard edge when he's singing loudly, which doesn't happen when he's singing quietly. . . . The problem with using static EQ to get rid of the hardness is that it might solve the problem of the vocal in the chorus, but when the quiet verse is being performed, it will lack presence, because you've taken out some 4 to 5kHz."[12] Delbert Bowers encountered a similar problem on the lead vocal of Lukas Graham's "7 Years," exacerbated by a cheap vocal mic: "A lot of it sounded great, but in certain registers Lukas can sometimes sound really powerful, and I guess the microphone made that sound harsh."[13]

With dynamic EQ, however, it's fairly straightforward to use a narrow peaking filter in compressor mode to control resonant peaks like these; the tricky bit is diagnosing the malaise in the first place and knowing exactly where the resonance lives so you can target it accurately. In fact, this particular balance problem is one of the slipperiest customers you're likely to encounter when mixing, because it takes a while to learn to recognize the specific piercing harshness (often on notes where the singer is stretching beyond her comfort zone in terms of pitch or volume) that is normally the first subjective clue to its existence. The second tip-off I typically get is that ordinary EQ draws a blank in terms of tackling the harshness effectively, and at that point I usually zoom in, loop one of the harsh-sounding vocal notes, pull out my highest-resolution spectrum analyzer, and scour the display for any spectral peaks that are shooting into orbit.

Sometimes, though, there may be four or five such peaks hopping up out of the mix at different points in the track, in which case manually configuring a separate dynamic EQ band for each one can become rather tedious. Fortunately, plug-ins such as Sknote's SoundBrigade and Oeksound's Soothe are pioneering a new kind of specialist processing that detects and reduces these sorts of narrowband resonances in a more automatic manner. While such resonance-suppressors can indeed save time, I would nonetheless advise against treating them totally as "set and forget" processors—as with de-essers, you'll only minimize unwanted processing side-effects (such as a loss of brightness and presence) if you automate the processing depth so that the plug-in only works on the syllables that need it.

Few other instruments present this particular problem because they don't change the way they resonate, but there are nonetheless still a few other

DEDICATED DE-ESSERS

A de-esser is a specialized processor designed to reduce sibilance levels in vocal parts. I've already mentioned how you can implement de-essing from first principles using a compressor (in conjunction with side-chain EQ), a multiband compressor, or a dynamic equalizer, and all of these methods can give good results. Almost all dedicated de-essers merely provide a more streamlined version of one of these methods, with an interface that's quicker and more convenient to use. My favorite de-essing algorithms these days, though, are a new generation of spectral sibilance processors that work on the vocal spectrum in a similar way that dynamic EQ does, but using a complicated adaptive filter curve that intelligently focuses its dynamic cuts on the harshest-sounding frequency peaks. There's also a much less common type of de-esser that operates in a totally different way, chopping out sibilant time regions from the audio and allowing you to process them separately, and while this design does give you maximum control over the processing of sibilant regions, I think it verges on overkill for everyday tasks. Whatever de-essers you have at your disposal, though, the main point is that each different design will respond differently, so

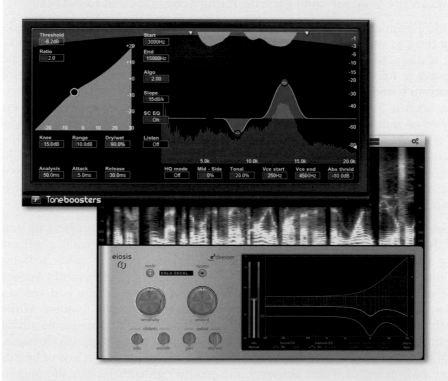

FIGURE 13.6
Although state-of-the-art de-essing plug-ins such as Toneboosters's TB_Sibalance (*top*) and Eiosis E^2 (*bottom*) will often make life easier, you can also achieve excellent sibilance reduction using less esoteric processing.

if you've got sibilance worries that cause one algorithm to break out in a sweat, then do a shootout of some other implementations.

Imogen Heap identifies the biggest reason why de-essing plays such an important role in many commercial mixes: "You can't bring up the level of the nice breathy part of the vocals if the esses are jutting out."[14] Perhaps as a result of this, it's quite common to find top producers using more than one de-esser in series. "I used two [de-essers]," says Ed Boyer, for instance, referring to his work with Pentatonix, "because if I had tried to take all the sibilance out with one de-esser, it would have affected the sound too much, so it was better to have two working less hard."[15] Tom Elmhirst put three on Adele's voice when mixing "Rolling In The Deep"![16] Other engineers use multiple de-essers for different tasks. "I'll usually have two different de-essers," says Mark Needham, for instance. "I'll set those [with] one wide and one narrow."[17] "I have two because one is doing high top sibilance, and the other lower stuff around 2kHz," reveals Rik Simpson,[18] echoing almost identical sentiments from Rich Costey.[19]

However many de-essers you use, I'd normally recommend putting them right at the end of the processing chain on vocal tracks, because otherwise the processing won't counteract any sibilance increase from down-the-line compression or EQ plug-ins. However, there's one important exception to this general rule: I'll always put de-essing before any type of high-frequency distortion enhancement, such as the Aphex Aural Exciter. This is because if the enhancer comes first, then the high-level sibilance will create lots of distortion harmonics, adding enough harshness to grill your eardrums to medium rare! Then even if the de-esser reduces the level of the sibilant regions, it won't soften their tone, so you can get into a situation where the de-essing starts making the vocal lisp well before the sibilants are actually sounding polite enough. De-essing before the enhancer means that the sibilants drive the distortion less hard, so they maintain a smoother sound. Leslie Brathwaite shares another general-purpose tip when layering vocal overdubs: "Every background vocal will have the esses at a slightly different place, so for them to be accurately de-essed I need to treat each track differently."[20] In other words, you can't just slap a de-esser on your backing-vocal group buss and expect great results.

Although de-essers are clearly designed to do what they say on the tin, you can often find other uses for them if you understand what they're doing behind the scenes. Most available de-essers work by compressing the high frequencies, which means you can often use them to rein in overprominent noise consonants such as "t" and "k," as long as the algorithm can be persuaded to respond quickly enough. The processing can also have applications for vowel sounds. "[I don't] actually use the de-esser as a de-esser," says Steve Fitzmaurice, for instance. "I instead use de-essers to get rid of harshness, kind of like an EQ. I will automate the threshold to start compressing around 2kHz, depending on the vocal, obviously. It may be higher, perhaps up to 3kHz, but not any higher."[21] And don't assume that de-essers have nothing to offer on instruments, either, because they do other useful jobs such as reducing acoustic-guitar fret noise or taming overbearing slap transients on close-miked hand-drum recordings.

FIGURE 13.7
Two specialist plug-in resonance suppressors: Oeksound's Soothe (*top*) and Sknote's SoundBrigade (*bottom*).

instances where dynamic EQ can save the day. The resonant sweep of a wah-wah guitar part might periodically coincide with a strong cabinet resonance in your guitar combo, for instance; the fundamental frequency of a certain upright bass note might vary considerably in level depending on whether that note has been stopped or played as an open string; or your pianist might be too heavy with his right thumb on one note of an important riff. Dynamic EQ may be fiddly, but it can also save your bacon when the going gets tough.

13.4 SPECTRAL DYNAMICS

The last type of frequency-selective dynamics processing to deal with is something I like to call spectral dynamics. In essence it's little different to the multiband processing we already talked about in Section 13.2, but it uses many more bands (typically more than a thousand) evenly spaced across the audible spectrum so that each band only handles an extremely small frequency range. Of course, setting up the dynamics controls for each of those processing bands individually would be prohibitively laborious, so spectral dynamics plug-ins have to be designed in such a way that you can adjust the myriad band parameters *en masse*. Probably the most popular interface for this is a spectrum analyzer display upon which are superimposed editable curves representing the values of important processing parameters. This setup lets you see at a glance, for example, how the signal level in each frequency region relates to the compression thresholds of its respective processing bands.

Despite the brain-melting complexity, spectral dynamics processing has many practical applications at mixdown. For a start, it's frequently at the heart of the specialist resonance suppressors, spectral de-essers, and noise-reduction algorithms I've already discussed earlier in this chapter. However, there are also more general-purpose spectral dynamics plug-ins such as Cockos ReaFir and ProAudioDSP Dynamic Spectrum Mapper that let you experiment more freely with the technology. One of my favorite applications of spectral dynamics is to compress the upper octaves of any sound's spectrum to increase the sense of brightness and detail while simultaneously smoothing potentially harsh-sounding high-frequency transients and resonant peaks. Gentle low-ratio compression across the whole frequency spectrum is also frequently worth a try on pitched instruments, because it'll tend to weaken the main pitched components (which normally dominate in level over other frequencies) in favor of the signal's noisy elements components and lower-level harmonics. This is a dodge that can work very well on vocals, adding extra character, grain, and breathiness to the timbre.

CUT TO THE CHASE

- Frequency-selective dynamics processing allows you to deal with balance problems that occur in both the time domain and the frequency domain at once and can therefore overcome some of the trickiest mixing obstacles. Despite the apparent complexity of such processing, using it is mostly just an extension of what we've already covered in previous chapters.
- Any normal dynamics processor can be made frequency-selective either by using it in a parallel setup with an equalized return channel or by equalizing the input to the processor's level-detection side chain. Side-chain equalization is particularly useful for refining a gate's triggering, as well as for reducing the sibilance emphasis and pumping/breathing side effects of heavy compression. Just be aware that any phase-shift in the parallel channel may impact on the timbre of the combined sound.
- If you imagine separate frequency regions within a given track as being separate tracks in their own right, the logical application of multiband dynamics processors becomes easier to grasp. Some tips: steer clear of presets; avoid automatic gain-compensation routines; spend enough time refining crossover settings; and don't feel you have to use every available band, or indeed all the processing available in each band.
- Dynamic equalizers and spectral dynamics processors can both mimic many of the effects of multiband dynamics and go further by dealing with dynamics problems that have very narrow bandwidth or that would require an impractically large number of processing bands.
- De-essers can operate in many ways, so if the first plug-in you try doesn't deliver the goods, make a point of auditioning some alternatives. De-essing normally works best at the end of a track's processing chain, although it's best to apply it before any distortion-based high-frequency enhancements.
- If you have to deal with a lot of noisy tracks, then specialist multiband noise-reduction processing can be extremely useful.

Assignment

- Check what de-essing and multiband dynamics plug-ins are available on your DAW system, and if the choice appears limited, then consider investing in third-party additions.
- Mute all the tracks in your mix and rebuild the balance, experimenting with frequency-selective dynamics processing wherever you feel it may help solidify the balance.
- Make a note of any faders that still feel unstable.

Web Resources

On this book's companion website you'll find a selection of resources to support this chapter, including:

- a special tutorial video to accompany this chapter's assignment, in which I show-case how some of the frequency-selective dynamics processes mentioned here can improve the balance of a simple multitrack session;
- dozens of audio examples demonstrating the effects of equalized parallel dynamics, side-chain equalization, multiband dynamics, dynamic EQ, and de-essing, as well as the action of specialist noise reduction and resonance-suppression software;
- links to a wide range of affordable plug-ins for frequency-selective processing: dynamics processors with side-chain EQ or external side-chain inputs; multiband dynamics and dynamic EQs; and specialized tools for de-essing, noise-reduction, and resonance-suppression.
- a number of mix case-studies from the *Sound on Sound* "Mix Rescue" column that focus specifically on practical applications of frequency-selective dynamics processing, along with links to more in-depth further reading on de-essing strategies.

www.cambridge-mt.com/ms-ch13.htm

CHAPTER 14

The Power of Side Chains

I introduced the idea of dynamics side chains in Chapter 13 but have so far assumed that the level-detection signal path is fed from the same source as the main processing path—albeit perhaps with different equalization. As it turns out, there are sometimes very good reasons for feeding a dynamics processor's side chain from a different track entirely, and many dynamics plug ins (most commonly full-band compressors, expanders, and gates) are now capable of accepting an external side-chain input.

The most basic reason for using an external side-chain signal is if you can't get the gain reduction to respond correctly based on the signal levels of the track you're processing. So, for example, imagine you have two mics on your rack tom: one on top and the other poked inside the drum from underneath. The top mic may well give the better sound, but it may also prove impossible to gate reliably because of snare spill. By using the inside mic's signal to "key" (via the side chain) the gate on the outside mic, you can achieve much better triggering, without any need to use the dodgy sound of the inside mic in the balance at all.

Side chains are about much more than this kind of troubleshooting, though, because they enable tracks to interact with each other to improve the overall mix balance. Probably the most commonplace application of this capability is in rock music where overdriven electric guitars are masking the lead vocal's high frequencies. Although you may be able to carve out space for the vocals with EQ in this scenario, the danger is that you'll simultaneously make the rest of the arrangement sound like *The Care Bears Movie*. So to take some of the pressure off the EQ processing, you can set up compression on the electric guitar parts that responds to a side-chain input from the vocal track. Whenever the vocalist sings, that triggers gain reduction on the guitars and therefore reduces the unwanted masking; but when the vocal is silent, the guitars return to their original level in the balance. Just a decibel or two of gain reduction can work wonders, although you do have to be careful that you don't go so far that the gain changes become distracting.

There are plenty of other situations where similar dodges can be of assistance:

- To get rid of a nasty snare-spill component on your hi-hat mic, use the snare close-mic to key a hi-hat compressor.
- To achieve the tonal benefits of strong drum ambience for individual instruments without making a mess of the mix, try keyed gating on your room mics. The most famous example of this, of course, was Hugh Padgham's trend-setting drum sound for Phil Collins's "In the Air Tonight," but before you write it off as just some cheesy 1980s gimmick, bear in mind that Chris Thomas and Bill Price also used the idea for the Sex Pistols' debut album *Never Mind the Bollocks*: "[It] involved me keying different ambience mics off the drums as they were being hit," recalls Price. "Chris Thomas's suggestion that we could shorten the ambience with gates, providing more without it sounding too distant, all made total sense to me."[1]
- To boost the snare sound in your overhead mics, use the snare close-mic to key a limited-range gate on the overhead channels and then fade them up to rebalance the cymbal levels. (Another time-honored trick, this one.)
- To avoid excessive low-end buildup, use the kick drum part to key a compressor on the bass. (Some people prefer to route the kick and bass through a single compressor to achieve a broadly similar result, but I don't find that method quite as controllable myself.) "I often do this when the kick is being stepped on by the bass," says Jason Goldstein. "So every time the kick hits, the bass ducks 2dB or so just for a moment. When you have a [prominent bass], you can't do too much."[2] You can hear a commercial example of this effect during the choruses of Coldplay's "Hymn For The Weekend," according to their mix engineer Rik Simpson.[3]
- To give a rhythm part or pad sound more rhythmic impetus, insert a limited-range gate on that channel and key it from an in-tempo metronome click or a simple repetitive audio track (perhaps created especially for the purpose). Work with the gate's attack and release times to get the most musical pulse.
- To add low-end welly to a kick drum, route a constant low-frequency sinewave tone through a gate, and then trigger it from the drum sound. This was a much-loved trick in the disco era and can still pay dividends nowadays in dance and urban styles. It doesn't really matter whether you create the tone using a simple test oscillator or a more fully featured synthesizer, but I'd suggest starting with a frequency of around 50Hz and then tweaking that by ear to find the best fit for your particular mix. If it's out of tune with your bass line, for instance, it can really destabilize the musical harmony.
- To prevent long echo and reverb effects from swamping a lead vocal, key a compressor on each effect's output from the vocal signal. You'll be surprised how much gain reduction you can get away with before anyone starts feeling seasick. "Dynamic delays . . . will duck out of the way, coming up only at the tail end of words and phrases," says Goldstein. "I almost never use a straight delay."[4]

DUCKING

Although a lot of engineers use keyed compressors to dip the levels of backing instruments in response to lead vocal levels (as discussed in the main text), I don't personally like this approach too much. In the guitars-versus-vocals case, for example, the quietest vocal notes will cause the least reduction in guitar levels, even though the quietest notes are the ones that need the most help. My preference, therefore, is to use a dedicated ducker instead. A ducker operates exactly like a gate, except that it opens when a gate would close, and vice versa—if you can already set up a gate, then operating a ducker should be a no-brainer. What's different about using a ducker for this kind of keyed mix-balancing task is that it will always introduce the same gain reduction for all the vocal notes, and though that's still not perfect, to my ears it's nevertheless a significant improvement over keyed compression.

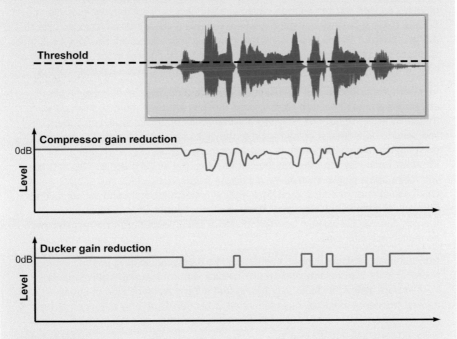

FIGURE 14.1
Comparing the gain-reduction actions of a compressor and a ducker. Notice that the compressor reduces the gain most when the vocal is highest in level, which isn't ideal for the purposes of side chain-triggered rebalancing of other competing tracks in the mix.

A stumbling block for some DAW users, though, is that they lack a proper side chain-enabled ducker, so it's fortunate that there's a clever little workaround you can do to arrive at an identical effect in pretty much any mixing system. Here's how it works within the context of the vocal/guitars scenario. Set up a gate as a send effect, send to the gate from the guitar channels, feed the gate's side chain from the vocal

(continued)

channel, and then invert the polarity of the gated return. What should happen with this setup is that every time the vocal is present, the gate will open to let through a burst of inverted-polarity guitar signal, and this will silence the main guitar channels in the mix by virtue of phase cancellation (assuming that the gate introduces no latency delay). All you now need to do is decide what level to set the fader of the return channel: the lower you have it, the less severe the phase cancellation, and the finer the ducking—as the Reverend Spooner might have put it.

This setup has a nifty application in dance music, too: it's a particularly elegant way of creating kick-triggered gain pumping across multiple tracks, because you can determine the amount of ducking for each instrument by adjusting its send level to the parallel gate. So if you wanted severe ducking of your synth pads and echo/reverb effects, you'd turn up the gate sends on those channels, while lead vocals or synth leads might have lower send levels to avoid the pumping making a nonsense of the main hooks. And you can go even more ninja with this concept too. For example, a favorite ducking technique of mine is where you add a linear-phase high-pass filter to the gate's return channel and trigger the gate from your lead vocal. Now any track that sends to the gate will be ducked whenever the singer is present in the mix, but only at high frequencies, and only to a depth determined by that track's send-level control.

In recent years, though, the availability of multiband compressors and dynamic equalizers with external side-chain access has brought even more sophisticated frequency-selective ducking techniques within reach of project-studio engineers. All you need to do is feed your lead vocal to the side-chain of a dynamic EQ inserted on your guitars buss, for example, and you can then configure the plug-in to cut whatever frequency you like from the guitars when the singer's active. Alternatively, you might use a plug-in such as Wavesfactory Trackspacer, which can analyze a side-chain signal from your lead-vocal track and automatically detect and cut conflicting frequencies from the guitar signal. The side-chain input of SoundRadix's pitch-tracking SurferEQ could also be used to similar effect, especially in conjunction with its multi-notch filter type.

Now although hardcore ducking maneuvers like these can achieve useful incremental improvements in mix clarity when you're battling with over-stuffed arrangements, I would warn against trying to use then systematically as some kind of panacea against frequency masking. As I mentioned in Section 11.4, only a fraction of the frequency masking in a mix will be problematic, so cutting all masking frequencies indiscriminately is a fool's errand.

14.1 KEYED MULTIBAND DYNAMICS

Once you get the idea of side-chain triggering for full-band dynamics, you don't have to extrapolate too much further to conceive the idea of keyed multiband dynamics, although you're unlikely to come across a balance problem that demands this kind of firepower more than once in a blue moon. If you do want to explore such avenues, however, a practical hitch may be that your choice of multiband dynamics processor offers no external side-chain access, in which case you may have to create your own do-it-yourself multiband setup (as discussed in Chapter 13) so you can key each band's individual processor independently. An alternative might also be to use an EQ'd parallel dynamics setup instead, thereby focusing a full-band processor's action into the desired frequency range. So, for example, if you wanted to boost just the mid-frequencies of the snare ambience in your overhead mics, you might set up a parallel gate channel fed from the overheads, trigger its gating action from the snare close mic, and then use an EQ to isolate the midrange frequencies.

CUT TO THE CHASE

- Some mix imbalances can only be resolved if dynamics processing on one channel is controlled by a signal from another channel. Full-band dynamics will be most straightforward to use in this way, but now that many multiband processors and dynamic equalizers also offer external access to their level-detection side chains, there's little to stop you creating keyed frequency-selective dynamics should the need arise.

- Duckers are often better suited to keyed mix-balancing tasks than compressors. However, if you can't get hold of one for your system, a keyed parallel gate channel can be polarity inverted to arrive at an identical ducking effect— a setup that has additional power-user applications.

Assignment

- Investigate the side-chain routing facilities within your own DAW system, and make a note of which plug-ins will accept an external side-chain feed.
- Mute all the tracks in your mix and rebuild the balance again, experimenting with keyed dynamics processing to see whether it can stabilize the remaining faders.

Web Resources

On this book's companion website you'll find a selection of resources to support this chapter, including:

- a special tutorial video to accompany this chapter's assignment, in which I showcase how side-chain triggering can help finesse the balance of a simple multitrack session;
- a selection of audio examples demonstrating various side-chain triggered dynamics processes;
- links to affordable dynamics plug-ins that offer external side-chain access;
- a number of mix case-studies from the *Sound on Sound* "Mix Rescue" column where I made extensive use of side-chain triggered dynamics.

 www.cambridge-mt.com/ms-ch14.htm

CHAPTER 15
Toward Fluent Balancing

You should now have enough tools and techniques at your disposal to create a good static mix balance. If you've worked through the end-of-chapter Assignment suggestions, you should also have gained some practical insight into which processes suit different balance problems within your musical style of choice. In other words, you'll have started to interpret the subtle whisperings of fader instability that are trying to tell you what the music needs. As I openly admitted back at the start of Chapter 8, the step-by-step procedure implied by the Assignments has no foundation in real-world mixing practice, but I make no apologies for that, because it makes the vastness of the subject easier to grasp initially. However, now that my overview of balancing procedures has run its course, it's high time we removed the didactic scaffolding and examined how all these methods fit into the more fluid workflow of the professionals.

15.1 REMOVING THE SCAFFOLDING

The key to fluent balancing is to let the faders set the agenda, so rather than rebuilding the balance multiple times and adding different processes with each rebuild, you only actually need to build the balance once. As you introduce each new track into the mix, you address its panning, filtering, and phase/polarity, and then try to set a balance. If you can't find a stable fader setting, then you keep experimenting with different processing until you can. Whether it takes a single subtle equalization tweak or unspeakable abuse at the hands of a whole gang of bloodthirsty plug-ins, you keep going until you get a solid balance without unacceptable processing artifacts. The goal is to get every aspect of the track to balance: every pitch and every noise; every transient and every sustain; every moment in time; and every region of the frequency spectrum. When you've succeeded (or else exhausted every processing option you have available), you move on to the next track and start again.

Of course, any given track may also feel subjectively unappealing, even at an appropriate balance, in which case you may also need to experiment with a certain amount of "suck it and see" creative tonal shaping. However, this kind of mix treatment is applied in a very different manner, so it has to be evaluated on

When push comes to shove, the needs of the balance should always be your first priority—if your mix doesn't have balance, no one's going to give a monkey's whether any individual instrument sounds stunning.

its own terms ("Do I like the sound better now?") and shouldn't be confused with balance processing ("Can I hear everything properly?"). Moreover, when push comes to shove, the needs of the balance should always be your first priority—if your mix doesn't have balance, no one's going to give a monkey's whether any individual instrument sounds stunning.

Although it's vital to have a well-developed plan of attack and to try your best to deal with each track's balance before adding in the next, it's also important to recognize that real-world mixing can never be a truly linear process, so no one's going to shoot you for tweaking the processing of your first track while balancing your twenty-seventh. On the contrary, it's often only when you're desperately trying to shoehorn your final tracks into the few remaining mix crannies that you get an accurate picture of what aspects of the more important tracks are truly indispensable. So if you become aware of a previously unnoticed fader instability on any earlier track, then go back and deal with it before introducing any further instruments. (And if you can manage to make this look completely erratic to the uninformed observer, you might also develop an aura of professional mystique into the bargain!)

Here's another point to be absolutely clear about: although fader instabilities can give you a hundred clues, all processing is one long experiment. "Real music is about experimentation," to quote DJ Premier.[1] You simply cannot know that a given process will work before you try it, because every mix is different. Only your ears can decide if any tone/balance improvement you've dialed in is worth the price you've paid for it in terms of processing side effects. So it's normal to make duff calls when it comes to choosing the right processor, and it's all right to strip away anything that isn't working and

CONFIRMATION BIAS AND ABX TESTING

I've already mentioned the importance of fighting against loudness bias while mixing—in other words the natural tendency for anything that's louder to sound better. Psychologists have discovered plenty of other ways in which we can deceive ourselves, though, and one in particular, usually called "confirmation bias," can be particularly insidious at mixdown. You see, by nature, it seems us humans generally prefer to perceive things as we expect them to be, not necessarily as they actually are. In a mixing situation, this means that every time you apply mix processing with an expectation of what that mix processing might do, confirmation bias will subtly alter your perception to fulfill that expectation, however misguided it may have been. Confirmation bias is the reason for an embarrassing mistake that every engineer I've ever met has made at least once in their career: spending some time subtly refining some mix processor's settings, and only then realizing that it's been bypassed the whole while, or else they've been tweaking the wrong channel! Whenever you do anything to your mix (or even just look at what your computer screen is showing you about it!) you're prone to confirmation bias, and although it only alters your perception quite mildly, it's enough to significantly skew some of the more subtle sonic mixdown decisions. Furthermore, it can waste your time and/or money by convincing you that some "pixie dust" process is significantly improving the sound when it isn't, or that the disparities between different designs/brands of processor are bigger than they are.

So what can you do against confirmation bias? Well, here's one simple trick. Any time you've set up some subtle sonic comparison and want to be sure you're not imagining things, first try to remove loudness bias from the equation as best you can, and then hover your finger or mouse pointer over whatever button makes the comparison—often a plug-in's bypass button. Then close your eyes and quickly toggle the button back and forth until you forget whether it's on or off. Now toggle the button more slowly (still without looking at the screen!) to ascertain whether you can really hear a difference between its two states, and if so which setting you prefer. Only once your ears have reached a firm conclusion should you look at the screen again to see which setting is which. In most cases, just doing this test once will give you a pretty good idea whether confirmation bias is playing a role, but if you're unsure, then just repeating the test a few times will provide more conclusive evidence. If even that won't allay your bias concerns, though, then check out dedicated software such as Lacinato's ABX/Shootouter, which offers the same kind of statistically rigorous testing procedure used for academic studies.

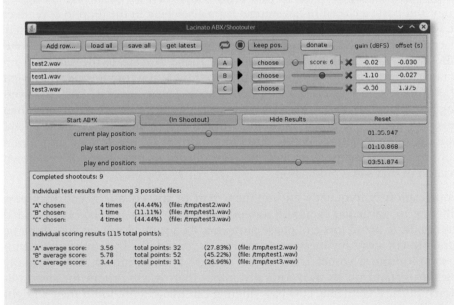

FIGURE 15.1
For maximum protection against confirmation bias, try Lacinato's cross-platform donationware ABX/Shootouter.

start afresh—even if that means zeroing absolutely everything and beginning the balance again from scratch. "I will often restart mixes three or four times," reveals Fabian Marasciullo. "Put everything back to zero and try again, reblend and EQ everything, and put the vocals back in."[2] Every time you flunk out, you'll have eliminated a setting that doesn't work, and that takes you a positive step closer to finding a setting that does. "I think a lot of the thing with music is pilot hours," muses Youth. "To go through all your mistakes and know not to go up those cul-de-sacs and fall into those elephant traps."[3]

15.2 REBALANCING PREMIXED AUDIO

Although most mixing involves taking multiple tracks and balancing them against each other, some styles of music rely quite heavily on premixed elements such as loops or samples, often taken from commercial music or multimedia releases. In such scenarios you may have to rebalance or otherwise adapt those samples at mixdown in order to fit them into their new arrangement context. This is one of the toughest balancing tasks there is, because it frequently requires a good deal of engineering ingenuity in addition to confident handling of intricate editing and processing techniques. So now that we've surveyed a wide variety of balancing tactics, let me pass on a few of my own personal tips for rebalancing premixed audio.

The first thing I'd suggest is not to neglect the power of editing. If you don't like a particular open hi-hat in your drum loop, then it's often easy enough to copy and paste a closed hi-hat snippet from elsewhere in the loop to cover over it, for instance. In fact, I've often found I can extend this concept to completely deconstruct a mixed drum-machine track by seeking out a handful of isolated hits in the song, and then using copies of those to recreate the pattern from scratch, relying on the waveform of the mixed track as a timing guide. But what if, say, there's no kick-drum hit without a snare layered over it? Well, try layering an isolated snare hit exactly in sync with the kick-plus-snare hit, and then invert its polarity to phase-cancel the sound from the mix. By the same token, you can often drastically phase-cancel important drum hits from a sample of a commercial release using isolated hits copied from an introduction/breakdown section in that same production.

Equalization is frequently an important part of the equation, and high-pass filtering in particular is usually essential if a new bass line has been added alongside, to avoid low-end phase problems. As a general rule, in fact, don't be afraid to cut away any inessential frequencies, especially if you're working with a sample from a commercial release, because by definition it was originally meant to stand on its own sonically, rather than leaving room for any additional musical elements. Low-pass filtering can make sense too if there's bright high-frequency background noise (hiss or vinyl crackle, say), since it may otherwise prove difficult to bring other lead vocals/instruments sufficiently up front by comparison, as discussed in Chapter 11. I'm also regularly surprised at how effectively you can adjust the apparent level of individual instruments or notes within a mixed sample using narrow peak filters (or their multi-peak cousins) trained on the relevant pitched components.

Full-band dynamics processes can achieve all sorts of useful results, such as compression to rein in peaks or bring up low-level detail, or transient processing to enhance drums. However, frequency-selective dynamics tend to have more to offer, because you can target the processing more critically. "I often do extensive work on the sample," says Young Guru, for instance. "I will go as far as splitting it up into frequency ranges as many times as I need . . . so I can carve out what aspects of the sample I want to cut through the track."[4] So let's say you'd licensed a sample of The Righteous Brothers' "You've Lost That Lovin' Feelin'" for your next EDM smash, but wanted to turn down that super-loud tambourine, a quick dose of high-frequency limiting could do that with negligible side effects, allowing you to feature the vocals, choir, and strings more in the mix. Given that the tambourine is panned hard to the left, you could also refine your results further by processing just the left-hand channel of the stereo signal. (In Chapter 18 we'll see that it's possible to process the center and the sides of a stereo mix independently too, which can be equally useful for rebalancing mixed signals.)

Occasionally, though, you may want to remove something from a mixed audio track completely, but simple audio edits won't do the trick for some reason. If it's just a note or two that needs muting, perhaps because of a harmonic clash, then multi-notch EQ is usually my first port of call, closely followed by the note-muting facility within Celemony Melodyne's polyphonic processing mode. Excising whole instruments can be fiendishly difficult, though, and is

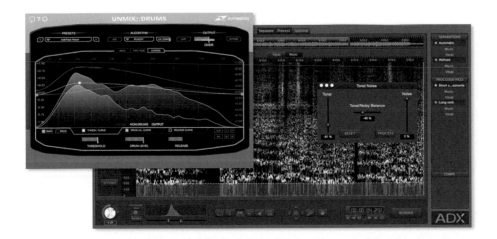

FIGURE 15.2
When it comes to rebalancing musical elements within premixed audio files, cutting-edge software such as Zynaptiq's Unmix::Drums and Audionamix's Trax Pro can now achieve feats that would have been unthinkable a decade ago.

usually well beyond the capabilities even of frequency-selective dynamics plug-ins. However, developers of offline mastering and audio-restoration software have steadily been pushing back boundaries in this field, and we're now beginning to see real-time mix processors such as Zynaptiq Unmix::Drums (for rebalancing drums) and Audionamix ADX Trax Pro (for rebalancing lead melodies) that are capable of pulling off jaw-dropping disappearing acts on occasion, especially if you're willing to put in the graft to finesse their more advanced settings.

The other situation where premixed audio may quickly exhaust your mixing mojo is when a particular instrument in the mix is too quiet, usually because there's a limit to how much you can fade it up without unacceptable knock-on effects on other instruments within the same mixed signal. Before you admit defeat, though, consider doubling that part with an additional MIDI instrument or audio overdub. You'd think this might be difficult, but the truth is that the sonics of the added part seldom need to match very exactly, because it doesn't normally need to be very loud in the mix to provide the necessary balance boost. I've been amazed how easy it is to get away with this in practice, in fact—I've used this dodge surreptitiously on dozens of occasions, and no client has ever yet spotted the sleight of hand. Do be careful with the timing of added parts, though, because any flamming will quickly give the game away, and also keep an ear open for phase-related tonal inconsistency, particularly if you're layering drum or bass sounds as discussed in Section 12.2.

SMALL IS BEAUTIFUL

There is a common misconception among small-studio mix engineers that it takes big, bold processing moves to bring about large sonic changes at mixdown, which is why so many amateur mixes suffer from overprocessing. "Mixing is just a lot of little subtle things," says Dave Pensado, for instance. If my experience with dozens of "Mix Rescue" remixes has taught me anything, it's that the most dramatic mix transformations are usually brought about primarily through lots of small mix tweaks. It doesn't matter if the benefit of any individual plug-in setting seems almost insignificant on its own, as long as its benefit outweighs any unwanted side effects, because even minuscule improvements add up surprisingly swiftly. So think like a sculptor: you won't get too far by clouting your piece of stone a couple of times with a sledgehammer; you need to chip away at it bit by bit to get anything worthwhile.

15.3 THE GEAR IS NO EXCUSE!

And finally, low budget is no excuse for bad balance. "There was a time when I fantasised about hardware gear," says Carlo "Illangelo" Montagnese, "and spent lots of money buying some pieces of hardware. But none of that is valuable to me any more at this point. I am over it . . . It's all about taste, it's all about ideas."[5] As I see it, if you can't get a balance with the bundled plug-ins in any mainstream DAW, then it's not the fault of the equipment, and a big budget will only make mixing quicker and more pleasurable if you can already get a great mix with no-frills gear. But don't take my word for it—ask Jake Gosling, who recorded and mixed Ed Sheeran's multiplatinum debut single "The A Team" on an old PC running Cubase SX3. "Funnily enough," he laughs, "I got a message from Abbey Road through the record label when we were mastering saying 'who mixed this record? This is amazing!'"[6] You might also ask Frank Filipetti, who won his 1998 Best Engineering Grammy for *Hourglass*, an album recorded in James Taylor's house using a first-generation Yamaha 02R and three Tascam DA88s. "In the end, your ears, your mind, your musical abilities are what it's all about. Put a George Massenburg, a Hugh Padgham, a Kevin Killen together with any kind of gear, and you'll get a great-sounding record. . . . If you have a sound in your head and you know what it is, you can get it. You may have to work harder on lower-quality gear, but you can get it."[7] Tony Visconti agrees: "I've heard people make very bad records on expensive gear. The gear does not dictate the quality. It's how you use it."[8]

You don't need every esoteric specialist processor to get the job done either, because there are so many workarounds when it comes to mix processing. If you don't have a de-esser, for example, then try a compressor with EQ in the side chain, or high-frequency compression from a multiband device or do-it-yourself equivalent, or an 8 kHz peaking cut from a dynamic EQ in compressor mode, or a polarity-inverted parallel gate with its side-chain filters set to isolate the sibilant frequencies, or a static EQ under the control of your DAW's mixer automation system. Or just mult the sibilant audio sections to another track and set its fader level separately! The gear is not the issue; it's you, your ears, and your monitoring that make a balance work. Or, as Dr Dre advised Derek "MixedByAli" Ali: "It's not what you're working on. It's who's pressing the buttons."[9]

CUT TO THE CHASE

- Fluent balancing technique is about reading the instabilities of your faders as they arise and drawing on the relevant tools in your processing arsenal as required. There is room within this process for subjectively enhancing sounds, as long as this doesn't undermine the success of the overall balance.
- Don't expect the mixing process to work in an entirely linear way, because it's normal to reassess early processing decisions later in the mix, once mix real estate begins to become more scarce.
- One of the toughest mixing challenges is rebalancing premixed audio, as it frequently relies on inventive combinations of editing, processing, MIDI programming, and audio restoration tools.
- All processing is experimentation. If any processing experiment doesn't work, then don't think twice about ditching it and starting over—you'll still have learned something useful about the mix problem you're grappling with and will be better equipped to solve it on the next attempt.
- Low-budget processing is no excuse for delivering a bad balance.

Assignment

- Save your mix as it currently stands, and then open a copy of it, resetting all the channel settings and stripping out all the plug-ins so that you're left with a completely blank slate. Now redo the whole balance in one go, introducing the tracks as before, reading each fader in turn, and applying whatever processing is required to nail down a good static balance.
- Once you've completed and saved your new balance, compare it with your first attempt. As likely as not, the second version will have been quicker and easier to do, because you didn't go down as many blind alleys, and it'll probably also sound better into the bargain.

Web Resources

On this book's companion website you'll find a selection of resources to support this chapter, including:

- a selection of audio examples demonstrating some of my favorite techniques for rebalancing premixed audio tracks;
- links to affordable audio-restoration and unmixing software for dealing with the thorniest of rebalancing tasks;
- a number of mix case-studies from the *Sound on Sound* "Mix Rescue" column where I made extensive use of side-chain triggered dynamics.

 www.cambridge-mt.com/ms-ch15.htm

PART 4
Sweetening to Taste

Once you've achieved a good balance, you're effectively on the home straight. Even if you do nothing else to the mix, your listener will at least hear everything on the multitrack clearly and at an appropriate level, and that should guarantee that the quality of the music comes across. Nonetheless, there are lots of bonus points to be gained in most productions by adding further "sweetening" effects to an already creditable balance, the aim being to present the recorded material in a more flattering light. As Lee DeCarlo explains, "Effects are makeup. It's cosmetic surgery. I can take a very great song, by a very great band, and mix it with no effects on it at all and it'll sound good, but I can take the same song and mix it with effects and it'll sound fantastic! That's what effects are for."[1] It's these additions that I want to focus on in the remaining chapters, as well as discussing the final stages a mix goes through before being signed off as finished product.

If you're wondering why we're only dealing with sweetening effects now, then think of this as an extension of the idea of building the balance in order of importance. Such effects are, pretty much by definition, less important than the tracks that they're designed to enhance, so it stands to reason that you should add them only once your

> Once you've achieved a good balance, you're effectively on the home straight. Even if you do nothing else, your listener will at least hear everything clearly and at an appropriate level, and that should guarantee that the quality of the music comes across.

raw tracks are balanced. That way you know how much space there is in the mix to accommodate the sweeteners, and are less likely to obscure important musical details by over-sugaring. Russ Elevado says, "I usually will not turn on any reverbs until halfway through the mix. . . . I try and achieve a big sound without resorting to reverb. So I just keep chipping away until the track is pumping. Then once I'm happy, I start to think about what I might want some room or ambience on."[2]

CHAPTER 16

Mixing with Reverb

The most widely used sweetening effect in record production has got to be artificial reverberation (or reverb for short), an effect that can generate complex patterns of echoes in response to an input signal. Artificial reverberators were originally designed to simulate the sonic reflections you get from boundaries in a real acoustic space, allowing you to add realism to unnatural-sounding close-miked recordings. However, reverb has a much broader brief in record production these days, and artificial reverberators are now used just as much for creative purposes. Indeed, many classic reverberation devices don't produce anything like the sound of a real room, but that hasn't diminished their enduring popularity!

16.1 FIVE ENHANCEMENTS AT ONCE

Part of the reason why reverb is incredibly useful at mixdown is that it can enhance several aspects of your sonics simultaneously. As I see it, it has the power to enhance the following significant elements:

- *Blend.* Reverb can increase the extent to which any individual track blends with the rest of the production, making disconnected instruments sound as if they belong together and giving the mix as a whole a more cohesive sound. An instrument that isn't at all blended sounds upfront and close to the listener, whereas an instrument that blends well is sucked more into the background, away from the listener. Therefore, the idea of blend is also closely related to the idea of front–back "depth" in a mix: less blend brings a track toward you, whereas more blend pushes it away from you.
- *Size.* Artificial reverb can increase the apparent dimensions of your mix's acoustic environment, making it sound as if your tracks were recorded in a larger (and maybe better-sounding) room than they actually were—which is one way to make low-budget projects appear more expensively produced. In addition, if any given instrument excites the simulated reverberation particularly audibly, then it creates the illusion that this instrument is large and powerful in itself, even if it's low in level and well-blended with the overall balance. So, in a nutshell, reverb can increase the size both of the whole mix and of individual instruments.

- *Tone*. The echoes that make up a reverberation effect have the capacity to phase-cancel with the dry track once they are added into the mix. The resulting comb filtering will alter the subjective tone of the instrument. Irregularities in tonal quality of each of the echoes will also contribute to this effect.
- *Sustain*. Because echoes are essentially delayed versions of the effect's input, any reverberation effectively increases the sustain of a dry sound it's added to. However, reverbs never sustain equally across all frequencies, so the tonal characteristics of the added sustain are a key parameter to be controlled.
- *Spread*. The simulated echoes in most artificial reverberators are distributed across much of the stereo image. This spreads information more evenly across the picture and may also increase the apparent stereo width of individual processed tracks, or indeed the whole mix.

Although reverb's multifaceted nature is good news in principle, the big catch is that a lot of reverbs tend to apply all these enhancements at once, whether or not you want them to! Furthermore, within a specific reverb effect, the amount and nature of each individual enhancement can be ferociously difficult to adjust independently of the others. The result is that less experienced engineers usually find that they can't get enough of one enhancement without overdoing another: they can't get enough blend without muddying the mix tone, perhaps, or they can't enrich an instrument's sustain enough without apparently transporting it to the Taj Mahal!

Whether you fall foul of these mix traps has little to do with how much you know about the physics of natural reverb or how well you understand all the

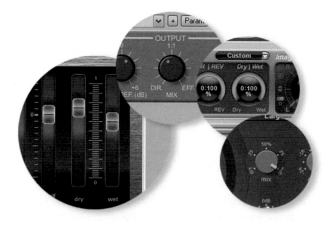

FIGURE 16.1
Almost every reverb plug-in has a control somewhere that sets how much signal with effects and how much signal without effects it sends to its outputs. These screens show a selection of different configurations.

inscrutable-looking widgets on a typical reverb plug-in. The main secret to getting decent results swiftly is learning how to create reverb effects that provide only a subset of the possible enhancements. Once you have a set of more specialized effects at your fingertips, it's more straightforward to apply them in combination to achieve precisely the reverb sweetening your mix demands.

16.2 ESSENTIAL REVERB CONTROLS AND SETUP

To create these kinds of reverb effects, you need to understand a handful of essential truths about reverb plug-ins. The first thing to realize is that reverb is almost always best applied via a send–return effect configuration, so that a single effect can be accessed from every channel of your mix, and can itself be processed independently if required. For that to work properly, you need to check two factors:

- that your reverb plug-in is only outputting processed effect ("wet" signal) and not any unprocessed sound ("dry" or "direct" signal), otherwise sending to the effect from a channel will also alter that track's level in your finely poised mix balance. Some plug-ins may have a single Wet/Dry or Direct/Effect Mix control, whereas others might have independent level controls. Whatever the interface, though, make sure to turn the Dry/Direct signal all the way down or switch it off completely;

- that the individual sends that feed the reverb are taken from the channel signal path post-fader. This means that the balance of wet versus dry sound will remain constant if you adjust the sending channel's main fader, and you won't have any ghostly reverb shadows remaining in the mix if you fade out the channel completely at any point.

Because reverb is such an important studio effect, most DAWs tend to provide more than one algorithm, and numerous third-party plug-ins are available too, including lots of freeware and shareware. If you don't already have a preference for one or the other, then don't worry too much about which you choose to start with, because a lot of reverbs these days are pretty usable and it's often tricky to know how well a given reverb suits a particular task until you hear it in action. One little tip, though: good-sounding reverbs normally require a fair bit of CPU power, so be wary of any that are slimline in this regard, no matter how flashy their graphics look.

As far as all those reverb controls are concerned, the only ones that you absolutely have to know about are those that adjust how fast the echoes die away. On simple plug-ins there might be a single control for this, usually called Length, Decay Time, Reverb Time, or RT60. Where there isn't anything like that, then sometimes this parameter is adjusted by manhandling the decay "tail" on a graphical representation of the reverb envelope. Alternatively, you may find that Room

FIGURE 16.2
The most important controls on a reverb are those that set how fast it decays, but different reverb plug-ins have different names for these, including Size, Damping, Decay, Time, and Length.

REVERB DESIGNS: A BRIEF FIELD GUIDE

You don't need to know the full history of how reverb effects were developed to use them effectively at mixdown. However, a little background knowledge about different reverb designs can nonetheless speed up choosing an appropriate plug-in or preset for each particular task, so here's a quick survey of the main options:

■ *Chambers*. The earliest type of added reverb, created by sending signals to speakers in an unused room and then capturing the reflected sound with mics.

■ *Plates and springs*. Two early types of electromechanical reverb, which were widely used during the 1960s and 1970s. The audio signal is used to set off vibrations in a bit of metal, and then the reflections of these vibrations are captured using pickups. Neither design is good at simulating realistic spaces, but both are nonetheless still highly regarded in the industry today, being well-suited to tonal and sustain enhancements.

■ *Digital algorithmic processors*. A type of reverb that rose to prominence during the 1980s and had a huge impact on the sounds of that era. Because the reverb reflections are created using mathematical models, the user has a lot of control over them. Although some algorithmic processors are not very realistic (especially early models), a number of these are nonetheless prized for their unusual spatial and tonal qualities.

■ *Digital convolution processors*. This recent technological development allows you to record the exact pattern of reflections from a real acoustic space as an "impulse response" file, which can then be used to recreate that reverb with unparalleled realism within a mix situation. The sounds of a range of different artificial reverberators can also be mimicked in this way, although any dynamically varying qualities of the original units cannot satisfactorily be emulated with standard convolution processing. (There is a technology called dynamic convolution that can emulate such intricacies, but it's quite CPU-intensive so most convolution reverbs don't implement it.)

Size and Damping are supplied instead, and although these fulfill a broadly similar function, they are also more likely to change other aspects of the reverb character as well. If you see separate level controls for the reverb's initial echoes (Early Reflections or ER) and the remainder of the reverb decay (Reverb or Tail), then those can also be helpful, but they aren't essential.

There'll be a slew of other controls on a lot of plug-ins, but for typical small-studio operators—and indeed a lot of professionals—there simply aren't enough hours in the day to worry about what all of those do. (They do serve a useful social purpose, however, in distracting computer anoraks from ever getting any music finished.) As long as you choose a reverb plug-in that has a load of programmed presets, you can get along just fine without any other parameters, because the balance processing you already know about from Part 3 provides ample scope for further molding the sound of each reverb if necessary. So, without further ado, let's see how you go about designing and applying the first (and probably most important) type of reverb patch: blending reverb.

16.3 REVERB FOR BLEND

The reason almost all small-studio productions need blend reverb is that a lot of their tracks are recorded as separate overdubs. Overdubbing avoids spill between different instruments in the arrangement, and though this causes fewer phase-cancellation complications, it also means that you don't benefit from the natural blending effects of sound leakage between different microphones. In addition, most modern recording sessions involve a lot of close-miking, and a close-mic usually picks up quite low levels of natural room reflections in most situations—and whatever reflections it does pick up will change a good deal as the mic is repositioned for recording different instruments. As a result, many tracks won't blend well enough with the mix, and sounds that should be in the background feel too close for comfort. Adding a common reverb to the unduly upfront tracks helps blend them together, pulling them further away from the listener and making them sound more as if they were recorded in the same place at the same time.

Play back the mix section you balanced for Part 3, and listen for any tracks that are having trouble blending—either they don't seem to connect with the mix as a whole, or they seem too close to you. Now select one of those tracks that has a comparatively full frequency range and preferably some transients too.

FIGURE 16.3
A number of reverb plug-ins provide the option to control a reverb's early reflections separately from its decay tail. Although it's nice to have access to such controls, you can actually get by fine without them most of the time.

(I often start with a mixed drum loop or a live drum kit's overheads, for example, but only if those parts don't seem to blend properly in their raw state.) Send at a decent level from this track to the reverb, and then solo the reverb return so that you only hear the reverb's added echoes, without any dry signal. Why kill the dry signal? Because that lets you really concentrate on the nature of the effect. Phil Ramone says, "One of the hardest things to teach somebody is to listen to the device itself. Take out the source and listen to the device. You'd be amazed how crummy some of these things sound! They flange, they phase, there's cancellation all over the place."[1]

16.3.1 Picking a Preset

You're now ready to start hunting for a promising-sounding preset that feels like it fits the sound you want for the mix. This is probably the most critical part of creating a reverb patch, but because it's subjective and genre specific it's also difficult to advise on. The main thing to realize is that the part of a reverb that is primarily responsible for its blending effect is roughly its first half-second. So whenever you try a new preset, reduce its length straight away to home in on the blending characteristics. If you have independent level controls for Early Reflections and Reverb Tail, pulling down the latter should help too. It doesn't matter exactly how short you make the reverb for the moment—just shorten it enough to make it into a brief, well-defined burst of reflections rather than an echoey decay tail.

> Take preset names with a large pinch of salt. Just because something says "Epic Snare Boosh!" doesn't mean that it won't serve as a good general-purpose blending reverb in your situation.

Beyond that, here are a few other general-purpose tips that may assist you in arriving at a good choice of preset:

- Take any preset names with a large pinch of salt. Just because something says "Epic Snare Boosh!" doesn't mean that it won't serve as a good general-purpose blending reverb in your situation. Nevertheless, do keep an eye out for any preset with "ambience," "early reflections," "short," or "dry" in its title, as there's a greater likelihood that it will be what you're looking for.
- Natural-sounding presets tend to do a better job for blending purposes than obviously synthetic reverbs (in other words, ones that don't really bear much relation to the sound of a real acoustic space). For example, any preset name containing references to plates or springs probably won't be that suitable.
- It can sometimes help to close your eyes and visualize the kind of space you want the blended sounds to inhabit. Picturing a real environment can help focus the mind here, although this may not help much if you're after a more otherworldly sound.
- Don't worry too much if the frequency-balance of the reverb doesn't seem right, because you can do a lot about that with equalization. What's most

important is that the overall acoustic signature feels right. That said, patches where either of the frequency extremes are very pronounced are unlikely to be useful for blending purposes.

- If possible, steer clear of plug-ins or presets that produce a kind of metallic sound with clearly audible pitched resonances, especially in response to transients.
- Try to find a patch that is evenly spread across the stereo image and doesn't swerve drunkenly off to one side or the other as it decays. You'll be better able to hear the reverb decay if you mute and unmute the track send a few times.
- Don't rush! It's not uncommon to trawl through more than a dozen presets and change plug-ins a couple of times before you find a reverb sound that really seems in keeping with the production you're working on.
- Rely heavily on your nearfield monitors here, because they'll give the best overall tone and quality judgments, but be sure to check the effect in mono as well to ensure that you don't get any nasty tonal changes. Mono compatibility is one of the areas in which expensive reverbs tend to outclass cheap ones, so be especially careful here if you're working with budget or bundled plug-ins. "You really should check in mono," stresses Phil Ramone, "to make sure that what you're hearing is what you get. That's why a lot of television shows get so screwed up. They come in with these great effects, but when they check in mono . . . half the reverb goes away."[2]

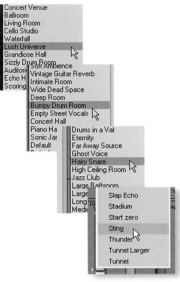

FIGURE 16.4
Just because a reverb preset's name is a bit silly, that doesn't mean it won't be useful. The only way to tell is to have a listen.

Once you've found a likely contender, unsolo and mute the return channel to remind yourself for a few seconds how the production sounds without the reverb. Then unmute the reverb in the mix, fade it up so you can hear it clearly, and confirm that it's really worth pursuing. If you think it is, then the first thing to do is tweak the reverb's length so that it's short enough to tuck behind the dry sound in the mix without creating any audible reverb tail, but also long enough that it can deliver sufficient blend as you fade it up. In doing this, you're minimizing the degree to which the reverb affects the size and sustain of tracks it's applied to, which gives you more power to adjust blend independently.

16.3.2 Tone and Spread Adjustments

Next you want to minimize undesirable tonal changes caused by adding the reverb to the dry signal, and this is probably easiest to do if you isolate the dry track and reverb return together. The tools and techniques for doing this are exactly the same as you used to adjust the combined tone of multi-miked instrument recordings in Sections 8.3 and 11.3: timing and phase shifts, polarity inversion, and EQ. Clearly, though, you want to process only the effect signal in this case, otherwise you'll upset the balance you've already achieved for your dry track. Although some reverb plug-ins may include tools for adjusting any of these aspects of their output, it's usually just as easy to insert a favorite delay, phase-adjustment, or EQ plug-in from your collection into the reverb's return channel. Mostly these processing choices should need very little further explanation—you just have to listen for any unwanted tonal colorations introduced when you fade up the reverb return, and then remodel the effect's timbre to counteract them. However, the idea of delaying the effect signal, usually referred to as adding "predelay," does warrant a little extra discussion. "I think the pre-delay is the most important part of any reverb," says Alan Parsons, "and it's the first button I reach for."[3]

The thing about predelay is that it doesn't just adjust phase-cancellation effects between the wet and dry sounds; it also influences how far back into the mix the blend reverb appears to pull the treated track, subjectively speaking. The less pre-delay your blend reverb uses, the further this reverb can push sounds away from the listener. If you use no predelay at all, then the implication is that the treated instrument is stapled to the rear wall of the reverb's virtual room, whereas adding predelay rescues is from that uncomfortable predicament and moves it closer to the listener. My advice here is to start off with about 10 to 20ms of predelay, aiming on the shorter side for more smaller-sounding spaces and on the longer side for more spacious virtual acoustics, and then to refine the setting up to a few milliseconds either side of that initial setting for tonal reasons. What this means is that you can apply the blend reverb across all your tracks to some extent if necessary, without distancing the mix as a whole unduly from the listener. However, if you later discover that

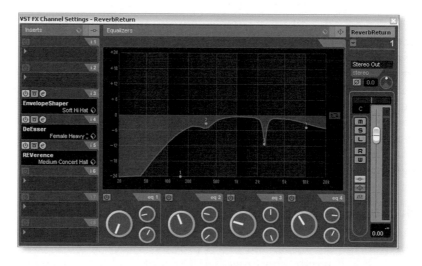

FIGURE 16.5
A lot of reverb plug-ins have balance-processing features built into them, but it's usually easier and more flexible to use the ones you already know. All you have to do is insert your normal mixing plug-ins into the reverb return, as shown here in Cubase.

you're unable to blend or distance any track far enough, you may wish to set up a second blend reverb with less pre-delay specifically to treat that—this eventuality isn't common, in my experience, but it's useful to recognize it as a possibility. One other issue to bear in mind with predelay is that it reduces masking of the wet signal by the dry signal, so you can typically use lower reverb levels when you've dialed in some predelay, and this can improve the clarity of your mix.

The final characteristic of the reverb you may wish to tweak a little is the stereo spread. Having a wide stereo spread on a blending reverb is usually a desirable side effect, because it helps the tracks in your mix feel as though they belong together by encompassing them in their acoustic environment, but there are some situations where you may want to narrow the spread. For example, you might wish to blend drum spot mics with stereo room mics that either don't have a wide stereo picture or have been deliberately narrowed by your choice of panning—a wide-sounding blend reverb could seem rather incongruous in that case.

16.3.3 Balancing Blend Reverb

Once you've designed your basic blend patch, you need to balance it with everything else that's in the mix. You can't do this with solo buttons down, though, so deactivate those in order that you can listen to the reverb in its proper context. For the moment, fade down the single send you've added so far, and if you've been messing with the return channel's fader, reset it to unity gain. Without the reverb in the mix, reacquaint yourself with the blend problem for a few seconds to regain a bit of perspective, and then slowly raise the send level control as far as necessary to connect the dry track more strongly with the mix as a whole—or, to look at it another way, to distance the instrument far enough from the listener. The more reverb you add, the more the instrument will blend, and the further into the background it will move. As Dave Pensado notes, "If you want the singer to sound like she's standing behind the snare drum, leave the snare drum dry and wet down the singer."[4]

Once you've got the most satisfactory setting you can, make a point of shifting your listening perspective (a little like we did when evaluating EQ settings back in Chapter 11) to check that the reverb isn't masking any other instruments. In other words, switch the effect in and out of the mix by toggling the return's mute button, and if some important frequency range is being obscured by the added reverb, then toss an EQ plug-in onto the return channel to carve away at it further. (You may then need to fade up the remaining reverb frequencies to regain the former degree of blend.) Pay particular attention to the region below about 300Hz or so, as this usually benefits from some degree of cut in most styles to preserve headroom and clarity for bass instruments at the low end and also to avoid muddy-sounding energy buildup in the lower midrange frequencies—the latter being one of the most common problems in small-studio mixes. "Often I'll EQ the reverb to attenuate lows," says Mick Guzauski,

WHAT IF THERE'S TOO MUCH BLEND?

Lack of blend isn't the only blend problem that you might be presented with: you may also encounter tracks that blend too well. In other words, they've been recorded with too much spill/reverb so you can't actually bring them far enough forward in the mix. This can be tricky to fix, but there are nevertheless a few processing tricks that can yield some success in "unblending," especially if used in combination:

- brighten the instrument's tone with equalization or distortion processing, because (as I explained in Chapter 11) we're psychologically programmed to interpret brightness as a clue that a sound source is close at hand;
- narrow the stereo image of stereo files, or even collapse them to mono, as the reverb will often phase-cancel more than the instrument itself;
- increase the instrument's dynamic range in some way to dip the low-level background reverb tails between notes and at the ends of phrases. Simple expansion or limited-range gating may work to some extent, whereas transient processors can sometimes deliver the goods on percussive tracks if they offer the option of negative "sustain" gain. Multiband expansion or even dedicated noise-reduction may offer further improvements, but it'll usually be fairly slim pickings;
- investigate some of the new generation of spectral processors that have been specifically designed to address this issue, such as Zynaptiq's Unveil, Sonible's Proximity:EQ+, and Izotope's RX De-reverb. Husky Hoskulds has found that such tools work better in combination. "After some experimentation I found that using Zynaptiq first and RX after it gave best results. Somehow a little of each sounded more natural than going heavy-handed with just one or the other."[5]
- you might make a last-ditch attempt at bringing the sound forward by layering in a dry MIDI instrument or sample alongside the recorded part;
- kidnap the client's ornamental fern, and demand that the part be rerecorded as a ransom. For every day's delay, send the client a frond in the post.

"so that . . . it doesn't cloud any instruments. Usually I'll roll off some low mids, around 200 to 300Hz."[6] The amount of EQ thinning that's suitable here, however, is quite genre-specific, so be prepared to revisit this setting after you've referenced your first draft mix against some commercial competitors.

Equalizing the reverb return also has other purposes, particularly if you'd prefer your reverb effect not to draw attention to itself—for example, in hard-edged or acoustic styles where audible artificial treatments might undermine the music's sense of rawness and emotional integrity. What most commonly gives the game away that you're using reverb is excessive high frequencies in the reverb return, because most musically useful natural acoustics rarely have lots of bright-sounding reflections. "If I'm going to use a reverb," says Elliot Scheiner, "I want it to be inconspicuous, so I don't use bright reverbs any-more; my reverbs are generally darker."[7] A low-pass filter or high-shelving cut in the reverb return is, for this reason, very common when using any reverb, but it's almost a matter of course when the reverb's specifically targeted at providing blend. Exactly how you set the filter must, of course, be judged by ear in order to find the frequency contouring that best reduces the reverb's audibility while at the same time allowing it to fulfill its blending role. The trick I suggested for EQing synth pads in Chapter 12 is equally useful in this context: try turning the reverb up a bit too loud, and then EQing it back to an appropriate subjective level from there. If you can't hear the artificial reverb as an effect in its own right, but the mix seems to thin out and fall apart when you mute the reverb return, then you're probably heading in the right direction. Ed Seay says, "You don't even have to hear it, but you can sense it when it goes away—it's just not quite as friendly sounding, not quite as warm."[8]

FIGURE 16.6
It's common to reduce the amount of low-end content in your reverb return (as shown in SSL's X-Verb here) to keep the low frequencies of your mix clear sounding.

Sometimes equalization doesn't go far enough and dynamics also need to be employed. The most common circumstance that calls for this adjustment is where transients are causing the reverb to produce distracting flams or stereo ricochets, in which case it's worth trying out a dedicated transient processor in the return channel, placed before the reverb plug-in, to see whether you can tone those down. Threshold-based transient processors are not usually effective in this role if more than one track is feeding the reverb, so threshold-independent designs are a much better choice. Sometimes frequency-selective dynamics prove necessary too, most commonly where vocal sibilance hits the reverb and sprays unnaturally across the stereo picture. A de-esser inserted into the return channel before the reverb is the obvious solution here, but a potential drawback is that threshold-independent de-essing isn't easy to find. If this causes a problem, one option might be to send vocal signals to the reverb via an extravagantly de-essed intermediary channel. (Just don't send this channel direct to the mix or it'll sound like the singers have swallowed their dentures.) Alternatively, you could use a trick Imogen Heap credited to Guy Sigsworth,[9] where you create a copy of the vocal track with all the troublesome consonants edited out, and then use that to feed your send effects.

Once you've applied your blend effect across all the channels that seem to require it, make sure to confirm the balances on your different monitoring systems at different playback volumes. It's also not a bad little reality check to bypass the reverb return for 10 seconds, envision how you want the track to blend, and then pop the reverb back in. Oftentimes this simple process will highlight some small balance/masking problem or a previously unnoticed tonal coloration that still needs attending to with further tweaks of the return EQ. You can also improve your ability to hear internal effect balances within a mix if you listen once or twice with the main tracks (perhaps bass, drums, and lead vocals) bypassed in various combinations, so that the foreground parts don't distract your concentration from background details—Geoff Emerick[10] and Alan Parsons[11] have both independently mentioned using this technique for their own mixes, so you'll be in good company.

All that said, it could well be that applying blend reverb to the first track sits it into the mix just fine without the need for any additional balance processing on the return channel. However, as you continue to identify insufficiently blended instruments and apply the same effect to those as well, the situation might easily change, so it's quite normal to keep tweaking your return channel balance processing right up until the moment you declare the mix finished.

BLENDING BY OTHER MEANS

Reverb isn't the only means to blend a production. Short delays will also work admirably (as explained in the next chapter) and many modulation effects will distance instruments by making them more diffuse sounding. Double-tracking can blend an instrument or voice better too, even if you keep the double-track at such a low level that it's not really audible in its own right. Adding background noise to your mix will often improve the overall sense of blend, and if you don't fancy sourcing your own noise recordings, then get hold of a media sound-effects library and look for what are usually called "room tone" or "room ambience" files. Room tone is the sound of nothing happening in a room—not the most interesting thing to listen to, but in a background role it can really help make all the tracks in your mix feel as if they belong together. Tape hiss and vinyl noise have similar blending effects, especially when they're in stereo.

16.4 REVERB FOR SIZE

The biggest difference between reverbs designed for blend and size is that where the former is best provided by the earlier reverb reflections, the latter is best created by focusing the effect sound on the remainder of the reverb tail. As such, you can leave the reverb length of each preset unchanged during auditioning and turn down any separate level control you have for early reflections. Presets where the reverb onset has a clear attack tend to work less well for this application than presets that start more softly and slowly. Beyond these suggestions, much of the previous advice about choosing a preset still holds:

- Beware of unnatural-sounding presets, as these will have trouble creating the sound of a larger space convincingly. CPU-light plug-ins will typically sound less natural than more computationally hungry algorithms.
- Feel free to ignore the preset names with impunity—the main goal is to try to imagine the space implied by each preset and decide whether it's the right kind of space for your mix to exist within.

- Don't be too concerned about tonal imbalances as long as there aren't nasty metallic resonances.
- Check that the stereo picture is fairly evenly spread, and assess the mono compatibility.
- Don't hurry the selection process, and make sure you ratify your choice properly with your different monitoring systems.
- When you've got a promising patch, mute it, recalibrate your ears to the mix as is, and then fade it up to confirm that it's actually what you're looking for.

Any reverb you create for size enhancement will inevitably also create a small amount of blend too (much as any blend reverb will also give hints of size), but you can reduce this "mission creep" to a useful degree by giving your size reverb a good dose of predelay—anything from 50ms upward. It's because of this longer predelay that reverbs with an attack "bump" won't usually work in this role, because the hard reverb onset can potentially make transients sound like they're flamming. It's partly for this reason that some engineers opt to use a tempo-related predelay time, because then the reverb attack following any percussive hit will be masked by the following one, making the bump less audible. (Tempo-related predelays can also subtly reinforce the song's groove.) Glen Ballard provides another good reason for increasing the predelay for any longer reverb: "I'm always interested in hearing the vocalist's words . . . so I like to have the dry signal a little bit clear of the effect, just for the articulation. If it's a really slow song, say a ballad, then the predelay will probably be longer."[12]

Size reverb will also inevitably add sustain, and though this may in general be a good thing, it's wise to try to make the tone of that sustain fairly neutral if

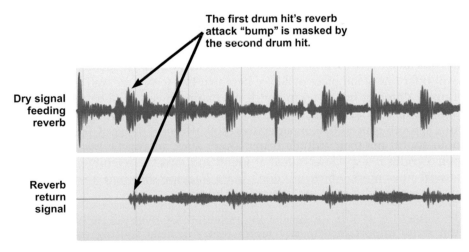

FIGURE 16.7
Reducing the audibility of a reverb's attack "bump" using a tempo-matched predelay time.

possible, because you may wish to apply this effect to many sounds in your arrangement if you're after any kind of "all in one room" feel to the production. Given the long predelay time, phase cancellation between the wet and dry sounds should be minimal, so any tonal adjustment can be carried out simply with EQ—solo a full-range instrument along with the reverb return, and then fade the reverb up and down to highlight tonal changes in the instrument's sustain that you might want to address. You may find notch cuts to be helpful here if the reverb turns out to emphasize any dissonant pitched resonances. As with blend reverb, the stereo spread can normally be left as is.

16.4.1 Balancing Size Reverb

With a suitable reverb patch on the go, deactivate those solo buttons again so that you can get on with balancing the effect within the context of the full mix. The procedure is similar to that described for blend reverb, the main difference being that fading up the reverb here gives the impression that an instrument is within a bigger space than it was recorded in. The more you apply, the clearer the impression of that larger acoustic becomes. The more you fade up the reverb for any specific instrument, the more audible the reverb becomes and the greater the additional illusion of relative size and power. With the size enhancement will inevitably come a certain amount of extra distancing and sustain, but these should be secondary effects that aren't as strong as the size illusion.

There is an additional subjective decision to make, though, with size reverb: Should you bring all the instruments within a single space, for a more realistic presentation, or be more creative and place individual instruments or groups of instruments into contrasting acoustic environments? The latter is far from natural or realistic, of course, but nonetheless provides excellent opportunities for musical contrast: your lead vocal might be dry as a bone so that it's right up at the front of the mix, the drums might be in what sounds like a large wood-paneled room, and the strings and backing vocals might float ethereally within a chapel-like environment—the sky's the limit, really. Creative choices like this aren't really within the scope of a general-purpose mixing primer, but I will offer one small piece of advice here that has served me well in the course of many different mixes: try to match the nature of each different space to the sonic character and artistic intent of the sounds it will be applied to. So don't put a smooth concert hall on aggressive punk drums if a ragged "in the garage" room sound fits their attitude better. Don't stick an angelic boys' choir in a long corridor if a spacious cathedral sound supports the serenity of the choir's musical lines more. Above all, don't put a bagpipe solo into a concert arena when it really belongs in an underground bunker.

The longer the reverb, the greater the potential for masking problems, so it's even more important with size reverb that you keep shifting your listening perspective to assess the impact of the reverb on the clarity of your dry tracks, especially at low frequencies. Within most mainstream productions, reverb tails on the low end of bass instruments are usually counterproductive—what extra impression of size you get from the seismic rumblings is more than offset by a loss of headroom and note definition. My recommendation is to high-pass filter the reverb return to keep the low frequencies tight and controlled, and also to be conservative with the amount of size reverb on your bass instruments in general. Another thing to be careful of is adding lots of this kind of reverb to sustained stereo chordal parts such as synth pads. The nature of such parts makes it difficult for any long reverb to enhance them much, so all the reverb tail actually ends up doing is blurring the chord changes, making it feel like your keyboard player can't keep time. (And after all that work you did quantizing him too.)

SPECIAL EFFECT REVERBS

This chapter is primarily concerned with reverbs that serve more technical purposes, but in some productions a reverb might also be used as a feature in its own right, a special effect designed to make a statement. "A lot of the guys I work with just hate reverb," says Billy Bush. "They just feel it puts things [to the] back in a mix . . . I use reverb more as an effect, not so much trying to create a space. I'm just trying to create some sonic element that'd find interesting."[13] If you're using a reverb in this way, there's not a tremendous amount of useful advice I can offer, simply because everything depends on your personal idea of what kind of effect you'd like to hear. My main recommendation is to consider incorporating that particular effect at the balancing stage, treating it exactly like any of the normal recorded tracks in your mix project and balancing it with just the same care. Bear in mind, though, that the effect might not rank nearly as highly in terms of "sonic importance" as the dry track that is driving it, so it may only be appropriate to add it to the mix further down the rank order. In addition, you'd be well advised to ask yourself tough questions about whether all the aspects of the reverb in question are absolutely essential to its artistic impact. If you don't, then your mix clarity may suffer at the hands of unnecessary blend, size, tone, sustain, or spread side effects.

Also bear in mind that the impact of any special effect diminishes with familiarity, so it may be counterproductive to leave it in the mix all the time. "I'd rather flick it on and off throughout the song in two or three places," recommends Al Stone. "That's what an effect should be. If it's on all the time it cancels itself out."[14] Guy Sigsworth shares similar views: "I'm very fond of these electronic pieces Stockhausen did in the 1950s. . . . One of the few effects he had to play with was reverb, and he did a lot of things where the first note's completely dry, the second one as reverbed as possible, the next one somewhere in the middle. The reverbs are very dynamic, they're not like just some vat you dip the whole thing in, they move around, they're very agile, they're like a nervous animal. I really like that, so I try to get that hyperactive approach to space."[15] Josh Gudwin is another big fan of these kinds of momentary spot effects, and has a neat method of implementing them: he puts his spot effects onto a set of empty tracks so that he can quickly drag fragments of trigger audio to them as required.[16]

Whether you want your size reverb to be audible as an effect in its own right is another judgment call that depends on the purpose of that reverb in the production. Where a single reverb is operating over most of the tracks in the arrangement to enhance the apparent size of the virtual venue, then keeping it understated is usually sensible, in which case you'll want to follow similar return-channel EQ and dynamics policies as when using blend reverb. However, where multiple size reverbs are being used more creatively, then making the sound of some of these conjured-up spaces less bashful might help with the musical contrasts within the arrangement—just be doubly careful of undesirable masking consequences from the reverbs in this case.

Whichever tack you adopt, do still go through the usual objectivity test with each reverb once most of the sends to it are operational: bypass the return for a few seconds, recreate in your mind the size illusion you're after, and then reinstate the reverb to see how it compares to that ideal. If dropping out some of the more important instruments for a while helps you refine the subtler reverb balances, then give that a razz as well. One matter to devote particular concentration to while finessing the overall sound of your size reverb is whether you've chosen the best combination of reverb length and overall reverb level. If the reverb is too long, then you may not be able to use enough of it to create a believable illusion of a larger space without it washing out the whole mix and obscuring all the nice mix details; if the reverb's too short, then you may not be able to enhance the apparent size of your production as effectively as you'd like to.

16.5 REVERB FOR TONE

Where it can frequently be appropriate to have sends to a blend or size reverb coming from most of the channels in your mix, reverb patches dedicated to tone or sustain enhancements will usually be applied only to small groups of tracks. This is because these two reverb types have more in common with EQ than they do with reverb, and it should be clear from Part 3 of this book that an equalization setting that works for one instrument is unlikely to work well for another. Indeed, the similarity of such processing to EQ was the reason I introduced the idea of tonal/sustain reverb back in Chapter 12—it's much more sensible to do any subjective timbral shaping of a sound early on in the balancing process, rather than leaving it until the balance is complete, whereupon any tonal changes could easily upset the whole applecart. Now that we've discussed reverb in more detail, however, let's look further at how to get the best from these more coloristic enhancements.

A reverb effect primarily for the purpose of altering the tone of a treated track is best shortened to avoid obvious reverb tail, much as it would be in a blend reverb, because this immediately minimizes size and sustain side effects. So when you're looking through your plug-in presets you'll want to keep the effect's length under control to get a decent impression of each reverb's tonal qualities. The similarity with blend reverb ends there, though, because the most usable kinds of presets will typically be those with an unnatural

FIGURE 16.8
Although quirky little plug-ins such as (*top to bottom*) Voxengo OldSkoolVerb, TAL Reverb 4, LongSound Microverb VST, or Viper ITB Vee Spring Verb may not sound natural for enhancing blend or size, they can really add character when you're looking for tone or sustain alterations.

sound—the less they sound like real acoustic spaces, the less they'll blend the track as a side effect. Hideous frequency-response imbalances and horribly clangorous resonances can also be an advantage, because those characteristics will be more likely to bring about truly meaningful changes in the timbre of treated sounds. Any rubbishy old plug-in with a wafer-thin CPU load is also right back on the menu for similar reasons. Jack Douglas observes, "Sometimes the weirdest things—like old spring reverbs—can sound really phenomenal in the mix. By itself it's going to sound awful . . . but use it right—color it a little bit, filter it, EQ it—and it's going to sound cool."[17] Manny Marroquin shares his taste for springs: "They're cool-sounding. You can make [the reverb] short and tight with a gate and it adds tone and depth to the sound without washing it out. . . . When you have one thing in a crowded mix you really want to bring out, but adding EQ would make it sound harsh, put it through a spring reverb."[18]

What you also need to realize is that the sound of a particular preset's echoes on its own is only part of what you're looking for, because a big proportion of the final tonal change available here will actually result from comb filtering between the wet and dry sounds. For this reason, you should always make sure that the dry sound is mixed in alongside the effect when auditioning presets for a tonal reverb patch, and bear in mind that the relative levels of the dry and wet signals will impact on the severity of the phase cancellation. Moreover, once a preset has been found that flatters the dry signal, you may be able to refine it further by applying predelay and phase/polarity adjustment to the return channel—note that predelays under about 10ms will usually give you more powerful comb-filtering effects. Equalization of the return channel also has a lot to offer. Manny Marroquin states, "I listen for the frequency where the reverb matches the input signal and tweak that."[19]

REVERB MODULATION

Because convolution plug-ins can capture the acoustic characteristics of real spaces so accurately, they might appear to be the most obvious choice for blend and size reverbs. However, impulse responses can't recreate the small dynamic changes in reflection patterns that derive from performers moving around or from naturally fluctuating air currents, so they typically have more clearly defined pitched resonances than the natural spaces they derive from. Fortunately, this is a pretty subtle difference, and one that you can usually disregard within the context of mainstream music styles, where many different effects are usually active at once. In more exposed situations, though, the extra resonances that develop within a convolution reverb can add an unwelcome metallic edge or harshness to the overall sound. Many algorithmic reverbs, on the other hand, modulate some of their internal parameters in real time to avoid this peccadillo, and can thereby produce a smoother-sounding end result. That said, once you realize this, you can often achieve a similar improvement for convolution plug-ins by applying subtle 100 percent-wet chorusing to the reverb's input signal. The trick here is to keep the modulation rate below one second, and to listen carefully for any audible detuning artefacts (especially on things such as pianos and acoustic guitars) while setting the modulation depth.

Ideally, the goal here is for the reverb not to be heard, so that it fuses with the dry sound in the listener's perception to provide the greatest tonal changes with the fewest blending side effects. One of the challenges to achieving this objective is finding the right combination of reverb level and reverb length—as the level of the reverb increases, its length will often have to be decreased if you want to maintain the same level of audibility. It's also wise to be careful of the stereo width and placement of your tonal reverb return, because if the effect is substantially wider than the image created by the dry signals, or in a different location in the stereo image, then the timbral change may be diluted and the treated sound will appear less upfront in the mix. By now, you're probably already firmly in the habit of switching monitoring systems to ratify tonal mix decisions like this, but do ensure you check mono compatibility in particular to avoid any nasty surprises—as I mentioned earlier, cheap and nasty reverbs can really fall on their asses in mono.

The danger with tonal reverb is that adding it will usually increase the overall level of the treated instrument, especially if you're using similar levels of wet and dry signal to exploit the full possibilities of comb filtering. For this reason, it's possible to mislead yourself if you simply switch off the effect in an attempt to judge its value. A better approach is to find a reasonable fader level for the processed sound in the balance and then mute the dry and wet sounds together for a few seconds. When you've had a chance to appreciate the state of the mix without those channels, try to bring to mind what you want the composite to sound like and only then reintroduce them to the mix. You're more likely to be dispassionate about the reverb's merits that way.

FIGURE 16.9
A typical tone reverb setting, running in Christian Knufinke's SIR2 convolution reverb plug-in: a characterful impulse response with a short decay time, a sub-10ms predelay, narrowed stereo image, and heavily sculpted frequency response.

Once you're confident that you've arrived at a combination of dry and wet signals that truly improves the character of selected instruments in the mix, you're still faced with the task of balancing these rejuvenated sounds within the mix as a whole. Because the tonal reverb effectively becomes an integral part of the sound, it stops functioning as a traditional send–return reverb effect, and it makes much more sense to treat the dry and wet channels in the same way as you would the individual microphone channels in a multimiked instrument recording. So I normally end up routing them both to a communal group channel for global control, which allows me to process the complete sound for balance reasons without altering the fastidiously honed phase relationships between the constituent channels. As with multimiked instrument recordings, however, there may be further creative potential in processing the individual channels, but you can run into difficulties here if you add the tonal reverb to more than one track, because if you process the effect channel for the benefit of one of the dry tracks, it may not suit the others. It's not much more effort to gain extra control here, though, by creating separate instances of the same effect for each treated track. You may also be able to reduce the audibility of the effect in this way, by replacing a stereo reverb patch with several mono instances, each panned to the location of the dry

REAL REVERB

Most of the discussion in this chapter is about using artificial digital reverb of one kind or another, but there are plenty of producers who prefer the real thing. "Digital reverb is the most ironic phrase you can come up with," says Jack White, for instance. "It doesn't make sense, because reverb is a natural, real thing. It needs springs, or a cave, to be there in real life. A plug-in emulation of that may sound OK to many people, but when you pile several emulations on top of each other, with nothing natural in it, you don't get results that are very interesting."[22] One reason real recorded reverb often sounds fuller and more satisfying is because it's fed from the entire sonic output of the instrument, whereas digital reverb can only respond to that subset of the instrument's sound that's actually been captured by the microphones. But there's clearly more to it than that, because there are many engineers who like to generate it at mixdown too, by using any nearby rooms as impromptu reverb chambers. "I like to use the [live] rooms while I'm mixing," says S "Husky" Hoskulds. "I'll usually have two small PAs going and usually a couple of speakers as well, so I often have all three rooms at the [studio] miked up . . . People will ask me how I get my room sound, and the simple answer is: by using the room! There's a lot more character to real rooms, and they don't have that flag on them that shouts 'Reverb!'. Room sound has dimension to it, because a real room has real dimensions."[23] "I think people don't use the room that they have enough," adds Young Guru, "because they're focused on looking into a screen. Even your bedroom. The outside. Your bathroom is one of the best places for reverb. Play with that stuff! I think guys that have always been in the box don't realise how good it is to push air."[24] Jack Douglas suggests another reason for giving it a whirl: "Find some kind of analog reverb in your house . . . Just the characteristics of that will make it stand out in your mix, simply because it's not going to sound like anything else."[25]

When it comes to balancing real reverb in a mix, you can treat it very much as you would artificial reverb, using the same predelay, phase, equalization, and dynamics tools to find a place for it in the mix. Many rock producers like heavily compressing room sounds to increase the sense of movement and excitement, especially on drums, but gating can also come in handy if you're trying to rein in the apparent reverb length, a trick Tony Platt mentioned using while working with industrial band Die Krupps. "Galaxy Studios has this massive recording room," he recalls, "so I put a huge PA up in there and I used it as an echo chamber, with mics the other end and a couple of gates on it. Depending on how quickly I shut the gates down I could vary the perceived size of the room."[26]

track that feeds it—a tactic that both Al Schmitt[20] and Elliot Scheiner[21] have employed successfully. Of course, this does inevitably multiply your plug-in count, so this is an additional reason to favor algorithms with a small CPU appetite in tonal roles.

16.6 REVERB FOR SUSTAIN

What tonal reverb is to blend reverb, sustain reverb is to size reverb—in other words, a longer reverb designed for creative timbral control rather than the simulation of natural acoustic reflections. As such, setting up a sustain reverb involves many of the same steps already discussed with relation to tonal reverb in Section 16.5:

- You'll usually only apply dedicated sustain reverb to small groups of instruments, and it's best to do this at the balancing stage if possible.
- You can only choose presets effectively while listening to the wet signal alongside the dry.
- Reverbs that sound too natural will tend to have more unwanted side effects, in this case primarily adding size enhancement, so don't be afraid to wheel out plug-ins that would, under normal circumstances, curdle milk at 20 paces.
- Your aim is to get the dry and wet sounds to meld into one perceptually. If they don't, the effect will be more likely to muddle the mix, increasing size-enhancement side effects and masking the details of other instrument sounds. Within this context, the reverb's level, length, and stereo spread/position all require due consideration.
- Resist evaluating the success of sustain reverb by simply muting/unmuting it, otherwise you'll fall foul of loudness bias. Instead, drop the whole sound out of the mix, give yourself a few seconds to reacclimatize and rebuild your mental image of how the treated sound should appear in the mix, and then switch the instrument (and its effect) back in.
- Once you've achieved the best sustain enhancement you can manage, you can effectively pretend that the dry track and wet return are the individual mics of a multimiked instrument recording for balancing purposes.

There are a few further considerations with sustain reverb, though. For a start, it's a smart idea to use a good dose of predelay (I'd suggest 25ms or more) to remove phase cancellation from the equation. That way, you'll minimize direct

---

GATED REVERB

It's a shame that the excesses of the 1980s have given gated reverbs such a dreadful reputation, because many engineers now instinctively shy away from them even when they might be useful. A couple of specific situations come to mind: when you want to achieve the maximum sense of size in a production, but without ridiculously long reverb tails obscuring all the mix details or blurring chord changes; and when you want to add a short, dense tonal or sustain reverb that doesn't gradually decay, an effect that's particularly good for bulking out thin-sounding drums/percussion. Although traditional gated reverb was generated using a gate acting on a normal reverb, in my experience the later "fake" gated-reverb effects (constituting a short burst of echoes that don't decay in level) are much more usable for mixing purposes because their additions are threshold independent—a quiet drum hit will receive the same length and density of reverb burst as a loud one. This effect is easy to emulate in a modern DAW by truncating the impulse response files within a convolution reverb.

EQ-style tonal changes, which gives you more independent control over the sustain enhancement as well as making any balance processing of the reverb return respond more predictably. However, as with size reverb, that increase in predelay brings with it the possibility that a clearly defined reverb onset may draw unwarranted attention to itself in the mix, so you either have to avoid bumpy presets or use a tempo-related predelay time.

16.7 REVERB FOR SPREAD

The stereo enhancement aspect of reverb is difficult to separate from its other functions, so most of the time it's best to think of it more as a bonus free gift rather than the main purpose of the effect. There are situations, however, where the stereo aspect of the reverb may be the main appeal. Here are a few examples:

- A classic effect for vocals involves a short fizzy reverb that adds density and stereo width to the upper frequency range. The lack of tail minimizes size and sustain contributions, whereas the combination of predelay and high-frequency bias avoids any sense of blend and restricts the tonal impact to a generic high-frequency lift.
- Any effect with a reverb tail can be used to "paint to the edges" of the stereo field in a generic way. Reducing early reflections and adding substantial predelay avoids adjustments to blend or tone; a smooth, bland, but nonetheless unnatural preset choice reduces size-related side effects; and balance processing of the return channel minimizes any undue sustain emphasis of individual frequency ranges and reduces the overall audibility of the reverb as an effect in its own right.
- Short reverbs can operate in a similar way if you use characterless but synthetic-sounding presets and increase predelay to scale down blend and tonal enhancements. Be careful of flams, though, if your predelay goes beyond about 20ms.

16.8 JUGGLING REVERB ENHANCEMENTS

Learning how to create and apply each of the different specialized reverbs I've discussed here serves two purposes. In the first instance, it means that you can use a combination of these effects to add exactly the right amount and quality of each enhancement without drowning in unwanted side effects. Whenever small-studio mix engineers tell me that they're having trouble finding the right reverb levels in their mix, it's invariably because their choice of plug-in isn't providing a suitable combination of enhancements. By the time there's enough blend, the tone warps; when you've managed

> Whenever small-studio mix engineers tell me that they're having trouble finding the right reverb levels in their mix, it's invariably because their choice of plug-in isn't providing a suitable combination of enhancements.

UNMASKING WITH REVERB

Although reverb can often cause additional masking problems in a mix, it's worth realizing that it can sometimes actually counteract the effects of masking. One way it can achieve this outcome is by increasing a masked instrument's sustain such that the wet signal trails on in time beyond the end of the specific musical event that is masking the dry signal. The stereo width of a reverb patch can also unmask instruments by virtue of the wet signal emerging from the "shadow" of an instrument that is masking the dry signal. For instance, if a central shaker in your arrangement were masked behind simultaneous central kick drum hits, you could use reverb to unmask it, either by extending the shaker's sustain past that of the kick or by widening the shaker sound on either side of it in the stereo field. In addition, tonal reverb might also unmask an instrument by adjusting its frequency balance, in much the same way equalization could.

to add some flattering sustain to your piano, it sounds like it's in a railway tunnel; and in the process of adding a sense of space, you blend things too much, so everything sounds a million miles away. Dedicating specialized reverbs to specialized tasks combats this all-too-common frustration. And, just to clarify, I'm not saying that you have to use all five reverbs (or indeed any of them!) on any specific mix, nor that you shouldn't use more than one instance of each type. You might decide, for instance, that you want to use no blend or size reverb at all, in order to retain maximum separation of the parts, but then use three or four tone/sustain patches to make individual sounds more characterful and unique. Or you might use three different blend reverbs, simply because each has tonal/sustain side effects that suit certain sounds more than others. The main thing is to think in terms of what you need to do, and try to use your reverb as efficiently as you can while still ticking the boxes you need ticked. That way you should still be able to keep the mix clear-sounding no matter what you use.

A worry that some people have is that combining different reverb patches on one mix will cause a muddled and contradictory acoustic impression. Strictly speaking, this is correct, but in practice it seldom causes any difficulties, simply because the spatial illusion in most music styles is already such an artificial construct as it is. Listeners are well accustomed to this on a subconscious level, and to a large degree suspend disbelief, in much the same way we do when we accept that camera angles and positions keep shifting in a film. The public are happy to imagine that your voice is soaring in a vast echoing space even when your drums are clearly sounding like they're in a studio live room and your bass DI is totally anechoic—it's all part of the confection! Of course, in more purist acoustic genres such as classical music, it usually makes more sense to minimize the number of reverb effects you use, and indeed to design each one either to complement or match any existing recorded room ambience, but even in that realm there's a lot of variation in the tastes and approaches of different high-profile engineers.

The second advantage of knowing which reverb characteristics affect the prominence of which enhancements is that it enables you to adopt a less regimented approach if the practical limitations of your studio demand it. Working in a hardware setup, for example, might limit you to only a small number of reverb processors, in which case it may be necessary to use each individual effect to cover more bases in terms of the possible enhancements. Once you understand how to minimize and maximize each individual reverb enhancement, it isn't a huge step to create reverb patches with desirable proportions of several enhancements simultaneously. If you have personal aesthetic reasons for using an idiosyncratic selection of effects, perhaps if you're after a retro-tinged sound, then an awareness of whether/how a specific processor's enhancements can be controlled is also extremely useful in determining how to get the maximum value out of its unique contribution to your production. The same applies if there are any recorded reverbs in the mix, whether in the form of ambient microphone recordings or "print it now because we'll never find it again!" effects bounces.

To finish up this discussion of mix reverb, my final piece of advice is this: don't use reverb as some kind of involuntary reflex. To quote Steve Albini, "[Reverb] is a thing that's done *pro forma* a lot of the time. [Engineers] put it on because they feel they're supposed to. I've never had that response. I'll wait until someone says, 'That sounds weird,' and then I'll try reverb."[27] "If I can, I'll mix completely dry," adds Bob Clearmountain, "[but it] changes radically depending on the piece of music."[28] Numerous commercial records have precious little reverb on them at all, because they already have sufficient blend, size, sustain, tone, and spread as it is. Many other modern productions apply reverb only selectively because a lack of some of reverb's enhancements can support the overall intent of the music—for example, a lot of music in pop, urban, and electronica styles benefits from some elements deliberately being poorly blended, so that they're right in your face at the front of the mix. "[I was] putting a little more reverb on something," says Jimmy Douglass, recalling his work with Justin Timberlake, "and he goes 'You know what? When you put too much of that stuff on there it seems like you're trying to hide something!'"[29]

CUT TO THE CHASE

- Reverb enhances several aspects of a treated sound at once, the most important for mixing purposes being blend, size, tone, sustain, and spread. To use reverb effectively at mixdown, you need to gain some independent control over each of these enhancements so that you can achieve exactly the right combination.
- To create a blend reverb, shorten a natural-sounding stereo preset to remove its decay tail, and add around 10 to 20ms of predelay. Address unwanted tonal colorations by adjusting the predelay, phase, polarity, and equalization of the effect.
- To create a size reverb, choose a natural-sounding stereo preset preferably with a soft reverb onset, and add around 50ms or more of predelay. Address unwanted sustain colorations by equalizing the effect.

- To create a tonal reverb, shorten an unnatural-sounding preset to remove its decay tail and refine the sound with predelay, phase/polarity, and EQ adjustments. Pay careful attention to the length and overall level of the effect, and consider whether its stereo width and positioning suit that of the dry signal.
- To create a sustain reverb, choose an unnatural-sounding preset, preferably with a soft reverb onset, add at least 25ms of predelay, and refine its sound with EQ adjustments. Pay careful attention to the length and overall level of the effect, and consider whether its stereo width and positioning suit that of the dry signal.
- To create a spread reverb, try an unnatural preset with a bland, uncolored tonality, and do your best to use predelay and balance processing to minimize blend/size/tone/sustain side effects.
- Blend and size reverbs will often be added to many different tracks in an arrangement and are usually best applied after the complete balance is in place, whereas tonal and sustain reverbs (and indeed more ostentatious reverb special effects) tend to be more useful for very small numbers of tracks and are often better applied during the initial balancing process itself.
- Blend and size reverbs can be balanced in the mix almost as if their returns were independent instruments. Equalization can reduce unwanted masking, avoid low-end clutter, and reduce the apparent audibility of the effect, whereas de-essing and transient processing applied to the reverb's input signal can also help make the effect more understated. Tonal and sustain reverbs are usually better treated as an integral part of the dry tracks they are treating—almost as if the dry and wet signals were individual mic signals within a multimiked recording.
- Despite their checkered history, gated reverbs can still be useful, especially the "fake" designs that create an unnatural burst of dense, consistent echoes.
- Don't disregard using nearby rooms as impromptu reverb chambers at mixdown. All you need is an extra speaker and microphone, and you'll likely get more characterful and organic results than you would from plug-in effects.
- Although reverb has the potential to mask elements of a mix, it can also help unmask instruments in certain situations by lengthening or spreading the dry sound or by altering its tone.

Assignment

- Do a survey of all the reverb plug-ins on your DAW system, and ensure that you have at least one fairly realistic reverb as well as a couple of character devices. If you feel restricted here, check out third-party alternatives.
- Return to your balanced mix section from Part 3 and see whether you can get extra sonic mileage out of any of the tracks using tonal or sustain reverb patches. Try at least a couple of each to get an idea of the possibilities. If they produce something worthwhile, you may need to rebalance to accommodate the additions.
- Set up at least one blend reverb and one size reverb in order to experiment with how these enhancements affect the character of your mix. Don't forget to spend time balancing the return channels, with processing if necessary.

Web Resources

On this book's companion website you'll find a selection of resources to support this chapter, including:

- a special tutorial video to accompany this chapter's assignment, in which I show how the reverb effects discussed in this chapter might enhance various aspects of a simple multitrack mix;
- a large selection of audio examples dealing with choosing reverb presets for different purposes; adjusting reverb-time, predelay, and EQ parameters; applying blend, size, tone, sustain, and spread reverb in practice; and blending mixes without using reverb at all;
- links to affordable algorithmic and convolution plug-ins, as well as to sources of useful impulse responses;
- a number of mix case-studies from the *Sound on Sound* "Mix Rescue" column where I dissect real-world applications of reverb in detail.

 www.cambridge-mt.com/ms-ch16.htm

CHAPTER 17
Mixing with Delays

A delay effect creates patterns of echoes that are typically much simpler than those of a reverb. Although most small-studio mix engineers seem to favor reverbs for mixing purposes, delays can actually be more useful because of the way they deliver reverb-style enhancements in more precisely targeted ways—think of delay as a sniper rifle compared with reverb's shotgun! For many upfront modern productions, in fact, delays may be more suitable than reverbs, simply because they take up less space in the mix and can be made less audible in their own right. "I use delays a lot more than reverb," says Spike Stent,[1] for example, and Jacquire King echoes the sentiment: "I use reverb, but I like delay more."[2]

17.1 ESSENTIAL DELAY CONTROLS AND SETUP

When faced with a delay plug-in, the two main controls to worry about are Delay Time, which sets the time-delay between the dry sound and the first echo, and Feedback Level (sometimes called Repeats or Regeneration), which determines how many subsequent echoes follow the first and how quickly they decay over time. If you're comfortable with these two controls, then you can get good use out of plug-in presets and pretty much ignore other more detailed parameters unless you feel particularly inclined to pamper your inner nerd. Delay and reverb are both echo-based effects, so it should be little surprise that using each of them at mixdown is similar in many respects:

- A send–return effect configuration will give you more control over the sound of the delay effect than an insert setup.
- You can use delay effects to enhance blend, size, tone, sustain, and spread, although the effects in each case typically take up less space in the mix than those created by reverb.
- Simple, clean echoes tend to work best for blend and size enhancements, whereas more characterful algorithms (such as digital emulations of analog echo devices) deliver more varied tone and sustain changes.

FIGURE 17.1
Emulations of old-fashioned delay devices are ten-a-penny, and their inherently unnatural (though appealing) sonic character makes them well-suited for delivering tonal and sustain enhancements.

- Clearly audible creative delays are best introduced while balancing, rather than trying to fit them in once the balance is complete.
- Don't forget to pay sufficient attention to balancing your delay returns, both in terms of level and in terms of equalization/dynamics processing.

If you're going to create a delay for blend purposes, then a single 50 to 100ms short "slapback" delay with little or no feedback is a good choice. This is a long enough delay time to avoid the tonal side effects of comb filtering, but it's still usually short enough to gel with the dry sound fairly well, provided that the effect levels are quite low and balance processing is used to camouflage the wet signal as much as possible. Don Smith is just one fan of this kind of patch: "It adds a natural slap like in a room, so to speak, that maybe you won't hear, but you feel."[3] For size purposes, a longer delay with some feedback makes more sense, but otherwise the role of balance processing remains much the same.

We dealt with the idea of using delays for tone and sustain back in Chapter 12, because their use is often better dealt with at the balancing stage. However, now that we know more about echo-based effects in general, let's be a bit more specific. A tonal delay will produce the most dramatic effects when its delay time is well within the comb-filtering zone (under 20ms), although unnatural-sounding analog style delays will also inevitably color the tone even beyond that range. You can add a little feedback if you want to emphasize the resonant "ringing" of the delay patch, in which case you may wish to match the pitch

THE DISTORTED SLAPBACK TRICK

From time to time, I'm asked to create an obviously distorted lead-vocal sound at mixdown, and this always presents something of a practical problem: by the time the distortion effect sounds suitably ostentatious, the intelligibility of the lyrics is suffering unacceptably. One way of working around this is to apply less distortion to the vocal itself, thereby maintaining sufficient lyric transmission, but then to add a slapback delay that has much more aggressive distortion in order to retain the required overdriven character in the mix. The distorted delay will, of course, pull the vocal a little backwards in the depth perspective, but not as much as a clean delay would, and in any case this residual blending action usually turns out to be a benefit on lead vocals most of the time, in my experience.

of that resonance to the track you're processing by finessing the delay time. (If you want more than one pitch, though, it's probably easier to get hold of a specialist resonator effect, such as Lexicon's classic Resonant Chords processor.) Delays for sustain purposes are more likely to use substantial feedback, as well as longer delay times.

The processing on a blend or size delay's return channel will usually focus on balance issues, much as with reverb patches fulfilling similar functions, although many engineers also send from their delay returns to reverb effects in order to push the delay repeats back behind the dry signals in the mix, making them sound more like natural echoes. It's also common for delay effects to be ducked in response to lead vocals/instruments (as I mentioned in Chapter 14) so that the delay repeats don't interfere with note definition and lyric intelligibility. With delays for tone and sustain, there's much to be gained by processing the effect returns quite severely with EQ or the other tonal effects discussed in Chapter 12—remember that it's not what they sound like on their own that matters as long as they enhance the dry track. Making such effects sonically quite different to the dry tracks has an additional advantage that it reduces blend and size enhancement side effects, which can be useful if you want your production to feel tight, compact, and right in the listener's face—a favorite trick of Spike Stent's, which you can hear all over his mixes for Massive Attack.[4]

With size or sustain delays, the issue of how their longer delay times lock in with the song's tempo bears a little further thought, because matching a delay to tempo effectively makes it less audible—the echo of any given rhythmic event will be masked to some extent by the next rhythmic

> Making delays sonically different to the dry tracks has the advantage that it reduces blend and size enhancement side effects, which can be useful if you want your production to feel tight, compact, and right in the listener's face.

event. A tempo synced delay can therefore be faded up much higher in the mix before it begins to become obviously noticeable, which frequently makes this kind of patch preferable for workhorse size/sustain-enhancement duties at mixdown. Most DAW delay effects now include tempo-matching functions as standard, allowing you to specify delay times in terms of musical note lengths, but if not then you can still work out tempo-related delay times in milliseconds very easily: divide 60,000 by the beats per minute (BPM) tempo value of your production, and then try different multiples and fractions of that value. So a tempo of 120bpm will give you a value of 500ms, for instance, and times of 1000ms, 250ms, or 125ms may also be useful—as indeed may triplet values of 167ms and 83ms.

It's not uncommon to have several tempo-synced delays at mixdown, as each delay-time note value may suit some dry tracks more than others. "If it's a very continuous [vocal] phrase, you've got to make shorter delay decisions," says Miles Walker, for instance, "because you certainly don't want it to get washy or overly filled up."[5] You may also find that a slight lengthening of the delay time beyond the strict tempo-matched value gives a more musical result in

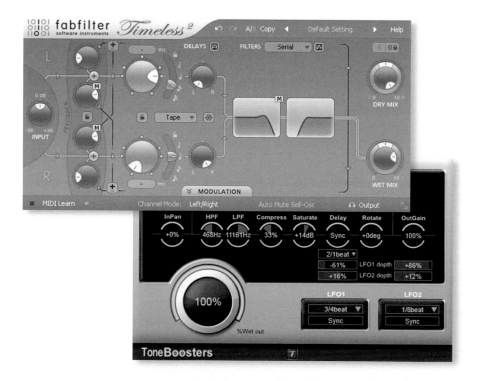

FIGURE 17.2
Plug-in delay effects such as Fabfilter's Timeless and Toneboosters's Module make it easy to synchronize your delay times to your project's master tempo. This is useful if you want the delay effects to sink into the mix and be less audible in their own right.

some cases. "I either set delays right on the beat, or . . . a little slow," says Neal Avron. "I don't like delays that rush. In more vibey songs, a slightly behind-the-beat delay can make things sound a little bit more laid-back."[6] Marginally lazy echoes can also be necessary where a performer is wandering a little ahead of the beat from time to time.

Delay effects that aren't tempo-synced can be much lower in the mix while still remaining clearly audible, so they're better for more creative applications where they enable the listener to appreciate the timing complexities without the effect trampling over everything else in the arrangement. The practical problem with unmatched delays, though, is that they quickly start to undermine the groove of your track if you use them too much, so if you're using delays as a statement within rhythmic styles, then it's usually better to take a kind of fence-sitting hybrid approach by choosing polyrhythmic delay times. These are tempo related but don't correspond to simple note lengths. Common examples are three-eighth-note and three-sixteenth-note delays. Polyrhythmic delays are the stock in trade of many dance styles, because they allow you to blow chemically addled minds with lots of trippy-sounding audible delay repeats while safeguarding the essential rhythmic clarity of the groove.

DELAY FOR RHYTHMIC ENERGY

Although tempo-synchronized delays are good for adding size and sustain, they can also add complexity to the rhythmic groove, especially when applied to percussive instruments. One of my favorite tricks, in fact, is to feed a subtle single-tap eighth-note delay from my drums submix. Mixing in a little of that echo subtly enhances the kit's sense of rhythmic energy, almost as if the drummer is adding more ghost notes, and I find it very effective giving a song's second verse a lift in relation to its first verse, for example. Be careful not to overdo the level, though, because overemphasizing a beat's minor subdivisions can also make it seem to drag perceptually.

17.2 USING DELAYS IN STEREO

Another reason why delays are often more useful than reverbs for modern styles is that they don't smudge the stereo field in the way a reverb does. As a result, delays typically leave a mix more open sounding and are preferred by many engineers.

There are different ways you might want to position echoes in the stereo field for different purposes. Blend and size delays tend to be most effective when you use a true stereo setup, in other words a delay effect where the stereo positioning/spread of echoes in the delay return matches that of the mono and stereo dry signals that are feeding the effect. In this scenario the delay repeats reinforce the stereo positioning of the mix and therefore draw less attention to themselves, which is ideal for subtle enhancement applications. In my experience, implementing a true-stereo setup in some DAWs is about as entertaining as brushing your teeth with an angle grinder, but try to persevere with it, because it's definitely worth the effort. (The stereo delay setup, that is, not the angle grinding.) Mono-input, stereo-output delay effects that add a sense of stereo width tend to draw more attention to themselves as artificial effects and may also confuse the imaging of stereo tracks, whereas simple mono-in, mono-out delays are less successful as global send effects because the mono return effectively narrows the mix and clutters the center of the stereo image. Mono delays do have their uses, but it's with tonal and sustain processing of individual instruments where they really come into their own, especially where the delay return is panned to match the dry signal being treated.

Of course, delay effects can also enhance the stereo spread, just as reverb can, if you can get your hands on a delay plug-in that gives you control over the positioning of the different echoes it generates. There are a couple of things to be aware of when doing this, though. First, panning short delays can disguise comb-filtering effects when listening in stereo, so you should confirm the mono compatibility of any such patch carefully. Second, there's a danger that panning individual echoes will skew the overall stereo image of the delay effect to one side, which might not be desirable. One response to this dilemma is to have alternate delay repeats opposition-panned so that they balance the stereo effect to an extent—a setup called "ping-pong" delay, which can be extremely effective for polyrhythmic patches in particular. Another tactic is to use a single echo panned to the opposite side of the stereo panorama than the dry signal. In either case, though, you should always ask yourself whether the stereo character of your delays is distracting too much attention from more important elements of the production as a whole, especially any lead instruments in the center of the stereo panorama.

You can, of course, also enhance the stereo spread of any echo (and indeed your dry signals too), by applying other stereo effects to it. That's another whole can of worms, though, so we'll focus on that specifically in the next chapter.

MULTITAP DELAY

Multitap delay effects occupy the middle ground between simple delay effects and reverb. What they allow you to do is independently specify the timing, level, and (usually) stereo placement of a handful of separate echoes. On the one hand, you can use these facilities to implement normal delay effects, but on the other hand, you can simulate something akin to a reverb's early reflections pattern. Where any specific multitap delay patch resides on this continuum will dictate how it's best dealt with from a mixdown perspective.

CUT TO THE CHASE

■ Delay effects offer similar mixdown enhancements as reverb, while taking up less space in the mix. Although the echo patterns they create are usually simpler, you can treat them in much the same way as reverb from the perspective of setup and balance processing.

■ Delay times that are matched to tempo will tend to sink into the mix, making them ideal for mixdown enhancements. Unmatched delay times will pop out of the mix at lower levels, so they are better suited to more ostentatious creative applications. However, if unmatched delay times are undermining the rhythmic flow of your production, then try out polyrhythmic delays instead.

■ A true-stereo effect configuration will work best for blend and size delays. Mono-in, stereo-out and simple mono delay effects tend to suit tonal and sustain roles better, but they may also be applied for the sake of deliberate stereo enhancement. If so, keep a careful ear out for mono compatibility and stereo balance.

Assignment

■ Do a survey of all the delay plug-ins on your DAW system, and check that you have at least one fairly clean digital algorithm as well as a couple of characterful analog-modeled devices. If you feel restricted here, check out third-party alternatives.

■ Return to your mix, save it, and open a copy. Mute any reverb effects you've used, and experiment with delay patches instead to see whether they can achieve similar enhancements with fewer unwanted side effects. Once you've seen what different delay patches might be able to offer in each of the roles, save your mix again, open another copy of it, and remove all the effects sends, leaving the delay and reverb returns in place. Now try to achieve the best balance of delay and reverb effects for each channel. Once you've achieved this, save the mix once more, and then compare it with both of the previous versions to gain a full appreciation of what delays and reverbs can do at mixdown.

Web Resources

On this book's companion website you'll find a selection of resources to support this chapter, including:

■ a special tutorial video to accompany this chapter's assignment, in which I show how the delay effects discussed in this chapter might enhance various aspects of a simple multitrack mix;

■ a selection of audio examples showcasing different types of delay effects in action, and showing how delays and reverbs work in tandem in real-world mix situations;

■ links to affordable delay and pitched-resonator plug-ins.

www.cambridge-mt.com/ms-ch17.htm

CHAPTER 18
Stereo Enhancements

During the discussion of panning in Section 8.2.2, I argued that stereo width is far more important to many mainstream productions than exact stereo positioning. "Modern mixing is all about being wide," says Delbert Bowers, "with different styles requiring different types of width."[1] On account of this, audio engineers have come up with an enormous variety of tricks for adjusting image spread, and this chapter details a selection of these techniques in order to clarify their respective pros and cons.

18.1 ARRANGEMENT TWEAKS

Some of the most effective stereo enhancements at mixdown can actually be achieved surreptitiously without loading a single plug-in. My most common stunt for inflating the sonics of Mix Rescue submissions, for example, is to generate fake double-tracks, that can then be opposition-panned for instant width and extra arrangement density. Tracks that offer this possibility will be those that include the same musical material performed more than once—not an uncommon occurrence in a lot of rhythm-section, riff, and hook parts. If you duplicate the track in question and use your audio editing tools to shuffle its repeated musical sections around, you can generate a version that combines with the original just like a separately overdubbed double-track. So let's say your song's chorus is eight bars long, but the chord pattern repeats twice within that. (If it doesn't, then it's your own silly fault for mixing prog rock.) It'll frequently be plain sailing to generate a double-track from a single strummed acoustic guitar part within that chorus, simply by copying the part and switching the order of the copy's two four-bar sections. When you mix the original and duplicate tracks together, they'll still sound like independent performances, because the original part will be playing bars 1 through 4 while the "double-track" plays bars 5 through 8, and vice versa. Andy Wallace talks about using this technique regularly: "Normally I like to create [double-tracks] from the performances that are already there by taking the double from

FIGURE 18.1
Generating a "fake" double-track with audio editing.

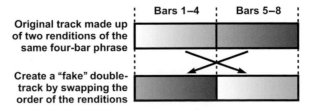

Original track made up of two renditions of the same four-bar phrase

Create a "fake" double-track by swapping the order of the renditions

Bars 1–4 | Bars 5–8

another moment in the song where the part is played, so I actually really get two different performances."[2]

Where sections don't repeat quite as closely, though, you have to think a bit more laterally. I regularly resort to pitch and time manipulation to square this particular circle, and we covered all the techniques you're likely to need for this back in Chapter 6. In other cases, I may build the double-track more as a kind of patchwork quilt by stealing individual chords and notes from all over the place to achieve the illusion of a fresh overdub, especially in those cases where a particular sound (overdriven electric guitar, say) throws its toys out of the pram if subjected to pitch shifting or time stretching.

Some small-studio users try to create this effect by simply delaying a copy of a single track, but it's important to realize that this won't create nearly the same result and may cause serious phase-cancellation problems in mono. (There are some useful applications of that idea, though, which I'll get to in a moment.) More usable results are possible with this concept, however, if you generate

ADDING ATMOSPHERE

Although this chapter is primarily about enhancing the width of a mix, the concept of adding stereo background textures is also tremendously helpful when a mix seems generally two-dimensional, emotionally unengaging, or lacking in atmosphere—shortcomings that commonly afflict electronic productions or multitrack recordings built entirely from overdubs. Under such circumstances, I often fall back on sample libraries or virtual instruments offering sounds aimed at film and television composers, and try layering in a few of those, either as background "beds" to increase differentiation between musical sections, or as swells to emphasize section transitions and musical accents. These multimedia-targeted sounds are often specifically designed to conjure emotionally evocative images to the mind's eye, and can really help intensify a production's mood if chosen sensitively. I'm also a big fan of breakbeat samples and abstract rhythmic loops, because they not only serve a similar atmospheric function, but can underpin or enliven the groove too. If this idea intrigues you, here are a couple of recent commercial productions worth searching out in this regard: Zayn's "Pillowtalk" and Lukas Graham's "7 Years." The thing to do while listening to them is to disregard anything that might be traditionally called a musical part (you know, the drums, bass, guitars, keyboards, strings, vocals . . .) and listen instead to all those other elements in the mix that don't have any obvious instrument names (the swooshes, bloops, and sound effects), because it's those extras I'm talking about here. "I added things to take the listener by the hand and walk him or her through this lyric and enhance the feelings that are in the song," explains Morten "Pilo" Pilegaard, the experienced film sound-designer who co-produced that Lukas Graham record. "It was like making a film. My first idea was to add a film projector sound . . . then I added some ticking clocks, a vinyl record sound, and some audience noises where the lyrics refer to people singing Lukas's song in 20 years."[4]

some kind of pseudo-random variation in the double-track, and one method of doing this that's gained popularity in recent years is applying different real-time pitch-correction settings to the double-track—a technique Dylan Dresdow has used when working with will.i.am, for intance.[3]

Another plug-in-free trick worth considering is adding some kind of wide-screen stereo background texture. A synth pad is a frequent choice here, but unpitched sounds can be equally handy. My stalwarts for this technique are things such as tape hiss, vinyl noise, and Foley room tones. The side effects of the pitched and unpitched additions will inevitably be a little different, though, so choose carefully. A pad will enrich pitched sustains, unpitched textures will to some extent improve perceived blend and obscure low-level details, and Foley ambiences may also clearly imply a sense of size and space. A particular advantage of using supplementary, but musically inessential, parts such as these to add stereo width is that mono compatibility rarely becomes an awkward issue. So what if your pad or room tone lose definition, or even disappear completely, in mono? Most mono listening environments are so lo-fi anyway that the typically subtle effects are unlikely to be missed.

18.2 ADJUSTING STEREO SPREAD

It's easy to adjust the spread of a collection of mono files in the stereo image using panning, but stereo recordings can be a bit tougher to deal with when it comes to adjusting stereo width. Granted, narrowing a stereo track isn't too tricky. All you do is adjust the individual panning of the left and right channels, as we saw in Section 8.2.5. Widening a stereo recording, on the other hand, requires you to look at stereophony in a slightly different way. Let me explain.

We're used to thinking of stereo recordings as being made up of one signal for the left-hand speaker and one signal for the right-hand speaker, but it turns out that you can also encode a stereo recording as two channels in a different way: one channel containing the mono elements at the center of the stereo image (the Middle signal) and one channel containing the elements that appear off-center (the Sides signal). As a sound moves off-center in a Middle and Sides (MS) recording, its level reduces in the Middle signal and increases in the Sides signal, its polarity in the Sides signal indicating which side of the stereo field it happens to be on. You can represent an identical stereo signal in both left–right and MS formats, and because each represents exactly the same information it's easy to convert between them.

But if MS encoding doesn't make any difference to the sound of a stereo file, what's the point of it? Well, the reason it's so useful is that it allows us greater control over the mono and stereo aspects of a recording and in particular gives us

> The crucial thing to remember is that a stereo recording's Sides component isn't audible in mono, so if you heavily emphasize that in the mix, the level drop when you move to mono may be quite dramatic.

the ability to adjust the stereo width by changing the level of the Sides signal. Fade the Sides level down and the stereo picture narrows, all the way to mono if you like; increase the Sides level and the stereo picture widens, and if you push things far enough you can even make the soundstage appear to extend beyond the positions of the physical speakers. However, it's crucial to remember that a stereo recording's Sides component isn't audible in mono, so if you start to heavily emphasize that component in the mix, the apparent level drop when you move to mono may be quite dramatic—in extreme cases a sound could disappear entirely on some listening systems! So when increasing stereo width in this way, you should be sure to compare your mix balance in both stereo and mono, and it's also sensible to relegate such widening treatments to musically unimportant parts, so that any level drops your widening does incur don't make a nonsense of the production.

Another way I like to look at MS widening, though, is as a means of clearing space in the center of the panorama. This is particularly useful for stereo synth pads, where pulling down their Middle level typically leaves more space for a song's drum, bass, and lead parts. I also frequently find it handy for reverb returns when I'm working with multimiked ensemble recordings. You see, it's not uncommon for things such as drum kits, pianos, and acoustic guitars to be recorded with stereo mic techniques, but I regularly find myself wanting to narrow the captured stereo image, the downside of which is that any recorded room reverb is also narrowed. By adding stereo reverb with a reduced Middle component, I can avoid too much reverb buildup in the center of the stereo picture, and make the instrument sound less like it was recorded in a corridor!

There's no shortage of stereo enhancer plug-ins that operate on MS principles, although you can easily implement similar widening without any of them simply by duplicating your stereo track, panning the duplicate's left and right sides centrally (thereby summing it to mono), and then inverting the duplicate's polarity. As you add the duplicate into the mix, it will progressively phase-cancel more of the stereo recording's Middle component, thus increasing the relative level of the Sides component. Notwithstanding, I personally find it more convenient to use a separate MS encoder/decoder, and my firm favorite for this is Voxengo's cross-platform MSED—not least because it's freeware, and I'm a cheapskate! This plug-in encodes a stereo left–right signal into MS format within the plug-in, gives you separate level control over the Middle and Sides levels, and then reencodes the stereo signal to left–right format at its outputs. MSED also gives you the option of just encoding to or decoding from MS, so you can process the Middle or Sides signals of a stereo file independently using other plug-ins in your DAW. This significantly increases the power of the tools discussed in Section 15.2 when it comes to rebalancing premixed audio, allowing you, for example, to EQ or compress a center-panned snare sound in a stereo drum loop without impacting as heavily on hard-panned percussion elements. (Indeed, quite a few

FIGURE 18.2
A great little freeware MS manipulation plug-in: Voxengo's MSED.

EQ and dynamics plug-ins these days actually have this kind of MS processing option built in, although it's hardly much of a selling point given that MSED adds MS capability to any stereo plug-in for free.)

Another application of MS equalization is to provide frequency-selective width control, because any frequency you boost in the Sides signal will be widened, whereas any frequency you cut will be narrowed. However, this kind of frequency-selective operation is also often implemented by the equally viable approach of splitting the audible spectrum into bands and then applying separate MS processing to each—exactly in the mold of a multiband dynamics processor. The reason frequency-selective stereo width processing is handy is that it allows you to reduce mono-incompatibility with the minimum overall impact on the apparent overall stereo width. So, for example, if a phase misalignment between the low frequencies in the left and right channels of a stereo recording is causing a loss of bass when auditioning in mono, you could encode the stereo to MS format and high-pass-filter the Sides signal to reduce the problem, without affecting the stereo impression delivered by the upper frequencies—a very common tactic in mainstream chart music. In a similar vein, you might feel you can get away with boosting the very high frequencies of a stereo recording's Sides signal to give it an enhanced impression of width on high-quality stereo listening systems, knowing that the loss of this mono-incompatible high end will probably go unnoticed on lower-quality mono playback equipment.

STEREO METERING TOOLS

As with most other mix decisions, the best way to judge the appropriate stereo width for each of your tracks is by comparing your mix against the commercial releases you're hoping to compete with. However, there are some specialized stereo metering tools that can provide you with useful supporting information. The one I find most helpful is called a stereo vectorscope, and it's great for highlighting subtly lop-sided stereo mixes and gauging overall mono-compatibility once you've acclimatized to how your specific plug-in responds. However, be aware that it's unlikely to highlight per-track mono-incompatibility problems within a full mix. For example, it'll show Maroon 5's song :Harder To Breathe," for instance, as having a fairly reasonable overall stereo width, without alerting you that the main rhythm guitars collapse under single-speaker playback conditions. So it's worth having the vectorscope open while you're working with individual tracks and effects returns to get the most value out of it at mixdown.

Another thing simple vectorscopes won't show you is frequency-selective information, such as the overzealous 5kHz and 16kHz Sides-signal boosts on Kanye West's "Breath In, Breath Out" that make mincemeat of the mono mix tone. Fortunately, a stereo spectrum analyzer can help out in two ways here: first, by giving the left-channel and right-channel curves simultaneously, it'll identify any seriously off-center

(continued)

frequency regions; and, second, if you precede the analyzer with an MS encoder, the display's two curves will show the Middle and Sides spectra instead, immediately highlighting frequency regions where the Sides signal dominates and mono-compatibility may be a concern. A number of software developers now offer other ways of visualizing the same fundamental data, though, so if the spectrum-analyzer approach doesn't feel intuitive enough to you, there are plenty of other options to choose from. For instance, I personally prefer using spectral panning and spectral correlation displays (which respectively show left–right and Middle–Sides level ratios plotted against frequency), because I find those easier to read.

FIGURE 18.3
Voxengo's freeware SPAN spectrum analyzer operating in MS mode, clearly showing the upper-spectrum mono-compatibility problems within Kanye West's track "Breathe In, Breathe Out."

18.3 STATIC ENHANCEMENTS

While MS processing can do a lot to manipulate the stereo information that's already present in a signal, you can only really use it to enhance stereo width information that's already present. What if you want to add stereo width to mono tracks?

18.3.1 EQ-Based Widening

Simple EQ provides probably the most straightforward method. Here's the deal:

- duplicate your mono track;
- pan the original and duplicate tracks to opposite sides of the stereo image;

- insert an equalizer into the original channel and make any old selection of cuts and boosts;
- copy that equalizer setting to the duplicate channel, but with the gain settings of the equalizer inverted so that cuts become boosts, and boosts become cuts.

The effect of this setup will essentially be to pan different frequency regions of the sound to different positions in the stereo field, smearing the sound's spectrum across the panorama to produce a wider image. Although the advantage of this processing is that it's comparatively mono-friendly (the EQ changes pretty much just cancel each other out when the left and right channels are summed), it can produce rather an odd listening sensation in stereo, especially if you only use a few broad-band EQ changes. An EQ setting with lots of small peaks and troughs usually creates a more even feeling of frequency spread, which is why this technique is often associated with graphic EQ—if you set up an alternating pattern of boosts and cuts on the sliders for one channel, even a roadie should be able to invert that pattern by eye on the other channel. (As long as he understands long words such as "alternating.") Nonetheless, I've never had particularly good results with basic EQ-based widening for important music parts, because of the way it destabilizes the phantom image and seems to make instruments sound less solid in the mix. Better to save it for incidental parts that suit a more diffuse and background role.

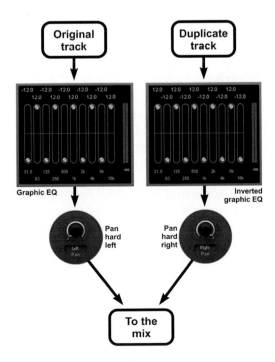

FIGURE 18.4
An EQ-based stereo widening effect.

18.3.2 Comb Filtering

There is an alternative way of implementing EQ-based widening, however, that works better as a general purpose treatment. The trick is to use comb filtering to generate many more frequency peaks and troughs than any EQ plug-in would be capable of. To set it up, first create a stereo delay send effect with around 20ms delay time (well within the comb-filtering zone) and no feedback (so there's only a single echo), and then invert the polarity of the return signal's left channel only. If you send to this delay from any channel in your mix, you'll cause that channel's sound to be comb filtered, but because of the effect's one-sided polarity inversion, you'll get a frequency peak in the left channel wherever there's a frequency trough in the right channel (resulting in leftwards panning), and vice versa (resulting in rightwards panning). The more level you send to the delay effect, the deeper the comb filtering, and hence the stronger the EQ widening effect.

The main side-effect to watch out for here is that the comb filtering can impart a slight sense of metallic pitched resonance, even without any delay feedback, and if this feels obtrusive you may wish to tweak the exact delay time to find a less noticeable pitch. Be aware, though, that shorter delay times reduce the number of peaks/troughs per comb filter, increasing the chance that lower frequencies in particular may start to veer undesirably to one side or the other of the stereo picture. If this happens, then either choose a longer delay time (if you're able to do so while remaining within the comb-filtering zone), or consider high-pass filtering the delay return to restrict the comb filtering to the upper spectrum.

18.3.3 Haas Delays

Another interesting effect for widening involves panning your dry track to one point in the stereo image and then panning a single echo to a different position, often on the other side of the field. The delay time here needs to be short enough that the echo can't be perceived separately from the dry signal and is therefore treated as if it were an acoustic reflection of the dry sound coming from another direction. In other words, it needs to be what is sometimes called a "Haas delay," after the German scientist who first explored this psychological effect experimentally. (Maybe it's just my schoolboy sense of humor, but I've always wished he'd been French.) If the delay signal can be perceived as a distinct echo in its own right, then you'll still get widening, but it'll sound much more noticeable as an artificial effect.

Typically a Haas delay will follow the dry sound within about 30ms, but this depends on the signal that's being delayed, so you always have to judge whether the delay is audible by ear—sharp transients can appear to flam at times well below 30ms, for example. You might be tempted simply to set the delay nice and short for safety's sake, but that brings problems of its own, because the further your delay time heads below 30ms, the more destructive the phase cancellation is likely to be between the wet and dry signals. You might not notice this if you've panned them hard left and right, but there'll be no escape in mono. The trick to getting best results here, then, is to try to set as high a delay time as possible without the echo breaking away perceptually from the dry signal. Some further processing of the delay return may help in this regard, along similar lines as when reducing the audibility of a blend reverb—high-frequency EQ cut and transient softening especially.

> Typically a Haas delay will follow the dry sound within about 30ms, but this depends on the signal that's being delayed, so you always have to judge whether the delay is audible by ear—sharp transients can appear to flam at times well below 30ms.

For the widest stereo effect, you should pan the wet and dry sounds to opposite stereo extremes, but I find that most of the best applications for this Haas delay effect are where the wet and dry signals aren't pushed that far apart. Haas delays are great for gently expanding mono signals in sparser mixes where there's a lot of space to fill, for example. They're also useful where you want to avoid a big gap in the stereo field between hard-panned double-tracked parts—pan a Haas delay for each track somewhere closer to the center, or even overlap the delays slightly, to stretch the instruments across as much space as you want to cover. When it comes to choosing a level for the Haas delay, it's tempting to use quite a lot, especially if you fall for the loudness bias by judging the impact of the effect just by switching the return on and off. However, Haas delays tend to rather homogenize instrument sounds they're applied to, smoothing over the internal details so that they sound a bit bland, and they also push sounds farther away from the listener, as you'd expect if you consider the effect's close relationship to blend delay/reverb. So tread a bit carefully here—listening in mono can really help clarify your decisions, because it takes the stereo widening out of the equation, thereby highlighting any potentially unwanted side effects.

One inherent drawback of this Haas-delay widener is that it almost always feels lopsided subjectively, even if you put the dry and wet signals at the same level and pan them to exactly opposing stereo locations. This is because the Precedence Effect (which I mentioned back in Chapter 8) identifies the dry signal as the sound source, and the delay as a reflection, so the listener's attention is drawn more strongly towards the dry signal. To a certain extent you can mitigate this bias by pushing the delayed signal a few decibels louder than the dry signal, but even then the stereo balance will likely feel pretty unstable. As such, I don't think simple Haas delays are the most effective way of "stereoizing" a mono recorded signal.

There's one other neat variation on the Haas idea that's also well worth investigating. Set up a single-tap stereo echo as a send effect, again using a delay time somewhere below 30ms, then polarity-invert just one side of the delay, and finally swap over the return's left and right channels (as shown in Figure 18.5). You can now use this send effect to widen any mono or stereo channel in your

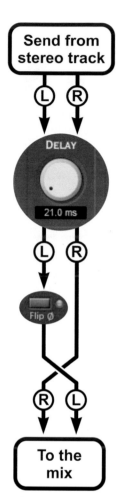

FIGURE 18.5
A stereo-widening send-effect based around Haas delays and polarity inversion: a sub-30ms stereo delay with one channel polarity-inverted and the left and right channels swapped.

mix. As with simple Haas-delay widening, though, be careful not to be misled by loudness bias and do listen for any loss of detail or flamming of transients. (Incidentally, if you remove the one-sided polarity inversion from this setup, it offers instead an interesting variation on the blend delay treatment discussed in Chapter 17, the Haas effect giving panned mono and stereo sounds a certain additional sense of ambient cohesion.)

18.3.4 Pitch Shifting

Pitch shifting offers a different way to widen mono recordings. Follow the same track-duplication scheme as for the EQ-based widening in Section 18.3.1, but instead of inserting EQ to differentiate the channels, use small pitch shifts—try shifts of five cents (in other words five hundredths of a semitone) upward in the hard-left-panned channel and five cents downward in the hard-right-panned channel. Although such small pitch shifts aren't quite enough to make the track feel out of tune, they do fool the ear into thinking that the signals it's hearing from the two speakers are actually independent sound sources, so you get the impression of a wide-sounding stereo image.

One problem with this particular implementation, though, is that pitch shifters that sound any good are also quite CPU hungry, so you'll rapidly suffer a serious processor hit if you want to widen several tracks this way. For this reason, a more practical approach is to set up the left-up/right-down pitch-shifting as a send–return effect, mixing that back in with the dry signal. That way you only need one pitch shifter for the whole project, and you can regulate how much it affects each instrument in the mix by adjusting individual track send levels. Of course, the send–return configuration brings with it the potential for comb filtering between the dry and wet signals, but if that causes unacceptable timbral side effects then try adding a short pre-delay (maybe 10ms or so) before the pitch shifter to avoid the worst of the phase cancellation.

18.3.5 A Classic Pitch-Shifted Delay Patch

One of the oldest studio effects tricks in the book can be seen as a kind of hybrid between the Haas delays of section 18.3.3 and the pitch-shifting send effect of section 18.3.4. Here's the way it works. If, instead of a single prede-lay in your pitch-shifter's return channel, you actually use slightly different delay values for each side of its stereo output, then the time-of-arrival difference between the two channels increases the perceived disparity between them and further increases the subjective wideness. All engineers seem to have their own preferences in terms of the exact pitch shift amounts and delay times, but so many people use this type of setup that it's got to be one of the all-time classic general-purpose mix treatments. Personally, I like to have a few milliseconds longer delay on the channel that is shifted upwards, because somehow the upwards shift seems to slightly counterbalance the Precedence Effect's tendency to skew the stereo spread towards the channel with the shorter delay.

Lead vocals are one of this effect's prime applications, because widening the vocalist's stereo image suggests that the singer is closer to the listener, but at the same time the effect also blends the vocals ever so slightly into the track as a side effect (much like the blend delay patch in the previous chapter), while maintaining an ostensibly "dry" sound. In addition, the presence of more vocal signal in the mix's Sides component decreases the disparity in perceived vocal level when moving between stereo and mono playback.

The biggest problem with this famous widener patch on lead vocals is that adding too much of it will make any band sound like Chicago, but even when you avoid that terrifying hazard, the pitch-shifted signal may still phase-cancel undesirably with the dry vocal sound. Although a little comb filtering may not matter on less important parts, on upfront lead vocals small timbral changes will be more noticeable. One solution to this is to increase the predelay by another 20ms or so, but this means that you'll then begin to unmask the effect by letting it trail slightly behind the vocal. You get less comb filtering, sure, but you also can't get away with using as much widening before the pitch-shifted delay begins to sound obviously synthetic. Another tactic I personally prefer is to subtly finesse your delay times so that the comb filtering is rendered more tonally benign—a millisecond either way can make a surprisingly big difference. I also frequently high-pass filter the effect return to some extent, which not only helps reduce the level of phase cancellation in the midrange (where it's usually at its most damaging in this instance) but also has the effect of supplying a little bit of high-frequency enhancement into the bargain.

Whichever way you set up this effect, though, you should make sure to check its mono-compatibility. One of the effect's strengths is that it's often fairly well-behaved in this respect, but if it isn't, then you can usually salvage the situation by massaging the pitch-shift or delay amounts while listening in mono.

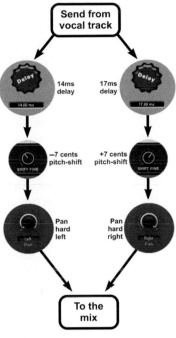

FIGURE 18.6
A classic stereo widening effect based around pitch-shifted delays, although every engineer has a favorite combination of pitch-shift amounts and delay times.

WIDENING DELAY RETURNS

In Chapters 16 and 17, I suggested a number of ways you could design reverb and delay effects to enhance stereo spread, but there's nothing stopping you applying this chapter's stereo-widening tactics to their return channels as well. In fact, this approach has the advantage that any timbral and mono-incompatibility side effects of the widener processing can't afflict the dry signal directly. This means that vocals and acoustic instruments can remain more natural-sounding in general, and summing to mono disproportionately weakens just the reverb/delay effect— the latter arguably no bad thing in complex arrangements where it's a struggle to maintain mix clarity once spatial separation is removed from the equation.

18.4 MODULATED ENHANCEMENTS

So far we've looked only at static effects that can alter and enhance stereo width, but there are further options once you get into the realm of modulated treatments.

18.4.1 Auto-panning and Rotary Speaker Emulation

Probably the simplest modulated stereo effect is auto-pan—an insert effect that sweeps a mono sound cyclically from one side of the stereo field to the other. Although this can be a nice effect on occasion, it's normally much too heavy handed for subtler stereo enhancements at mixdown and quickly becomes distracting for stereo listeners. However, there are still some useful widening patches to be had from an auto-panner if you keep the modulation depth low

FIGURE 18.7
Rotary-speaker emulations, such as Plug & Mix's LS-Rotator (*left*) and NuBi3's Spinner LE (*right*), can work well as stereo wideners at mixdown.

(so the panning isn't too wide), use a smooth modulation waveform such as sine or triangle (to avoid jerky stereo shifts), and push the modulation speed up into the 5 to 10Hz range.

Personally, I prefer to use auto-panning in a frequency-selective manner instead if possible, because it tends to feel less disconcerting if different frequency regions are panning in different ways—at least you don't get the whole sound charging rhythmically back and forth across the stereo field like some kind of audio tennis ball. Multiband auto-panners provide one solution here, but I find that a Leslie-style rotary speaker emulator often provides a nice alternative. At its simplest a plug-in like this will pan the low and high frequency ranges independently to simulate the speaker's two rotating horns. However, most models also modulate other aspects of the sound simultaneously, so you may wish to apply the simulation as a send effect to make the results more subtle, perhaps with some predelay in the return channel to avoid comb-filtering problems.

18.4.2 Dynamic Tone and Pitch Changes

Although the static EQ widener I mentioned in Section 18.3.1 has some applications, most of the time modulated tonal changes provide a more usable stereo enhancement. Chorusing, flanging, and phasing will all achieve this end when they're used as send effects, and while the exact mechanics of the way each works are slightly different (and not really essential to understand for mix purposes), their stereo widening tactics all essentially boil down to the same thing: modulating the tone of the left and right channels of the frequency response independently. Andy Wallace likes this kind of treatment on bass: "Not so much as 'Dig the flange on the bass!', because I kind of prefer that nobody even knows it's there. . . . I use that just to open things up a little bit so that everything is not kick, snare, bass right down the middle. Not that there's anything wrong with that, but sometimes it feels sonically a little more interesting to me."[5] Adam Moseley does something similar, but for different reasons: "Sometimes the emotional point of the song will be the warmth of it, and you need to have this surrounding warmth come out of the speakers when the first bass note happens. I want it to sound like it's just come out and put its arms around you, like it's warm water, or your parent with a warm towel. So even if it was a mono bass . . . I'll put it into some kind of stereo thing."[6]

Only some implementations of these plug-ins will allow the individual stereo channels to be processed separately like this, and others have this facility switchable, so your first setup step will be to choose an appropriate plug-in and preset. Beyond that, I normally feel that the most successful widening treatments for mixdown arise when the effect's Feedback and Depth controls are kept low, although the strength of the widening will, of course, depend on the send level from the dry track.

> ### WIDEN YOUR WIDENERS!
>
> Many of the widening tactics I discuss in this chapter involve setting up some kind of send effect, but a downside of many of these is that, if the mix is summed to mono, they can affect the timbre of tracks to which they're applied, usually by adding a flavor of comb filtering, phasing, or chorusing. If this change is undesirable, you can avoid it by using an MS adjustment plug-in to reduce (or even mute) the stereo Middle component on the widener's return channel.

Modulated pitch shifting, better known as vibrato, can yield widening effects if you can actually find a plug-in that offers it (audio vibrato processors aren't that common) and then persuade it to treat the stereo channels independently. If you have this option available, keep the vibrato's Depth control low again to avoid tuning vagueness creeping in, and consider introducing some predelay into the effect return to reduce unwanted tonal colorations.

Whatever modulation effect you use, it is still vital to check how your processing sounds in mono. One of the good things about dynamic widening effects is that the damage caused by undesirable phase cancellations will tend to be softened by virtue of the fact that the modulation will be varying them the whole time. Still, you can't know how your mono-compatibility stacks up without using your ears.

FIGURE 18.8
If you can find a modulation plug-in that allows its left and right channels to operate out of phase, then that may work well as a stereo widener. Here you can see a couple of freeware examples: Blue Cat's Flanger (*top*) and GVST's GChorus (*bottom*).

18.5 CHOICES, CHOICES . . .

I've covered a lot of options for stereo widening here, and you might be forgiven for wondering why you actually need more than one. The reason there
are so many approaches is that each one creates a different subjective sound,
exhibits different side effects, and affects mono-compatibility in different ways.
It's only by trying them all out and developing a memory of what they sound
like that you can choose the best one for any given widening task.

CUT TO THE CHASE

- Stereo width is at least as important to modern record production as exact
  stereo positioning, so it's important to have a selection of stereo width-
  manipulation techniques at your disposal. All of them will change the
  way stereo and mono listening affects your mix balance, so you should frequently check mono-compatibility as you're working with them.
- Generating "fake" double-tracks via audio editing is a very effective strategy
  for filling out many small-studio productions, both in terms of texture and
  stereo width. Mixing in a delayed copy of a track doesn't produce the same
  effect. Pads, tape hiss, vinyl noise, and Foley room tones can all add stereo
  width too, alongside note-sustain, blend, and size side effects.
- MS stereo encoding/decoding gives you a lot of control over the image
  width of stereo signals, particularly when applied in a frequency-selective
  fashion, but it can also lead to serious mono-compatibility problems if
  pushed too far.
- Static EQ, pitch shift, and delay effects can all deliver stereo widening artifacts, as can modulation processes such as auto-panning, rotary speaker
  emulation, chorusing, flanging, phasing, and vibrato—assuming that your
  particular DAW and plug-ins allow you to process the two sides of a stereo
  signal independently.

Assignment

- Do a survey of the plug-ins on your DAW system, and try to find at least one of
  each of the following: auto-panner, rotary speaker simulator, stereo chorus, stereo
  flanger, stereo phaser, and stereo pitch-shifter.
- If you can see any fruitful avenues for editing together "fake" double-tracks in your
  own example mix, then try some out to see whether they might fill out the arrangement texture and stereo field in a beneficial way.
- Return to your mix and experiment with a good selection of the different stereo
  adjustment effects covered in this chapter. Try at least MS-format rebalancing,
  static EQ-based widening, the classic pitch-shift delay patch, Haas delays, and a
  couple of different modulated stereo treatments. You shouldn't need to use all of
  those on every mix, but having a feel for the sonic characteristics and side effects
  of them all will better prepare you to choose the most suitable option for any
  given task.

Web Resources

On this book's companion website you'll find a selection of resources to support this chapter, including:

- a special tutorial video to accompany this chapter's assignment, in which I show how you might use a variety of different stereo enhancements to add width to a simple multitrack mix;
- a selection of audio examples demonstrating different stereo-widening techniques, including audio-editing, sound-layering, and effects-processing approaches;
- links to sources of stereo-widening audio layers and a variety of affordable stereo-widener plug-ins, including MS processors, pitch-shifters, and stereo modulation effects;
- some mix case-studies from the *Sound on Sound* "Mix Rescue" column where I focus particularly on stereo-manipulation methods.

 www.cambridge-mt.com/ms-ch18.htm

CHAPTER 19
Master-Buss Processing, Automation, and Endgame

You should now have the most important section of your mix balanced and sweetened to taste, which brings you to within striking distance of a complete first draft. This chapter explains the steps you need to follow to go the remaining distance.

19.1 MASTER-BUSS PROCESSING

At this point in the mix, you may also wish to try adding further processing to your master buss to refine the combined sound of the entire mixdown signal. This is one of the most complicated mixing tasks there is, because any buss-processing move you make will usually incur some undesirable artifacts, and mitigating these side effects can involve a good deal of finicky parameter tweakage, not only on the master buss itself, but also to per-channel mix settings. That alone already makes things intimidating for many small-studio users, and high-profile practitioners only add to the confusion by continually innovating in this area in their unending quest for improved sonic results, generating a blizzard of alternative approaches. So what I'd like to do is identify some of the most common master-buss processing tactics, explain why you might want to use each one, and pass on some of my own hands-on tips. Once you understand the basic building blocks, it becomes a lot easier to create your own bespoke setup that'll suit the specific music you're working on.

19.1.1 Compression

Probably the most common master-buss process in most music styles is compression, which can appeal for a variety of potential reasons:

- It introduces some level interaction between different parts in your mix by ducking quiet signals slightly in response to louder ones, and in doing so it gives a feeling that the mix coheres better. It's common to hear engineers talk of master-buss compression "gluing the mix together" on account of this.
- It can create pumping effects that add a subjective sense of loudness and aggression, which suits certain music genres, especially rock music.

- By boosting lower-level signals, it draws more attention to internal mix details and generally makes the music seem more emotionally engaging.
- If a characterful compressor is used, then it may subjectively enhance some aspects of the mix—for example, by adding subtle distortion artifacts at certain frequencies (of which more in Section 19.1.4).
- Some engineers feel that it reduces the negative side effects of down-the-line mastering or transmission compression. Here's Andy Wallace, for example: "A long time ago, I learned that [radio compression] was seriously changing how the low end sounded and the balance . . . That's when I really started experimenting with a substantial amount of stereo compression. And I found that if I had something compressed ahead of time and was happy with the sound of it, the additional compression from the radio station had less effect."[1]
- It evens out the dynamics of the entire mix signal, increasing the production's average levels so that it appears louder at a given metered peak level.

Of these, the last isn't actually a very good reason to apply master-buss compression during mixdown, because loudness processing is usually better carried out as a separate specialized processing stage once the mix is complete. (We'll focus on this in Section 19.2.). It's also important to remove loudness bias from your decision-making process when applying master-buss compression so that you're not misled into overprocessing. In other words, adjust your compressor's Makeup Gain control to compensate as best you can for any loudness increase, so that bypassing the compressor doesn't boost the subjective volume.

As far as which model and setting of compressor to use, my own research suggests that opinions in the professional community are many and varied, so master-buss compression isn't just some kind of preset you can slap over your mix without thinking. When a compressor is acting on a complex full-range mix signal, the sonic differences between compressor models and the effects of small adjustments to fine parameters begin to make much more profound sonic differences, and subjective judgments are often as much about stylistic conventions and the engineer's personal preferences as anything else. However, that insight isn't going to help you much in finding a setting that suits your particular mix, so let me at least suggest some of the practical factors you may need to consider.

Master-buss compression simply for the purposes of "mix glue" rarely uses more than about 2 to 3dB of gain reduction at most. Where you want to keep transient smoothing and pumping artifacts to a minimum, you'll likely want to rely on a slow attack time and automatic release time, whereas faster release-time settings will tend to draw out more of the mix details. Ratios over 2:1 are unlikely to be useful in these cases, and you may discover that ratios as low as 1.1:1 may be more appropriate, depending on the musical style. If you're looking for more aggressive pumping compression, however, then you might even see 8dB of gain reduction on peaks, with faster attack/release times and higher ratios. Lee DeCarlo also recommends one important refinement: "I'll play with the release and the attack times until I can actually make that limiter pump in time with the music."[2]

But no matter how much your band look like extras from *Lord of the Rings*, the common thread with any type of master-buss compression is that you don't want to compress the master buss beyond the point where unwanted processing side effects become overbearing. Much of the art of successful buss processing therefore lies in working out how to rein in unwanted side effects as far as possible, so that you can maximize the desirable aspects of the sound. We looked at these side effects in Chapters 9 and 10, but here's a quick refresher of the main offenders, with suggestions of how you might deal with them:

> No matter how much your band look like extras from *The Lord of the Rings*, you don't want to compress the master buss beyond the point where unwanted processing side effects become overbearing.

- *Loss of attack on prominent transients, such as kick and snare drums.* Suggested remedies: reduce gain reduction; increase attack time; reduce ratio; adjust knee softness; switch level-detection mode from peak to average (RMS); select a different compressor design.
- *Excessive gain pumping.* Suggested remedies: reduce gain reduction; adjust release time or switch to automatic mode; apply low-frequency EQ cut in the level-detection side-chain; select a different compressor design.
- *Unwanted distortion.* Suggested remedies: lengthen attack and release times; adjust knee softness; select a different compressor design.
- *Loss of weight on instruments with low-end transients, such as kick drums.* Suggested remedies: reduce gain reduction; lengthen attack time; apply low-frequency EQ cut in the level-detection side-chain; adjust knee softness; select a different compressor design.
- *Undesirable alterations in the subjective mix balance.* Suggested remedies: reduce gain reduction; lengthen attack and/or release times; adjust knee softness; select a different compressor design.
- *Unappealing tonal changes to the mix.* Suggested remedies: use a gain plug-in to adjust the signal level feeding the compressor; select a different compressor design.

If you're having trouble hearing the side effects of your master-buss compression, here's a tip from Andy Wallace: "It's not unusual for me to really make a compressor slam, to hit it pretty hard. I'll often put it on an extreme setting, to get a sense of what the ballpark is, and then back it off to what sounds right to me."[3] So crank the gain reduction to get a handle on what you're listening for, and then reduce the amount of compression to more sensible levels. Sometimes a few of the side effects may be impossible to dispel entirely with adjustments to the compression alone, without losing the desired degree of mix glue, pumping, or compression character. In this event, there are a number of different options. The first is simply to adjust the mix itself to compensate for the compressor's less delectable side effects. You might enhance the kick and snare transients with additional processing to keep them crisp despite the attack-dulling effects of heavy mix-pumping compression. Or you might tweak a few channel faders if some instruments (or particularly effects returns) have been knocked out of balance by detail-enhancing buss processing.

Another remedy is a parallel master-buss configuration, whereby different blends of tracks are sent to several independently compressed master busses, which are themselves summed together to feed your DAW's master outputs. The intricately interwoven parallel-processing schemes used by some well-known proponents (particularly Michael Brauer[4]) may initially seem like unfathomable voodoo, but the primary function of parallel master-buss compression is actually quite straightforward: it lets you side-step unwanted artifacts from traditional single-channel processing. So, for example, let's say you want to compress the living daylights out of a rock mix for hurricane-force mix-pumping, bucketfuls of juicy harmonic distortion, and maximum air-guitar potential. However, while cranking the compressor up to 11 in search of moshpit nirvana, you discover that you've thoroughly emasculated the kick and snare hits, and no amount of parameter twiddling seems able to salvage them. A workaround here would be to set up a second master buss in parallel, to which you send a bit of extra kick and snare. Because these signals will now effectively bypass the main master-buss compressor, they'll help restore some punch to the drums in the final mix. You might even compress the second master-buss too, perhaps using more conservative settings designed simply to keep the kick and snare additions at a consistent level.

> It may seem like unfathomable voodoo, but the primary function of parallel master-buss compression is actually quite straightforward: it lets you side-step unwanted artifacts from traditional single-channel processing.

Another helpful way to look at parallel master-buss compression is as a means of controlling the balance interaction that occurs when different instruments are compressed together. If you feed your band's drums, bass, guitars, keyboards, and lead vocals through a single compressor, the loudest signals will influence the compressor's gain reduction most, which likely means that level peaks from the bass and drums will duck the other instruments, whereas the guitar and keyboard backing parts won't affect the compression nearly as much, seeing as they're typically quieter and carry less low-frequency energy. But what if you want more level interaction between the guitar and keyboard parts, and less ducking of the lead vocal to improve lyric intelligibility? Well, you could justifiably set up three different parallel master busses: the first for the drums and bass, with assertive peak-oriented compression to control the peaks and increase sustain; the second for guitars and keyboards with slower low-ratio compression to enhance the musical interaction between those parts; and the third for the lead vocals, perhaps with no compression at all so that it doesn't fight your channel fader settings. There'd also be nothing preventing you from feeding, say, the guitars to the first master buss as well, if you wanted them still to retain some interaction with the drums and bass, or indeed from sending different blends of all your tracks to each master buss for even more subtly nuanced ensemble dynamics.

Now, although the logic of complicated parallel master-buss compression setups usually emerges once you start thinking in terms of (a) avoiding unappealing master-buss compression side effects and (b) managing a mix's internal level interactions, it should be obvious that such setups often spawn mightily tangled signal routing. So unless trawling DAW support forums is your idea of fun and frolics, I'd advise against running multiple master busses as a matter of course, and suggest saving parallel master-buss compression for clearly identified troubleshooting tasks, at least until you've got considerable mixing experience under your belt.

What about multiband compression, especially in cases where you're having problems with gain pumping? While in theory it's fair game, my own view is that multiband master-buss compression is almost always best left to the mastering engineer—for practical reasons as much as anything. Multiband compression is just too complicated and finely tuned to manage alongside all the other aspects of the mix, and every small-studio engineer I've ever seen

FIGURE 19.1
For master-buss processing, it's usually wise to use gentle, broad-band equalizers so as to avoid upsetting the internal balance of the mix unduly. A couple of my favorite affordable plug-ins for this are Tokyo Dawn's Slick EQ and Sonimus's SonEQ Pro.

using it on their master buss has fallen into at least one of the following traps: they put insufficient effort into getting the balance right, because the multi-band compressor's automatic frequency adjustments make spectral balance problems in the mix more difficult to spot and remedy; or they undercook their blend effects and then overcompress the buss in an attempt to compensate with additional "mix glue"; or they overprocess individual tracks to compensate for dynamic frequency-response side effects of the buss treatment. In short, full-band compression is more than enough to worry about while mixing, so I'd save yourself some grief and leave it at that.

19.1.2 Equalization

Because our ears are so good at adjusting to differences in tonality, it's common to discover towards the end of the mixdown process that your overall mix tone needs tweaking. Although you could deal with this on a track-by-track basis, it's normally more convenient to add an extra EQ plug-in on your master buss, and I'd suggest inserting it after any master-buss compression you may have on the go, so it doesn't interfere with those finely honed settings. But it's more than just convenience, because using a single high-quality EQ in the master-buss will typically sound better than applying lots of lower-quality EQ processing on individual channels, particularly where you want to increase brightness. Spike Stent explains the rationale here: "Basically what I do with every mix is put a GML EQ across the stereo buss of the SSL, and just lift the top end, above 10K, so I won't have to use as much high end from the SSL. The mid and high end of the SSL is a little harsh and cold. So I turn that down a bit, and then lift the high end with the Massenburg to make the mix sound warmer and add sheen."[5] Although your choice probably won't be between an SSL console and a GML EQ, the same reasoning applies when working on DAW systems, because adding high end in the digital domain with lots of low-CPU channel plug-ins tends to introduce more harshness than when you use a single high-CPU plug-in over the master outputs. "Digital EQ doesn't handle high frequencies very well," says Jacquire King.[6]

Whatever you do with master-buss EQ to fix your overall mix tonality, be aware that you'll probably upset some aspects of your mix balance into the bargain, so don't be surprised if you need to revisit a few of your individual channel settings in response. In my experience, it makes life easier in this regard if you restrict your master-buss EQ moves to broad brushstrokes, as wide-bandwidth frequency-response adjustments tend to leave internal mix-balance relationships more or less intact. If you find yourself wanting to boost or cut narrow frequency regions on your master buss, then it usually indicates that you're better off going back to your individual channel EQ settings instead.

19.1.3 Stereo Manipulation

Another attribute of your overall mix sound that's easy to misjudge while you're up to your elbows in per-channel mixing decisions is stereo width, and it's sometimes only once the job's nearing completion that it becomes apparent how comparable your imaging is with commercial releases. Again, this is something you could deal with on a channel-by-channel basis, but that risks messing with the relative positions and widths of the sounds, so it often makes sense to apply stereo-adjustment processing to your master buss instead. This could be as straightforward as a simple MS plug-in, but in practice it usually makes more sense to use multiband MS processing, or to equalize the Sides component of the stereo signal, so that you can adjust different frequency ranges independently. Any of the other stereo-enhancement techniques in Chapter 18 might also be used on occasion, but they're typically less generally applicable because of the greater likelihood of unwanted tonal side effects when processing the full mix signal. Whatever processing you use, though, the two most important things to remember are: firstly, to listen to the mix both in stereo and in mono before signing off on your changes; and, secondly, to recheck the balance of your stereo mix after the stereo-image processing, because adjusting the Sides level in any frequency range may also affect the apparent overall volume of any instruments whose energy is focused in that spectral region. By the same token, confirm that centrally panned instruments (e.g., bass, kick, snare, lead vocals, instrumental solos) haven't been knocked out of their desired positions in the level balance or depth perspective by any increases in Sides-signal energy.

MULTIBAND CROSSOVERS

Any multiband processing involves using crossover filters to split up the audio signal into different frequency bands, and it's worth realizing that this process may not be transparent. In other words, the crossover process itself can change the sound, even before any per-band processing is applied. Although crossover side effects are usually subtle, even minuscule sonic colorations can have a profound impact on your mixdown sound, so do make a point of first assessing any multiband plug-in's crossover character before you start fiddling with its per-band processing.

19.1.4 Special Sauces

Although there are often clear technical grounds for master-buss processing, it's fair to say that most top engineers also select master-buss processors for less tangible subjective reasons, as a means of adding their own unique "special sauce" to the sonics. In this context, clearly the specific processor selected for any dynamics, EQ, and stereo-manipulation work plays a role, but other mix tools that are purely coloristic in function are also very popular. The general-purpose distortion techniques I talked about in Section 12.1 are all on the menu, for instance, although parallel processing will often be essential to keep the added harmonics sufficiently subtle and carefully targeted for use on full-mix signals. There are also now heaps of hardware and software processors

> Distortion levels that sound acceptable on nearfield monitors may become overbearing under the forensic microscope of headphone listening or when compounded by the additional sonic grubbiness of your grotboxes.

on the market that promise the elusive magic of some kind of analogue circuitry, such as vintage tube outboard, classic mix consoles, and historic studio tape machines. Unfortunately, there's little direct advice I can give about using these kinds of treatments, because they're inherently rather "suck it and see" by nature. Clearly, if you like the work of particular mixing engineers, then you could do a lot worse than checking out interviews with them and investigating their preferences as a starting point. Just be super-careful to avoid loudness bias, since many of the "analogue warmth" plug-ins I've tried seem to subtly boost the overall level as well. Also, be sure to check the sound on all your different monitoring systems, because distortion levels that sound acceptable on nearfield monitors may become overbearing under the forensic microscope of headphone listening or when compounded by the additional sonic grubbiness of your grotboxes.

19.1.5 General Considerations for Master-buss Processing

Whatever master-buss processing strategy you go for, here are a few bits of general advice. First, if you're going to use it, never try to completely finalize all your mix settings before getting your choice of master-buss processing involved, because you'll almost always have to adjust the mix settings in response to how the master-buss processing reacts, whether you're using compression, EQ, stereo manipulation, distortion, or a combination of all those and more. But neither should you be afraid to keep readjusting your master-buss processor settings as the mix continues to develop, in order to keep each one operating in its most magical-sounding zone. It's because of the interactions between individual channel processing and master-buss processing that it's essential to get any master-buss plug-ins involved during the mixdown process itself, rather than leaving it until the mastering stage. "Why would you wait till the end?" asks Jaycen Joshua. "You want to hear what it's doing immediately. If you wait to the end, it's going to change your mix completely."[7]

A much thornier question, though, is this: how early should you get master-buss processors involved? To avoid overcomplicating the step-by-step mix-teaching approach in this book, I've deliberately waited until most of the balancing and sweetening work has been done for the production's most important section before broaching the topic of master-buss processing. Without master-buss processing, per-channel mix decisions can be taken on their own merits, rather than being affected (or even counteracted) by plug-ins over the whole mix, which I think makes it easier to develop a solid understanding of how fundamental mix techniques affect the sound, as well as encouraging students to target mix problems more precisely, thereby reducing unwanted processing side effects.

However, while this "bottom-up" mixing method (i.e., doing most of your channel processing before your buss processing) has advantages for learners, it's important to realize that there are plenty of mixing engineers who advocate a "top-down" approach instead. The thinking here is that it's much easier to make broad-brush changes to your mix with buss processing, so you'll get your final sound into the right ballpark much more quickly if you concentrate on the busses first, rather than immediately working one track at a time. But although the top-down method clearly has the edge when you're fighting the clock, it's not without its own weaknesses when quality is more of the essence than time. My main concern is that, once you get beyond that intoxicating first round of easy-win mix improvements, it can quickly get frustrating trying to compete with the less tangible characteristics of commercial releases ("clarity," "sustain," "depth," "groove," and so forth) which tend to be a cumulative product of many tiny mix tweaks rather than of larger-scale processing maneuvers, in my experience. If it feels like you've put together 80 percent of your mix in an hour using top-down methods, it's going to be pretty tough to discipline yourself to spend another, say, two days on detailed work to achieve the remaining 20 percent that separates your mix from the market competition.

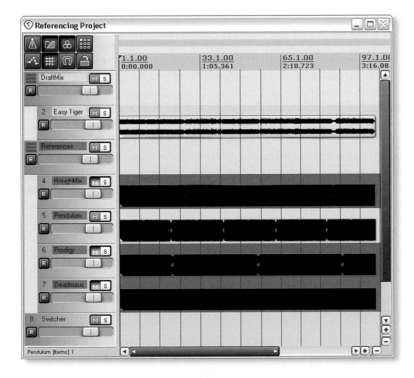

FIGURE 19.2
For referencing purposes, it's good practice to export your mix into a separate DAW session so that you can easily compare it to a selection of commercial productions.

As such, I've noticed that the top-down approach frequently encourages project-studio mixers to overprocess their busses, instead of engaging with painstaking but vital nitty-gritty tasks.

Fundamentally, though, the end justifies the means in mixing, and although both bottom-up and top-down mindsets have enthusiastic supporters, neither has a monopoly among world-class engineers. In my view, small-studio users get best results if they can develop a mixing workflow that balances benefits from both approaches. For example, I personally prefer to deal with most phase, balance, and masking issues on a per-track basis before activating any master-buss compression, but I might introduce a master-buss EQ or special-sauce processor very early on to swiftly shunt the overall tonality or character of the production in the right direction, and I'll often apply additional master-buss EQ and stereo width-adjustment during the very last stages of mix referencing too. Does that make me a bottom-up, top-down, or inside-out mixer? I don't think it really matters, as long as I can work out how to get the necessary results that way.

19.2 REFERENCING AND LOUDNESS PROCESSING

In an ideal world, once you've got your master-buss processing up and running and you've rebalanced individual tracks as necessary in response to it, you should have arrived at the final sound for your most important mix section. But, sadly, the world isn't ideal! The cold, hard, truth of the matter is that nothing you've done so far can be considered final until you've properly compared your mix to its commercial competition. Then why have I waited until now to return to a subject first broached in Chapter 4? Well, it's tricky to make sensible comparisons between a half-finished mix-in-progress and fully polished commercial releases, so mix referencing tends to be most revealing once your own first mix section is approaching completion. That said, my own initial balances frequently include a couple of fader levels that are less well judged than the contents of Noddy Holder's tie rack, so I do find that a brief bout of referencing early on can usefully nip those slip-ups in the bud, as well as indicating whether some initial master-buss EQ might be in order. Further mid-workflow referencing can also help answer specific mixing questions at any time during the process, but I think it's also just fine to go with your gut and leave the main referencing work until you've got as far as you can under your own steam. However, I wouldn't recommend trying to mix other sections of the production before you've properly referenced the first, otherwise you may discover later on that you've been building those new parts on shaky foundations.

Once you feel it's time to get properly stuck into referencing, export the section of the production you've mixed so far as a stereo audio file, and import it into a fresh DAW session along with some appropriate reference material, perhaps made up of specific target productions suggested by the client, as well as some tracks selected from your personal reference library. As I mentioned in Section 4.2.3, the most important things to do when making these comparisons are to switch between the mixes instantaneously, so that your hearing system has no time to adapt itself to differences in balance and tone, and to adjust the subjective levels (not the metered levels!), so that loudness bias doesn't interfere with your quality decisions. If you immediately discover that your overall mix tonality or stereo width is significantly off-target, then I find it's worth correcting those things straightaway with equalization or MS processing (which can subsequently be transferred into your mix project's master buss), because it's often hard to evaluate more subtle mix attributes before that's done. Furthermore, in styles of music that use loudness processing at the mastering stage (which is pretty much everything these days), you might feel you want similar processing at the end of the plug-in chain on your own unmastered mix section in order to make truly meaningful comparisons. However, like any other master-buss processing, loudness maximization has side effects, so you need to be aware of whether these might require compensatory tweaks to the mix itself.

REFERENCING THE ROUGH MIX

In addition to referencing commercial releases, it usually pays to have a good listen to any rough mixes that are available. At the very least, this ensures you've not inadvertently missed out anything important from your mix! "I make sure that all the elements are there," says Rich Costey, "and that I have replicated or maintained any effects that the client may have spent time creating, and that they care about."[8] A lot of engineers also use the rough mix at an early stage in the mix to confirm the validity of their basic balances. "[I] quickly reference where their balance is," says Bob Horn, "and then decide if I want to keep my version of that or vary it. That's the fastest way for me, because I know it's probably going to be what they're looking for level-wise."[9] "I like to hear the rough mixes," adds Randy Staub, "because people will have spent a lot of time on them, and they give a basic idea of the kind of balance and perspective that the artist and producer have in mind."[10]

The crucial question when presented with any rough mix, though, is how far to deviate from it for your "real" mixdown. "The rough mix is often more important than the final mix," cautions Josh Gudwin, "because it sets the direction for the song, and if the artist likes that, you don't want to lose that by redoing sounds and so on. So my final mixes almost always are refinements of the rough . . . For example, with [Justin Bieber's] 'What Do You Mean' we literally went with my rough mix."[11] One vital factor is how long the client has had that particular rough mix. "The first question I ask is 'How long have you guys had the rough mix?' and if it's three months or longer, they're getting a loud version of the rough mix!" says Miles Walker. "But if they've had it a month, or did it a week ago, then I know I have some leeway to try out ideas."[12] "When you mix something made by guys who have been working on a rough mix for a couple of months to a couple of years, it's a very delicate balance," agrees Leslie Brathwaite. "What I try to do in those situations is to simply enhance the mix and make sure that it's technically correct, and not to change what they have gotten used to."[13]

Now clearly a typical project-studio rough mix will probably be a whole lot rougher than what these A-listers are turning out, but regardless of how polished it sounds,

(continued)

it's important not to let yourself be terrorized by it, because the mix engineer is usually also expected to bring something new to the table. "Unless the [rough] mix is shockingly bad, you don't want to go miles away from it," says Tom Elmhirst. "On the other hand, while I will obviously have heard the [rough] mix at some stage, I try not to listen to it after that. Otherwise it'll affect the way I work too much."[14] "I'll reference the rough initially, but at some point I will end up flying on my own," adds Andy Wallace, "because when you're making decisions just based on somebody else's mix it can be self-defeating. There's this syndrome that if you listen to something often enough, it will start to sound right . . . I try to avoid that. If you allow the rough mix to cloud your own vision, you end up with nobody's mix."[15] Remember too, that the people who created the rough mix may have totally lost their objectivity after prolonged work on the project, so a dispassionate reassessment by the mix engineer may be exactly what the music needs. "Normally the reason why they're calling me," remarks Mark "Spike" Stent, "is because somewhere along the line maybe they've lost sight, or got to close to it, and that's where I come in."[16]

Increasingly, musicians are also expecting the mix engineer to have more creative input, and many top engineers now see this creative collaboration as their primary role. "There's certain mixers that just enhance the producer's idea," says Jaycen Joshua, for instance. "You'll get basically the rough [mix] enhanced. I don't work like that . . . I crash or burn! If I can make your record the best record in the world, I'm going to do everything possible. And sometimes I've got to dial it back. Sometimes they'll come in and be like 'I love everything except this, this, this, this, and this.' But I'd rather dial it back."[17] Of course, the danger here is that you go too far and alienate the client before they've even properly evaluated the sonic improvements. Miles Walker suggests a useful trick in this respect, though: submit a first mix that more closely matches their rough, and then follow that with your more creative version. "I won't ever submit [the more creative mix] first, because if they are so inundated with the rough they're going to be taken aback and won't judge me on the sonics or anything like that—they'll just be 'yeah, we don't like it.'"[18]

19.2.1 Basic Concepts of Loudness Maximization

The basic purpose of loudness maximization is to cram the maximum subjective level onto fixed-headroom digital delivery media such as CDs and MP3 downloads, but without unacceptably compromising other qualities of the production. There are lots of ways to achieve extra apparent loudness without increasing peak signal levels, each of which has its own cost in terms of potentially unwanted processing artifacts, so choosing between them (or indeed combining them) depends on how much loudness you want and what kinds of side effects you're willing to tolerate. These decisions must reflect the wishes of the client, the nature of the commercial competition, and the way the mix in question responds to the different processing options. Demacio Castellon suggests, "If you're making a jazz record you're not going to make it as loud as a pop record. You're not going to make a blues record louder than a heavy metal record."[19] While I can't make the choices for you (your client, target genre, and mix will be unique, after all), what I will do nonetheless is describe some common loudness processing options so that you're aware of the main side effects in each case. That way you should be able to come to reasonably well-informed conclusions when working with loudness maximization processing for reference purposes.

But before I get into specifics, I want to stress a crucial general principle that should be at the front of your mind when flirting with loudness processing: it's the side effects you should be concentrating on, not the loudness hike. This has two main ramifications, the first of which is that you should never, under pain of a visit from the Audio Engineering Society's ninja death squad, evaluate the success of a loudness processor by comparing it to a subjectively quieter unprocessed signal. If you don't match the subjective loudness of the processed and unprocessed sounds for comparison purposes, then the loudness bias will render any subjective quality evaluations worthless. Most loudness processors make it ridiculously easy for you to mug yourself in this way, so you have to be super vigilant. My preferred working method is to duplicate my mix to two separate stereo mixer channels, but to process only one of them for loudness purposes. I can then switch between the two for comparison and adjust the faders for the best subjective loudness match. If you don't trust yourself to be impartial here, then get hold of one of the new breed of affordable loudness metering plug-ins designed to meet the R128 Loudness Recommendation of the European Broadcasting Union (EBU), and use its Integrated Loudness measurement function to match the levels.

> A crucial general principle that should be at the front of your mind when flirting with loudness processing is this: it's the side effects you should be concentrating on, not the loudness hike.

The second ramification of focusing on subtle differences in processing side effects is that it demands that you push your monitoring technique to the limits. Some side effects aren't necessarily very audible on one type of listening system, for example, so you can only make well-rounded judgments about the

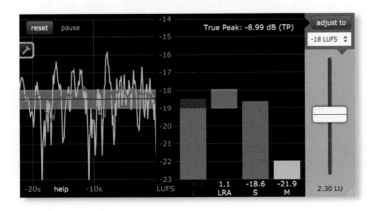

FIGURE 19.3
Klangfreund's affordable LUFS Meter is a great little cross-platform plug-in for carrying out loudness-matched listening comparisons.

relative merits of different loudness enhancements based on information from all quarters. Other side effects may only come to your attention at certain listening volumes, in certain places in your control room, or with the assistance of audio metering and analysis software. There are no shortcuts to achieving sensible loudness processing in this regard, so consider yourself warned.

19.2.2 Suggested Processing Strategies

With those stern provisos out of the way, let's have a look at some actual processing strategies and their side effects:

- *Full-band "top-down" squeeze.* This is where you attempt to gently squeeze a large section of a signal's dynamic range, using very low-ratio compression operating above a low threshold level, such that subtle compression is happening almost all the time. As long as gain reduction is kept within 3dB or so, it's usually possible to keep pumping artifacts relatively benign by adjusting attack/release times by ear. Potential side effects include undue emphasis of low-level details such as background sounds, reverb/delay effects, and incidental noises; unwanted overall level increases during sections with sparser arrangement; and reduction in transient definition.
- *Full-band "bottom-up" squeeze.* A similar approach as in the top-down squeeze, except that the dynamic range below the compressor's threshold is targeted. There are specialist processors for this purpose (called upward compressors or de-expanders), but you can also achieve much the same effect using subtle full-band parallel compression. Potential side effects are akin to those of top-down squeeze—transient definition tends to suffer less, but you get less peak control and an increased likelihood of unwanted overall level increases during sparser arrangement sections.

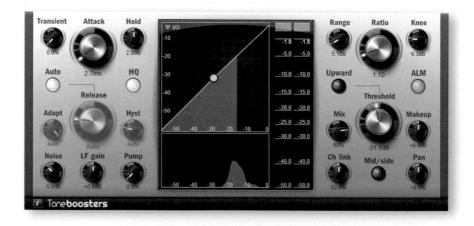

FIGURE 19.4
This screenshot of Toneboosters's TB_BusCompressor plug-in shows a typical gentle setting for full-band top-down squeeze.

- *Full-band limiting.* This is usually a much faster-acting gain reduction designed to stop signal peaks dead, while leaving the remainder of the dynamic range comparatively unscathed. In some types of music, an intelligently designed peak limiter can achieve a peak reduction of several decibels before audible problems start arising. Potential side effects include pumping, bass distortion, softening of transients, and reduction of apparent drum levels.
- *Multiband compression/limiting.* Using a multiband configuration for any of the three dynamics processes I've just described gives you additional scope for dynamic-range reduction before pumping artifacts impinge on your enjoyment of the music. However, pumping is only one of the potential side effects of compression-based loudness enhancement, and if you use multiband processing to increase the amount of gain reduction, then you can easily end up with greater side effects in terms of mix-balance alteration and transient softening. Any multiband approach also adds further problems of its own: because the amount of gain-reduction in each frequency band will depend on the overall level of frequencies within it, changes in the spectral content of the mix (perhaps as a result of alterations in the instrumentation) can trigger unwanted changes in the overall mix tonality as the compression/limiting adjusts its gain-reduction—it can be almost as if some maniac were randomly tweaking an EQ over your whole mix in real time. There are also engineers who feel that multiband gain reduction drains the life out of a production by ironing out the tonal contrasts between its musical sections.
- *Subtle distortion.* By adding distortion harmonics to a mixed signal, you can increase its harmonic density and apparent loudness with very little increase in its peak signal levels. Subtle valve, tape, and transformer distortions are all options here, and if you need greater control over the exact nature of the harmonics additions, then parallel or frequency-selective configurations may be appropriate. Potential side effects include fatiguing tonal harshness, emphasized vocal sibilance, increase in the apparent level of treble percussion instruments in the balance, veiling of midrange details, and unwanted overall changes to the mix tonality.

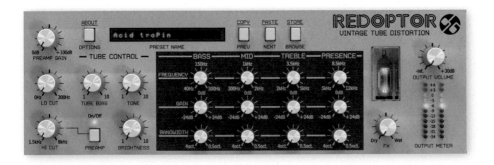

FIGURE 19.5
Subtle treatment with a multiband distortion plug-in such as D16's Redoptor can boost the subjective loudness of a mix.

- *Clipping.* Yes, clipping—whether it's straight digital-style flat topping of the waveform peaks or some kind of modeled analog saturation that rounds them off more smoothly. You can find any number of textbooks that threaten hellfire and brimstone should you dare abuse your full mix in this way, but such admonitions are at odds with widespread commercial practice—examine the waveforms of the top 40 singles any week of the year and you'll see clipping in abundance. The advantage of clipping as I see it is that it doesn't seem to affect the subjective attack or balance of prominent drum parts as much as peak limiting, so it tends to suit styles with hard-hitting rhythm parts. Potential side effects include subjective tonal change of clipped peaks and unwanted distortion on more steady-state signal waveforms, although this may be disguised to some extent by distorted instruments, such as electric guitars, within the music itself. (For a particularly striking example of this tactic, check out the Imagine Dragons single "Radioactive.")

Of course, each of these strategies will suit different productions, and each will only give you a certain range of loudness enhancement before it begins to reveal its weaknesses, so it's not unusual to find that several loudness processes in combination are required to push your mix as loud as your references. You may also quickly discover that it's still impossible to push your mix to the target loudness without unconscionably heinous side effects, and this can be a clue that some aspect of your mix itself needs to be tweaked to allow the loudness enhancement to work more effectively. "[I'm] always working with the question in mind of how the mix is going to sound once it's mastered," says Shawn Everett, "and I am adjusting what I'm doing for that. I know how loud people want things nowadays, and it is almost like an art form to see how loud I can get my mix and at the same time retain all the things I like about records that aren't loud."[20] For example, drum peaks commonly find themselves bludgeoned down into the mix balance by extreme loudness processing, so you may need to boost the apparent power of these instruments by increasing their sustain instead.

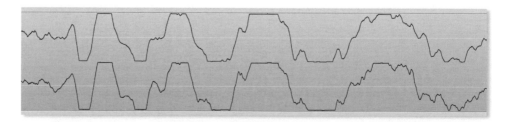

FIGURE 19.6
A lot of high-profile releases use clipping as a means to increase subjective loudness—multiple "flat tops" of 70–150 clipped samples (as seen here in Logic's hit "1-800-273-8255") are by no means unusual among current chart tracks.

19.2.3 Loudness Processing Beyond the Referencing Stage

Although I've thus far focused on loudness processing for referencing purposes, it's as well to be aware that most high-profile mix engineers also choose to loudness-process the mixes they send to their clients for approval. "When I send my mixes to the artists and management," says Matt Hyde, for instance, "I crank them up with the Massey L2007 [limiter], so they get a kind of fake mastering, just for them to have perspective."[21] Leslie Brathwaite gave a similar reason for doing this with mixes he sent to Pharell Williams: "[The Waves L2 limiter] I put on purely for Pharell, to give him some idea of what mastering would do to the mix."[22] But the biggest reason most mixers now deliver loudness-processed mixes is simply to avoid falling foul of loudness bias should the client compare their mix with a competitor's without proper loudness matching. "The biggest problem I'm having today," says Tom Elmhirst, "is being sent [rough] mixes that are overloaded with mastering plug-ins and limiters, because producers are having to compete for album cuts to such a degree. It's a battle for me as well, because people don't like it if my mix isn't as loud as the [rough] mix."[23] "So many records today just give me a headache," says Steve Fitzmaurice, "but basically I have to send everything to the labels loud-ish. If the producer has sent a rough mix to the label it will often be way louder than any mastering engineer will ever make it. So in order for me to even vaguely compete, I have to send a mix that has been limited to some degree."[24] Delbert Bowers puts it more bluntly: "When you do your auditions for labels and your song is not louder than the [rough mix], you won't get the job."[25]

When it comes to sending an approved mix to mastering, however, most engineers will usually provide a file without any loudness processing. "Pseudomastering gives me an idea of what elements potentially suffer during the mastering stage," says Dylan Dresdow, "but [I send] everything to the mastering engineer without any processing."[26] That said, it can also be useful for the mastering engineer to hear the loudness-processed version the client approved. "I send the mastering engineer both versions," says Rob Kirwan, "so he knows what the artist has been listening to, but also has the freedom to do what he wants."[27] However, it's very clear to me, having analyzed hundreds of chart singles for *Sound on Sound* magazine's "The Mix Review" column, that there are also some top-flight mix engineers who are delivering loudness-processed mixes to the mastering engineer. Delbert Bowers, for instance, tries "to get the mix as loud as it possibly can be, so it is ready to be released when it comes out of mixing, even before mastering . . . Part of the whole modern sound is how far we can go into the realm of distortion. That sound is part of what people want to hear on records now. You take that away and it just doesn't sound current . . . Loudness really is part of what people are listening to nowadays. There is distortion on everything, it is a matter of how much distortion."[28] Whatever you send to the client or the mastering engineer, though, I'd nonetheless suggest at least archiving an unprocessed version of the final mixdown file so that you can refine or redo any loudness processing at a later date if required.

LOUDNESS-NORMALIZED PLAYBACK

For decades, mastering engineers have been engaged in the so-called "loudness wars," competing to squeeze the maximum subjective loudness from a succession of limited-headroom playback formats, frequently to the detriment of sonic quality. In response to this, audiophiles have long been campaigning for the widespread adoption of loudness normalization (i.e., matching different music files in terms of subjective loudness, not peak signal level) within music broadcast systems and end-user playback devices. However, it's only now internet streaming has come to dominate music consumption that loudness normalization has really gained market traction, because most of the major streaming services appear to be implementing such normalization technology in one form or another behind the scenes. In principle, this promises an end to the loudness wars, although my own experience is that the amount of loudness processing applied to the latest commercial releases (at time of writing) remains largely unchanged despite this technological paradigm shift, so professionals are still clearly hedging their bets here. As such, loudness-processing skills look likely to remain relevant to small-studio mix engineers for some time yet. That said, with the tide now clearly on the turn, forward-looking mix engineers would do well to begin considering how their mixes compete under loudness-normalized conditions as a matter of routine.

19.3 REFERENCING CHECKLIST

Once you've managed to match the loudness of your draft mix section fairly well against your reference material, you need to take a deep breath, leave your ego at the door, and submit every aspect of your mix to merciless side-by-side comparison. Give all your monitoring systems a good workout, check at a range of different listening volumes, and don't stop until you've winkled out every last insight. Remember that this mix section should be the high point of the production, so it should shake the listener's socks off one way or another. Here's a quick rundown of issues I usually make a conscious point of assessing:

■ *How does the overall mix tonality compare?* This is something that can easily be out of line, because it's difficult to stop your ears from adapting to a skewed tonality while you're stuck into detailed mix activities. If the tone feels wrong, then first try inserting an equalizer over your mix to see whether you can correct the frequency imbalance by ear. However, if you're struggling to find the right frequencies to adjust, then by all means take advantage of the analysis routines in the EQ-matching software discussed in Section 11.4.2, as this may help highlight the most relevant spectral regions. Whatever you do, though, don't just try to match the spectral curve of your mix exactly onto that of a reference mix like you're using some kind of cookie-cutter, because that rarely works very well. Different mixes should have different spectral curves, because arrangements, keys, and mixing styles naturally vary between productions, even within the same musical subgenre. However, if a matching EQ shows that most of your reference tracks have less 300–500Hz energy than your in-progress mix, then it stands to reason that you should at least try EQing your mix in that zone, either on the master buss or on some individual channels.

■ *How does the balance compare?* Concentrate primarily on the most important instruments here, because they usually offer the least leeway within any given musical genre. I usually make a point of checking at least kick, snare, bass, and lead vocals, but the more parts of the mix you direct your attention toward specifically, the more representative you'll be able to make your own balance. Don't just listen for overall levels, but also consider how much dynamic range each instrument has.

■ *How does each instrument's tone compare?* Again, concentrate primarily on the most important instruments. Also, if the tone of any instrument in your mix appears to be less appealing, then ask yourself whether there's actually anything you can do about that without upsetting the balance.

■ *How does the use of reverb and delay effects compare?* Listen for how well the different instruments blend with each other, how close/distant they are in relation to each other, and how any different size reverbs have been used. Give a thought to overall reverb/delay lengths and levels as well.

■ *How does the stereo image compare?* This is partly a question of the apparent width and placement of each individual instrument, but you should also consider how wide the stereo image is overall in different regions of the frequency spectrum.

If you've selected good reference material, you're using all your monitoring systems effectively, and you're being brutally honest with yourself, then you should end up with a mix adjustment "to do" list as long as your arm. You may even feel like you're back at square one, but try not to be daunted. This is the make-or-break point of any mix as far as improving your skills is concerned, so if you can keep your nerve you'll be putting yourself on the fastest track to commercial results. "You're going to make mistakes," reassures Humberto Gatica. "The important thing is to learn from them, to tell yourself 'Oh my God, that sounds terrible, I will never do that again.' . . . You correct yourself and then you balance it so the good overrules the bad. And then you go on from there."[29] I can't stress enough that the difference between small-studio engineers who can mix and those who can't is that the latter fight shy of proper referencing because they can't handle a bit of disappointment in the short term. What are you: man or mouse? You can read all you like about processing techniques, but it's the referencing process that actually teaches you how to mix.

So have a sushi break, catch up on the latest YouTube kitten videos, and then steel yourself to head back to your mix project and deal with those revisions. When you're done, export that section of mixdown and run through the whole referencing process again, keeping all your different monitors in constant rotation. And keep repeating that process as long as you can stand it. "The idea is to make it work on all systems," says Andy Johns. "This is a lot of work, but it's the only way to go."[30] Worked your way through a dozen different versions? Pah! You're clearly not taking things seriously enough . . .

19.4 AUTOMATION FOR LONG-TERM MIX DYNAMICS

The goal of the recurring nightmare that is the referencing process is to arrive at a version of your climactic mix section that is not just earth-shatteringly magnificent on its own terms but also stands shoulder to shoulder with the market leaders. If your references are any good, you'll probably find that you never quite reach this goal, but that's fine—you're trying to clear an incredibly high bar, so even if you don't quite make it, you should at least end up with something that is solidly of a commercial standard. Your job now is to fill in the remainder of the mix sections. I described the general principle behind this step in Section 8.1.1, but there are a few additional points to raise based on what we've covered since then.

> The goal of the recurring nightmare that is the referencing process is to arrive at a version of your climactic mix section that stands shoulder to shoulder with the market leaders. If your references are any good, you'll probably never quite reach this goal, but that's fine— you should at least end up with something that is solidly of a commercial standard.

First of all, make sure to save a new version of your mix project before proceeding any further, so that you can always refer back to those settings should anything go awry during the rest of the mixing process. Once that's done, you can build up the rest of the musical structure in a fairly methodical way by muting all the currently active tracks, looping a new musical section, and then reintroducing the tracks in the appropriate rank order. Whenever a track needs adjustment to fit its new context, either try multing that section to a different track, or engage the DAW's mixer automation facilities to handle the change. Bob Clearmountain uses a similar section-based approach: "In the '70s, when we were recording on analog with no automation . . . we would mix sections and then edit them all together, and I tend to work pretty much the same way on the computer. . . . I'll try to get each section sounding right until I have a complete pass that is a pretty good basic mix."[31] Mults are best for situations where a track's processing and effects need to change in a variety of different ways simultaneously, because it's easier to deal with lots of changes from the mixer controls than by tweaking dozens of streams of automation data. "I might treat the vocal in the verse completely different to what I do in the chorus," says Tim Palmer, for instance. "[it] might be a completely different chain, channel, compressor, everything."[32] Where it's just a couple of parameters, though, automation is usually a more efficient solution and will save the CPU overhead of duplicating plug-ins.

Automation of practically any mixer or effect parameter is now a standard feature of almost all DAW systems, and if it's not on yours then sling that software into the trash and move to a different platform! Seriously, it's pretty much impossible to compete with commercial releases without the use of mix automation these days, so it should be considered a prerequisite. "Fader moves are even more important than EQ," says Chris Lord-Alge. "No matter what song you have, you need to help it build."[33] By using mults and automation, you can slowly assemble the overall production piece by piece, and as long as you pay enough attention to the order in which you tackle the sections, your long-term mix dynamics should make fairly good sense once all the gaps are filled. If the arrangement isn't particularly inspired, however, then you may need to reinforce the long-term dynamics, in which case here are a few tips to get you started.

One of the most hackneyed techniques (but no less useful) is to contrast the levels of reverb/delay between the verses and the choruses. Mark Endert says, "You can get a lot of contour out of a track that lacks dynamics just by changing the acoustic space around it."[34] The most common configuration is for the verses to be drier and closer, where the choruses are bigger and more live sounding. "The reverb that would sound plenty wet in a sparse part of the track will be too dry as the track builds dynamically," says Mick Guzauski. "So, often, I'll ride the reverb and delay as the track builds."[35] So when that first chorus hits, push up the room mics, or fade up the return from the parallel compressor on the overheads (like Tom Elmhirst did on Adele's "Rolling In The Deep"[36]), increase some of the sends to the size reverb, or switch on an epic tempo delay. When you reach the final chorus, you might go the whole

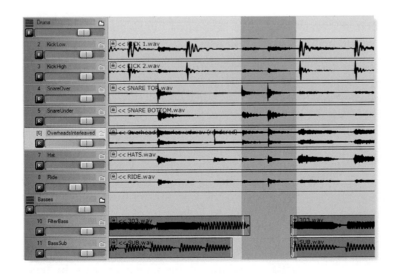

FIGURE 19.7
If you're using synth pads of any kind in your mix, you'll probably need to automate their levels so that they adequately fill out the lush mix moments, but don't also become too audible in sparser sections of the arrangement.

hog and do all of those things at once! And when you've got chorus 1 really opening out, you then have the option to catch the listener's attention at the start of verse 2 by abruptly cutting the reverb and delay tails dead just before the lead singer starts their first line. Bear in mind, too, that it often makes sense for verse 2 to be slightly less dry than verse 1, given that chorus 1 has by that point introduced the idea of a more reverberant sound.

Not everyone's a fan of the "more reverb in the chorus" stereotype, though. "I've found the opposite works so much better," says Tim Palmer. "In the verses, where it's open, that's where you use your room mics, and you shut them off completely in the choruses, because you want all that space that's taken up by room mics [to belong to] the guitar, with all that energy right into the front of the speakers."[37] Some pop and electronica productions also make a virtue of super-dry chorus textures (try Taylor Swift's "Look What You Made Me Do," for instance) so that the main hooks catch the ear by suddenly leaping forward in the depth perspective.

Also firmly in the "cliché for a reason" category is a similar approach toward stereo width. Widening out your choruses, or perhaps just packing their stereo picture more densely, is a time-honored crowd-pleaser. "When it hits the chorus I'll punch wide on the acoustic guitars," says Adam Moseley, for instance, "and suddenly . . . your stereo image has gone 'wow!'"[38] This is where having stereo enhancement send effects can really pay off, because you can easily ride their levels as in the previous reverb situation. Introducing opposition-panned "fake" double-tracks, Haas delays, or wide stereo pads only for the choruses would be alternative options to weigh up. Opening up the overall spectral bandwidth for important sections is another popular ruse. "Like low- and high-pass filters in the verses on certain keyboards," explains Manny Marroquin, "so that they would be fuller in the choruses."[39]

And, if all else fails, there's also the option of simply riding the level of your whole mix, either before or after your master-buss processing, to emphasize the section dynamics. "I'm very concerned with preserving the dynamics of each mix," says Andy Wallace, "as it is part of a greater picture that I try to paint. Riding the master fader during the mix is part of my emphasis on the sonic architecture of the mix and of maximising the feel that I get from it."[40] Of course, this kind of stunt may also affect the response of down-the-line master-buss or mastering processors, so keep your ears open for knock-on sonic consequences.

If one of the instruments in a packed-out chorus arrangement also appears in less crowded conditions elsewhere, don't forget to reassess its high-pass filter and EQ settings within the new context, particularly when the instrument in question is something acoustic. Piano and acoustic-guitar parts are regular cases in point here. They often require fairly severe low-frequency cuts in the context of a busy mix, but that same EQ will make the instrument sound like a tap-dancing vole if it ever moves further into the limelight. In a similar vein, Mike Shipley routinely automates compressor parameters in response to the needs of the musical arrangement. "This is very handy," he explains, "because you may need one section to be as dynamic as possible and another section fairly compressed."[41]

Pads regularly require automation adjustments if you want them to consistently enhance the mix without anyone noticing. A pad that plays across several different musical sections will almost certainly need at least level automation to retain the perfect enhancement/inaudibility equilibrium, although you may also find that the EQ has to change too if the synth is masked much more heavily in some arrangement textures than in others.

Automation can be a great boon when you've got to create any kind of buildup, whether that's a short ramp-up into a chorus or the slow development of an extended outro workout. There are two important points to realize when mixing buildups. First, the peak of any crescendo must inevitably be fixed to fit in with the production's long-term dynamics, so the only way to create an impressive buildup is to drop the overall excitement level as far down as you can in anticipation—the less you drop the energy levels, the less room you leave yourself for maneuver. The second crucial point is that you should try to rely as little as possible on sheer level increases to implement any buildup, because these won't survive downstream loudness processing particularly well. It's much better to intensify factors such as arrangement complexity, instrument timbre, effects levels, and stereo width to create the majority of the impression, as these illusions will tend to survive better than level increases in real-world listening situations.

19.5 DETAILED RIDES

If you mix each section of your production according to the principles I've discussed so far, you should eventually end up with a complete mix that balances sensibly throughout, while at the same time appearing to progress in a logical and musical way from start to finish. However, it will still need one final ingredient to bring it up to a professional level: detailed rides. These short-term automation-data wiggles are what take a mix beyond what plug-ins can achieve, adding in the essential human element of intelligent, dynamic musicality.

Although I briefly mentioned the idea of detailed rides when discussing lead vocals back in Chapter 9, I've otherwise deliberately glossed over their role until now because they're something of a two-edged sword. On the one hand, automation is the most powerful and musical mix process you can possibly apply, and it is capable of matching and exceeding the capabilities of even the most advanced dynamics processors. An automated fader will produce more natural results than any compressor, for example, because it's got a human being acting as its level-detection side chain and gain-reduction circuit. An automated EQ will remove sibilance more sensitively than any automatic de-esser, because your ears will be much better than any DSP code at judging the exact nature of the frequency-selective gain reduction required in each instance. An automated reverb send can respond to arrangement changes in a way that no plug-in can manage.

> An automated fader will produce more natural results than any compressor, because it's got a human being acting as its level-detection side chain and gain-reduction circuit.

But on the other hand, detailed rides are labor intensive too, and it can be difficult to implement broad-brush changes to automation data after it's been created. It might take you a couple of hours to compress a single bass line using automation, whereas compression will do most of the job for you in minutes and will let you read-just the attack time easily at a later date. There's no sense in de-essing everything manually if an automated de-esser will do the bulk of the work for you in a fraction of the time. If you can achieve useful improvements to your reverb levels by using a keyed ducker in the reverb return, it'll save you ages automating a bunch of sends. Although there's nothing stopping you using automation to construct a perfectly good mix without the help of a single dynamics plug-in, you'd have be an utter fruitcake to work that way voluntarily.

The upshot is that it's not usually a great idea to get rides involved too early in the mixing process if you can avoid it. The closer you can get to a great mix without them, the less time you'll have to spend fiddling with little graphical curves in your DAW system, and the more productive your limited mixing time is likely to be. Reserve rides only for those situations where multing and dynamics processing fall short, then you shouldn't waste too much time. As such, rides usually serve one of three main functions: they troubleshoot mix problems that mix processors are unable to isolate effectively, they improve the subjective mix balance beyond what processing can achieve, and they direct the listener's attention toward interesting arrangement features via momentary balance changes. Let's look at each of these roles in turn.

19.5.1 Intelligent Troubleshooting

Time-domain processing is all about prediction—the mix engineer's job is to listen to the moment-by-moment variations in a track, build a kind of generalized mental picture of what's going on, and then forecast the kind of gain-change processing that might be able to fit all those moments into the mix. Unfortunately, though, music recordings aren't always predictable. For example, your acoustic guitarist might normally be very good at keeping any fret noise under control, but just before the second chorus he accidentally plants a real humdinger. Dynamics processing won't supply the answer here, because the kind of drastic processing you'd need to target that lone fret squeak would also mash up the rest of the performance. Assuming that there's no option to patch over that transition with audio copied from elsewhere in the song, you could mult this squeak to a separate track to sort out the problem, but it's often more elegant to deal with this type of isolated blemish by using automation instead.

There are lots of other common applications for these kinds of rides:

- dipping vocal breath noises;
- adding attack to a fluffed guitar note by fading up the initial transient;
- rebalancing individual bass notes. Comments Mike Shipley, "I find that if the player's pickup is substandard, one note will 'out-boom' another, and though I'll first try to fix it with a dynamic EQ, it rarely works. So I'll just automate the track, dig in and find the frequency, and smooth the bass out";[42]
- zapping momentary feedback howlarounds on live recordings;
- catching overprominent "s" or "f" consonants that manage to bypass your de-esser. Steve Hodge states, "I routinely de-ess with automation now by [ducking] the level of the esses down without affecting the rest of the word. It can be that precise."[43] "As soon as you put the de-esser on you're taking away that sparkle that you might need, so why would you do that if you can actually hand-draw it out?" adds Jaycen Joshua;[44]
- reducing the low-frequency thump when a mic stand has been knocked during recording;
- automating a snare mic's gating threshold to achieve reliable triggering despite heavy hi-hat spill;
- compensating for changes to recording settings accidentally made during a take: a performer might have moved position in the height of artistic inspiration, or the engineer might just have nodded off onto some faders!

REGION-SPECIFIC PROCESSING

Alongside multing and automation, most DAWs now offer another option for implementing momentary processing changes: applying plug-in processing to specific audio regions. Although in essence this doesn't really offer any new processing possibilities (in other words, you could usually create an identical end result using either multing or automation instead if you preferred), it does have the advantage of keeping your projects a little more manageable visually, because it doesn't require you to create hordes of new automation curves or audio tracks.

19.5.2 Perfecting the Mix Balance

The problem with most dynamics processing for balance purposes is that it's only concerned with the track it's processing, so it can't deal with problems arising out of momentary masking between different parts in a musical arrangement. The side-chaining strategies I explored in Chapter 14 go some way toward tackling the simplest of these situations, but the masking scenarios in musical compositions are typically too complex to be successfully untangled solely using keyed dynamics plug-ins. Masking is a perceptual phenomenon, which means that the human hearing system is essential for detecting it, so it stands to reason that the only truly successful masking compensation comes from manual adjustments evaluated by ear. Andy Wallace provides one such example: "When there are many guitars in the track that have to work together with the bass, certain notes in the bass track will often stand out more than others. Compression just does not level that out for me . . . Instead, I'll spend a lot of time riding certain notes up that seem to be getting lost, and notes down that are jumping out too much."[45]

These kinds of balance refinements may be subtle, so it makes sense to switch to your Auratone-equivalent while working on them, because that'll cause them to pop out most clearly—they can be very difficult to resolve on stereo nearfields. Chris Lord-Alge also recommends undertaking such tweaks only when listening at low volume: "You can hear better what's going on. When you turn things up, after a while all your moves become a smear. So when you're doing really critical moves, do them at low level."[46] "In my experience, listening at low volume is the best way to fine-tune my balances," agrees Andy Wallace, "making sure I can hear that one small guitar part that plays an accent in one place, and that all the details of the reverbs work well."[47] Choose an important instrument and then listen to it for the duration of your mix. If it seems to duck down unduly at any point, edge it up in the mix using automation; if it draws too much attention to itself, then pull it down a fraction. In this kind of application a quarter of a decibel can make a significant difference, although it's also not unusual to see momentary rides of 6dB or more—for example, if there's a little low-level detail in a note's sustain tail that's disappearing from view.

One of the established masters of this type of automation is hit mixer Andy Wallace, and here's Jim Abbiss describing being a fly on the wall during one of his mixes: "He was constantly adjusting instruments that were playing with the vocals, so you could hear the vocal, and when the vocal stops you can hear the instruments—obvious things which people do, but he did them quite extremely. It meant that he did not need to effect things as much or EQ things as much, because he made space with his balance."[48] The man himself explains the impetus behind his riding of drum overheads: "I ride them a lot of times because sometimes a cymbal won't be as loud as another cymbal. . . . But, also . . . every cymbal crash will be ridden up maybe 5dB or more. Sometimes I will feel that I'm hearing more ambient stuff in the overheads than I want to hear in the mix. [When] the overheads [are] balanced where I want to hear the ambience . . . sometimes the cymbals won't be loud enough to have the impact that I want."[49] This is where it really pays off if your faders are set in their area of best control resolution, as I recommended in Section 8.2.3. If they're right down at the bottom of their travel, you'll struggle to achieve any kind of subtlety with your automation, because you'll only be able to adjust the gain in steps of one or more decibels.

And, of course, such balancing tactics may also extend to automating the parameters of processing plug-ins too. Here's Jason Suecof, for instance, talking about a perennial mixing problem in metal styles: "If your kick's just going "bu-da-bu-da-bu-da" and then starts going "dudldudldudldudl," the bass guitar is going to get crushed in two seconds, so you can either bring your kick drum 1.5db to 2dB down . . . but it's also cool to put on [an EQ plug-in] and roll up to like 44Hz and automate that so that the low, low end of the kick drum doesn't squash the bass . . . Sometimes you've got to do both, sometimes you can get away with just one."[50]

But that's only scratching the surface of what some people get up to. Rich Costey recalls, for example, the lengths to which über-producer Mutt Lange went on Muse's album *Drones*: "It seemed like he's not a fan of compression to restrain tones. Instead what he had done was apply volume automation and automated reductive EQ on almost everything—and I mean every single snare hit, piano chord, and bass note, and so on. You could see the EQ work, just for an instant, on every snare hit, at 1kHz or sometimes 700Hz, the moment the stick hit, so that the decay would sound as fat as possible. The depth he went into with the bass guitar was incredible. The first mix we did was 'Mercy', and Mutt had not only EQ-automated every eighth note, but had also volume-automated in between each eighth note to remove the sound of the fingers on the strings. As a mixer I've seen many people's sessions, but I'd never seen anything like it."[51]

It can take a little while to sensitize your ear to subtle balance changes, so don't abandon the listening exercise straight away. Give yourself a few minutes practice to really get into the zone. "The more you listen, the more you hear," says Steve Parr.[52] Nowhere is this more important than with lead vocal parts, of course, so you should listen particularly carefully there. (For more specialist advice on lead vocals, see Section 19.5.4.) What can sometimes help, especially when listening on an Auratone substitute, is to think in terms of distance rather than level. If the bass takes a step toward you or away from you from time to time, then it's a clue that rides might be called for. I also find that keeping my monitor level constant for this stage helps make subtle balance shifts easier to detect, because that way you can build up a kind of short-term "memory" for what the target level/distance should feel like, which makes the work quicker and more instinctive.

FIXING BAD COMPRESSION WITH AUTOMATION

Detailed automation provides a solution to one of the trickiest mix-salvage jobs. Tom Lord-Alge explains: "Often I want to uncompress things that are on the tape, when they are overcompressed or compressed in a nonmusical way. But I fix that by compressing even more and then creating new dynamics by using the faders. Automation comes in handy for that."[53] Dave Pensado has his own slant on this tactic for vocals: "I go into Pro Tools and I type in the level on every syllable . . . and I automate the EQ on every syllable . . . On a four-minute song, to fix a lead vocal with just the most horrible compression takes me about two hours."[54]

19.5.3 Directing the Listener's Attention

In any mix, the listener's ear is constantly "refocusing" on different parts as they all jockey for attention. To make a visual analogy, it's a bit like watching a soccer or basketball game on television. While you're focusing on the player in possession, the other players are only in your peripheral vision, and hence are seen in less detail. When the ball's passed to another player, you shift your focus to him instead, and it's then the first player's turn to be in your peripheral vision. The musical listening experience works a bit like this (although hopefully with less dribbling), in that most people will focus most of their attention on only one thing at a time. Your business as the mix engineer is to direct the listener's attention toward what is most important in the track at any given instant. Although the rhythm guitar part might not be nearly as important as the lead vocal, a little fill from that guitar might be much more interesting to listen to than the sound of the singer taking a breath. Fade up just that tiny moment, and it'll demand its own little slice of attention before the vocals return to take pride of place. Gus Dudgeon says, "When you're doing a mix there are slots that appear where you can crank something just enough to help it through so it still makes its point, but isn't blowing your head off and hasn't got lost."[55] "I apply quite a bit of volume automation to sections and individual parts in places where I want to emphasise or de-emphasise them," adds Kevin Savigar, "because I like the song to breathe as it goes from beginning to end."[56]

Doing a series of rides like this has several potential benefits. First, the production as a whole will usually appear more involving as a result—in other words, the listener's interest doesn't have as much chance to wane during the lull in vocal activities in our example. "Not every singer has perfect emotion all of the time," says Jon Gass. "So if you can build the mix and craft the textures around the vocal, that's what it's about. It helps sales."[57] Phillipe Zdar feels it can also breathe life into repetitive programmed parts. "I use it to make the music come alive. If I have a drum track that's a loop or that comes from an MPC or SP12 drum machine, it will never sound like a real drummer. So I do lots of volume rides to add a live feeling."[58]

Second, the main parts in the mix will be appreciated more deeply because the interludes continually cause you to refocus on them—so you'll concentrate better on the postfill vocal phrase in our example by virtue of coming back to it afresh. These kinds of balance adjustments are, of course, completely unavailable from any kind of mix processor, because they're based on musical judgments rather than on anything a DSP chip can measure, so automation is the only way to bring them about.

My strategy is usually to listen to the song from top to tail, again on my Auratone substitute, and try to think in terms of maintaining the unbroken attention of the listener throughout the time-line, much as Matt Serletic describes here: "A great record should be like you're pulling a string toward you constantly, and you never let up: there's a constant tension that maintains your interest. That can be the vocal dipping into the way the guitar line leads into the prechorus; there's always this constantly connected thread. Especially when I'm doing a final mix, I'm always looking for that point where the string breaks. Then I say 'OK, we've got to fix that.' If I'm not believing it past this point, I've lost it, it's not right."[59]

> Don't forget the possibility of riding up reverb or delay tails, because those can be surprisingly arresting in the right circumstances—and they're also something you can add at mixdown if there's nothing else happening in the arrangement!

In a typical song structure, the lead vocal might be responsible for that thread a lot of the time, but wherever this vocal becomes momentarily less interesting (perhaps between phrases, during long note sustains, or after a few repetitions of any simple riff), I'll hunt around among the other arrangement parts to see if there's anything that might provide a welcome diversion. It's in this context that I'll often call on any buried gems I may have turned up during Chapter 5's initial reconnaissance of the multitrack. If they don't appear at a suitable point in the arrangement, I'll consider moving them somewhere else where they're better suited to bridging a gap in the thread of attention. Don't forget the possibility of riding up reverb or delay tails either, because those can be surprisingly arresting in the right circumstances—and they're also something you can add at mixdown if there really is nothing else happening in the arrangement at a given moment!

What's interesting about tinkering with the listener's attention like this is that you can often get away with a surprisingly skewed balance without raising eyebrows, as long as that balance only lasts for a moment and normal service resumes fairly quickly thereafter. This gives you a way to raise the audience's awareness of background parts in your mix, even when they normally have to be balanced quite low. You also have the scope to clearly indicate to the listener that a new instrument has entered the arrangement, even when there's only actually room for the merest suggestion of it in the balance during the remainder of the timeline. If you're relying on the entry of a new part to help differentiate a given section as a way of improving the long-term mix dynamics, then that little dodge can be worth its weight in gold. Another time-honored trick here is to push up the first couple of syllables of a lead vocal part a bit too loud to draw the ear—by the time the listener has had time to think "blimey that vocal's loud!" the vocal has already returned to a less controversial level. Besides, in the words of Steven Tyler (recalls Warren Huart): "If it's worth hearing, it's worth hearing too loud!"[60]

Although little fader pushes can really work wonders on a mix, even to the extent of making the music itself appear more engaging, there will occasionally be times when you find you can't make a big enough statement in this way without making a complete mess of the balance. The way to get better results from your rides on these occasions is to simultaneously dip some of the competing background parts. This will decrease masking of the part you're riding up in the mix, while the unchanged levels of more important subsidiary parts in the mix will normally prevent the balance change feeling subjectively awkward. So let's say you really wanted to hear the details of that guitar fill in our vocal/guitar example, but pushing it up too far in the balance felt odd. You could draw more attention to it by pulling back the faders of the other background parts in the mix at that moment (say piano, synth pad, and backing vocal lines) to clear more space for the fill's qualities to shine through at a lower level. The beauty of doing it like this is that as long as the vocal, drums, and bass remain at their usual levels, the listeners probably won't even be conscious of what you've done at all, because their attention is targeted at the vocal and guitar lines. In some such cases, you may find that you can completely mute some background parts without anyone being the wiser—all people will notice is that the instruments they're supposed to be listening to sound clearer. Result!

19.5.4 Vocal Rides

"With the vocals you're chasing the faders to get them really in your face," says Chris Lord-Alge. "It's all about automation."[61] The amount of time top engineers spend riding lead vocal levels reflects the central importance of singing to most commercial music styles, and it's by no means uncommon for more automation to be applied to lead vocals than to everything else in the track put together—indeed, the data will often look messier than a graph of the Dow Jones. The primary reason is that most listeners remember lyrics and melodies more strongly than anything else, so making sure that both of these elements come through as strongly as possible is fundamental to a track's commercial prospects. "The vocal is probably the most important part of the mix," confirms Tony Visconti. "If the vocal is poorly placed in the mix, it's going to defeat the purpose, it's not going to sell the music. People will remember vocals more than they'll remember the guitar licks. . . . Don't ever be afraid of putting the vocal too high in the mix. . . . What you should do is ask someone not involved in the production if they can hear every word. That's really the acid test."[62]

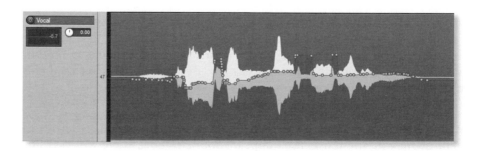

FIGURE 19.8
Pushing up the levels of weaker syllables, consonants, and vowel transitions with detailed fader rides can dramatically improve lyric intelligibility.

Although maximizing the audibility of lyrics with automation isn't very different from perfecting the balance of any other instrument in a mix, the level of detail required can initially take some getting used to. It's a world where the level of each note, each consonant, each vowel, and each inflection is carefully balanced to create the most solid subjective vocal level possible in the mix. In high-level commercial productions, this activity alone might easily take up several hours, and Bob Clearmountain has even mentioned how Mutt Lange would spend all day just riding one vocal part.[63] So if you think you're done in ten minutes, think again.

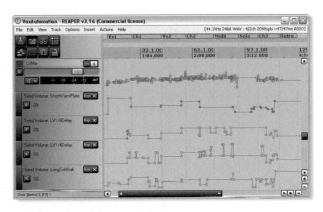

FIGURE 19.9
The vocal level and effect-send automation from a fairly typical Mix Rescue remix. It may look like a lot of data, but that's what it takes to compete in most mainstream musical styles.

Here are a few pieces of advice on automating for better vocal intelligibility. Dull-sounding consonants often benefit from being faded up quite a bit, because their tone by nature cuts through the mix less. I'm thinking of consonant sounds such as "n," "ng," "m," and "l" in particular here, but it depends a bit on the specific singer whether other consonants might also come across unclearly. You can also imply that those consonants are louder than they are by emphasizing the transitions to and from the surrounding vowels. In a similar fashion, consonant sounds "w" and "y" can benefit from an emphasis of their characteristic fast diphthong transition.

But vocal rides aren't just about intelligibility; they're also about maximizing emotional impact, and this is where automating vocals becomes more of an art. All singers have certain vocal characteristics that help to express their emotions, so if you can find and emphasize these moments, you can increase the power of their performance. "It's a great technique," says John Leckie, "because all the little secrets get revealed. At the ends of lines, a lot of singers will trail off, and if you lift the fader 10dB right at the end of the line, there's lots of things you haven't heard before . . . suddenly there are new things happening in the song."[64] As Leckie suggests, many of the most characterful aspects of vocal performances are to be found in the spaces around the main notes: the little moment just before the vocal properly revs up at the start of a note, the tiny crack in the tone as one pitch changes to another, an expressively extended consonant, a sassy little pitch glide, the unfocused noisy onset of a high-register wail, or the hint of extra breathiness and fall-off as a vulnerable note dies away. Often these kinds of details are hidden within the mix, but if you can unmask them, then you can significantly elevate the power of the vocal performance—and by association the perceived improvement that your mix has made to the overall production.

The other area where vocal rides can actually appear to improve the vocal performance is in rhythmic styles where the lead vocalist is aiming to deliver

HARDWARE FADER CONTROLLERS

Although different DAW systems have slightly different ways of implementing fader automation, they usually offer two basic methods of data entry: either you can use your keyboard and mouse to draw in automation curves, or you can record and modify the level rides in real time using a dedicated hardware controller equipped with touch-sensitive motorized faders. Technically speaking, though, there's nothing you can't achieve with the former method, and there are plenty of world-class engineers who work exclusively in that way. "I do all this with mouse and keyboard now," says Rich Costey, for instance. "I find that I've quite naturally moved into working with mouse and keyboard, and I've stopped thinking about it. I miss the fun of riding a fader, but I am so busy listening that I don't really care what I do with my hands."[69]

However, there are also numerous top-name engineers who continue to extol the virtues of hardware faders. "You can do everything electronically," says Phil Ramone, "but for me it takes the spirit out of it when I don't make the moves with my hands."[70] "When I am working with faders I feel like I am touching the music and am part of it," adds Derek Ali. "I don't like looking at a screen for hours. It makes me feel like I am not free. I want to feel free when I am working, I want to be like an artist in a booth who can move his hands and feel free and express himself."[71]

Steve Fitzmaurice stresses the advantages of being able to work on a mix without looking at the screen. "I don't sit there looking at things and going, 'I need to turn the drums 2dB up.' Instead, I just move the faders until it sounds right. Sometimes you look down and you think, 'Bloody hell, I just turned that up by 5dB!' By contrast, like many people, I find myself turning things up 0.3dB in the box, and then I remember what a waste of time that is, because no one is ever going to hear that. You seem to mix far more conservatively when you are looking at things than when you are just listening to them."[72] "Honestly, when you are looking at a screen, you are looking at numbers," continues Derek Ali. "You can turn things up or down a specific amount of decibels, or tune this or that frequency. But how useful is that? . . . It is just a number. Instead you have to train your ear, you have to learn to notice the different frequencies and sounds, and then let your own taste decide."[73] "Sometimes I switch the screen off," says Tom Elmhirst. "You have to forget the screen, because you end up thinking about the music, rather than feeling it."[74] "I am not looking at waveforms either," adds Rich Costey, "so I can focus on what I am hearing."[75]

However, I think it's unwise getting too dogmatic about it either way, because each of the two automation approaches actually suits different tasks. "I do rides in Pro Tools for very detailed work," says Mike Shipley, for example, "and the more intuitive fader rides on the board."[76] There's little benefit in trying to duck individual fret squeaks from a hardware controller, because even The Fastest Fader In The West is likely to be a bit haphazard with such fleeting nuisances. You'll get a much cleaner fix by just clicking in three or four perfectly placed automation points with the mouse. But if you're trying to automate a pop vocal to compete against the stiffest commercial competition, that might require literally hundreds of automation data points if you use the mouse, in which case working with a hardware fader instead may offer a substantial time saving—as well as potentially reducing the risk of repetitive strain injury! Certainly, the simple single-fader controller I bought a few years ago effectively paid for itself in this respect during its very first album-mixing project.

As the size and complexity of any prospective hardware controller increases, though, the question of whether it's good value for money becomes more complicated. How much more time would eight hardware faders (or indeed 48) save you compared with a single fader? Would hardware transport controls make your session navigation appreciably quicker? And could you work significantly faster by adjusting plug-in parameters from little illuminated rotary knobs? To be honest, if mixing is the only concern, then I think the "time is money" argument becomes pretty weak after the first fader, simply because it's the listening and thinking that takes most of the time at mixdown, not the actual manipulation of faders and plug-in parameters, so I don't think using the mouse presents that much of a workflow bottleneck. That's not to say, though, that other factors might not swing the argument, of course. Instant access to multiple faders and mute/solo buttons becomes much more valuable in high-pressure multimic recording sessions where real-time reaction speed is of the essence, for instance, and it's also important not to undervalue the image-boosting cachet of an imposingly space-aged control console if your studio regularly plays host to paying clients.

FIGURE 19.10
A touch-sensitive hardware controller can speed up level-automation tasks considerably, even if it only offers one fader, like the affordable Presonus Faderport.

a fast line with a good deal of punch and aggression—the most common example these days being the rapping in various up-tempo urban and electronica styles. This kind of performance is tremendously difficult to pull off successfully, especially by an untrained vocalist with underdeveloped diction and breath control. The result is that some syllables may come across well, whereas others will feel rather limp. The solution is to go through and automate a little gain pulse at the onset of each underwhelming syllable, reinstating the missing sense of energy and urgency. Tedious? Well, you'll may consider throttling yourself with the mouse cable before you're done, but the alternative is sending your rapper for elocution lessons, which is never going to be pretty . . .

It's worth pointing out that the subjective process of emphasizing those bits of the vocal that you like, and de-emphasizing those you don't, effectively makes your vocal *sound* more appealing too. This may seem like an odd assertion, given that fader automation only affects the signal level, but when you consider that engineers often apply mix processors for exactly the same reasons (i.e., to maximize those characteristics they like and minimize those they don't), this perceptual sleight of hand begins to make a lot more sense. In a similar vein, Steve Fitzmaurice has found that automation obviates the need for a lot of mix processing: "In general I use less and less EQ on lead vocals. I prefer simply turning them up or down . . . The fashion these days seems to be to compress the shit out of a vocal, then EQ it so it sounds really thin, put lots of midrange EQ on it so it cuts through, and then turn it down. I don't like that, and instead I prefer to have a good lead vocal sound on which I do loads of rides."[65]

One final point: vocal automation isn't necessarily just about levels. Many engineers talk about automating de-esser thresholds, for example, while Mike Shipley routinely automates the vocal EQ for his work with Mutt Lange: "With Mutt we always have programmable equalizers where we can EQ every word . . . every consonant of every word if we want—literally, every part of every word."[66] Delbert Bowers mentions automating multiband compressor thresholds: "The thresholds are moving, sometimes per word."[67] Automating effect sends is also commonplace. "I use faders for sends to the effects," remarks Tom Elmhirst, for instance, "especially on the vocal, on which I may have four effect sends . . . and I can ride these and play with them."[68]

19.6 NAILING DOWN THE FINAL VERSION

Once your automation work is done, you'll have arrived at a complete mix draft. Only two things now stand between this and the final mixdown file: another bout of mix referencing, and the process of incorporating any revisions suggested by the artist/client. (If it's your music, of course, then both these tasks may be rolled into one.) As far as the referencing stage goes, you should first of all proceed with the same kinds of A/B comparisons you performed with the first mix section in Section 19.3. However, there's one further referencing activity I'd recommend, which really helps to nail down the final subtle tweaks: building a "snag list." This involves listening to the whole mix on all your different monitors and making notes of any final nips and tucks required. Although that sounds similar to the mix referencing you've already done, it actually works a bit differently, and the devil's in the detail.

19.6.1 Effective Snagging

You should now make a point of listening to your mix (and any reference tracks) from top to tail, in order to cement your decisions regarding the long-term mix dynamics—a lot of your mixing time will inevitably have been spent playing back small snippets, so a good few complete play-throughs at this stage acts as a valuable counterweight. "My usual rule," says Andy Bradfield, "is that if I play a track three times in a row and something consistently bugs me, then I'm going to have to do something about it. That's how I know that something isn't right—my ear starts complaining and my brain keeps bringing it up."[77] Bob Clearmountain imparts some further words of wisdom: "Make sure you're listening to the whole mix, not just the vocal, the guitar, or the snare drum. . . . You have to train yourself to listen to the song and not the individual parts, and then things that are wrong will jump out at you—it'll be obvious that there's a part in the second verse that's pulling your ear away from the vocal when the lyric at that point is really important to the song."[78]

There's no better snagging time than first thing in the morning after a good night's sleep. "Generally, my process is to start work on a song and get it to sound great during the first day, leave it overnight, [and] listen with fresh ears the next morning," comments Randy Staub.[79] Keeping your concentration levels up throughout even a single complete mix playback is demanding, so it's vital that your ears and brain are as fresh as possible before you start. "The hardest thing is trying to stay fresh, to stay objective," says Nigel Godrich. "It's so hard to let it all go and become a listener."[80] I'd also encourage you not to play the mix from your DAW system, either, but instead to burn a CD (or export the mixdown file to your normal mobile music player) so that you can listen to it in exactly the same way you listen to your own record collection. This helps you trick yourself psychologically into listening more as a consumer than as a mix engineer, which is a remarkably powerful way of highlighting when a mix is truly fit for general consumption. Within this context, a couple of playbacks on your favorite day-to-day domestic systems can be a good idea. Although your hi-fi, car stereo, or earbuds won't actually provide you with much extra information about the technicalities of your mix if you've used your monitoring systems wisely thus far, they can make it easier to appreciate your production from Joe Public's perspective.

An additional benefit of playing from a CD or bounced file is that you're less likely to be distracted by visual feedback from your DAW's meters, waveform displays, and arrangement-window layout. "It's real easy to be objective when you're not staring at the speakers and looking at the meters," says Ed Seay.[81] If you can see your software's play cursor advancing to a busier-looking section of the onscreen arrangement labeled "chorus," you'll instinctively be inclined to think that this section sounds bigger when it arrives, even if it's actually falling a bit flat. Furthermore, because you're preventing yourself tweaking the DAW's mix settings while listening, you'll typically be much more disciplined in basing your decisions on the consensus of all your different monitoring systems—it's all too easy to start fiddling with mix settings simply to compensate for the vagaries of a particular monitoring system, when there's nothing actually wrong with the mix. Gareth Jones recalls making this mistake with a Depeche Mode mix early in his career: "Since the record company was so concerned about getting the song on the radio we spent a lot of time listening and mixing on a two- or three-inch speaker. . . . We probably spent too much time doing this, because while it sounded absolutely amazing on a small system, on bigger systems I can hear all the faults. You know, 'Oh God, I could have got more bass on it,' or 'It would have been great if the high end could have been a bit smoother.' Obvious things. If I'd been a more experienced engineer, I would have . . . also paid a bit more attention to the big monitors."[82] In my experience, there's little danger of falling into this kind of trap if you make a point of whittling down your list of snags on paper first, leaving only those that are truly justifiable across all your monitoring systems.

A snag list helps you wrap up a mix much more efficiently and decisively, because you always have a definite target in your sights while working and it's very difficult for niggly mix mistakes to slip through the net. When you've ticked off all the snags on your list, and referenced again to check they've been properly dealt with, then you'll be much more confident in consigning your mix to the out tray. If you still worry that your mix is unfinished after all that, then perhaps it's time to just resign yourself to it. Phil Tan notes, "If you were to give me the same song on different days, it would probably come out differently. There's no such thing as perfection, and I think music is more fun when things are a little inexact."[83] "You can always change a mix and not make it worse, but do the changes improve it?" asks Andy Wallace. "In my experience, a mix rarely gets better with endless changes and recalls. For me a mix is about trying to find something that works and makes the hairs on the back of my neck stand up, and believing in that. If you are rethinking and second-guessing that all the time, you risk losing that feeling."[84]

19.6.2 Revision Requests

Of course, if you're mixing someone else's music, then it's neither here nor there whether you're personally happy with the mix, because it's the client who decides what's finished and what isn't. No matter how good an engineer you are, there's simply no getting away from the fact that a lot of your mix decisions will effectively be guesswork—calculated guesses, admittedly, but guesses nonetheless. Without being Marvo the Incredible Mind Reader, you can't know exactly how the artist, manager, or A&R rep want the record to sound, even if you've taken the wise precaution of asking them for some representative reference tracks for guidance. Therefore, even at the highest levels of the industry, mix engineers expect (and plan for) the eventuality that they may need to revise their first submitted mix in response to suggestions from the client. "I did 42 mixes of this song. Forty-two!" recalls Carlo "Illangelo" Montagnese of his work on The Weeknd's "The Hills." "All of them were about finding the perfect balance."[85] Bruce Swedien can trump that, though: "I did 91 mixes of [Michael Jackson's] 'Billie Jean', and finally Quincy [Jones] said 'Let's go back and listen to mix number two.' And we did, and it blew us away! I had overmixed that song right into the pooper, so the mix that went onto the record was mix number two."[86] As you might imagine, this process can be inconvenient or vexing at times, but it's inevitable, so there's no sense in getting upset about it (grrr!), even if you find you have to go right back to the drawing board.

Mixing isn't just a technical exercise; it's a process of discovering what the client's looking for. "It is during mixing that the tracks really come alive," says Jesse E. String, "and at that point you gain new perspectives and you often want to change things."[87] Sometimes you may also need to try a number of blind alleys before you can find exactly the right direction. Marcella Araica, like most high-profile mix engineers, has been there and bought the T-shirt: "When you're recording, people are usually satisfied just hearing the sound come back from the monitor speakers, but when it comes to the mix it's suddenly the real deal . . . you get a lot of cooks in the kitchen, which can be frustrating. But it's part of the whole mix thing. I find that I can learn a lot when people have comments. They may be hearing something that I don't, and even when the critique makes no sense, it may lead me to bring something out that I hadn't heard before."[88] Daniel Lanois picks up on the same theme: "Just go with the other person's idea. The thing about ideas is, if you don't chase them up and see them through to their conclusion, you may have frustration living in a corner of the room. And I've never had a disagreement regarding result. Somehow it always works out that everybody agrees on the approach that sounds best."[89]

> There's simply no getting away from the fact that a lot of your mix decisions will effectively be guesswork—calculated guesses, but guesses nonetheless. Without being Marvo the Incredible Mind Reader, you can't know exactly how the artist, manager, or A&R rep want the record to sound.

Making the process of revisions as painless as possible is as much psychology as anything, so it will always depend heavily on the kinds of people you're dealing with. However, Kyle Lehning takes an approach that has proved its worth for me on many occasions: "I tell the artist very early on in the process that there will never be a tie. 'Whatever we find ourselves disagreeing on, you win!' But because there won't be a tie, I have the freedom to tell you exactly what I think. At the end of the day, it's your career, so it will be your responsibility to decide whether what I have to say is helpful or not. . . . Sometimes people are afraid of losing their job or offending the artist, but the more successful an artist gets, the more important it is for them to find somebody who will actually tell them what they really think."[90]

Beyond that general advice, though, I do have a few further suggestions that have helped me navigate this particular minefield in the past. The main one is to ask your clients to describe the changes they're after with relation to other productions (as suggested in Section 4.1.4). So if the vocal reverb isn't right, ask them to play you something more suitable from their record collection. "If you're a producer and you say 'I want this record to sound like this, this, and this' . . . it's like a cheat sheet," says Leslie Brathwaite.[91] This strategy heads off two frequent problems: the first is that opinions may differ regarding what a "silky vocal reverb" is, and the second is that what they call "reverb" might actually be delay, chorusing, double-tracking, or any number of other things. To misquote a well-known adage: an audio example is worth a thousand words! Also, if you're working for a band (or indeed any group of clients) and you value your sanity, do your best to nominate a single person as mix revisions spokesperson, their job being to collate everyone's alteration requests and then hammer out a unanimous agreement among all parties as to which of them to implement. Failing that, try to make sure you always copy all revision requests to everyone, to avoid suspicions that things are being stitched up in any underhand manner. Interpersonal tensions can run pretty high towards the end of a band project, and the atmosphere can quickly grow prickly if you find yourself trying to mediate conflicting revision requests from bandmembers on the war path.

With the best will in the world, though, tempers do sometimes get frayed, and it's when goodwill evaporates that it's useful to have some written record, however informal, that sets some boundaries in case of dispute. Some mixers codify this into a formal legal contract, but most small-studio owners shouldn't need anything that formal. "I have an email that I send to potential clients," says Justin Meldal-Johnson, for instance, "that outlines how I like things to go. It says I'd like a list of all the people involved in the decision-making process on the mixes; how I like the files delivered; how many recalls; how much it costs when it goes over. All that sort of stuff."[92]

> Ask your clients to describe the changes they're after with relation to other productions. So if the vocal reverb isn't right, ask them to play you something more suitable from their record collection. An audio example is worth a thousand words!

Fundamentally, though Mark "Spike" Stent speaks for most high-profile mix engineers when he remarks, "I always say to every artist . . . it's your record, not mine."[93] "You might put all your creative effort as an individual into a mix," elaborates Miles Walker, "and you might try all these different ideas that *you* think are amazing, but if the client doesn't like it, you have to be able to get away from that, because you're getting paid to drive the bus: sometimes you've just got to drive them where they want to go."[94] "I can't assume that my sensibility is the right sensibility," adds Justin Niebank. "That's where going back and listening to rough mixes can be especially useful—they can tell you what people were thinking."[95] In some cases, you can find that a specific effect is vital to the client's vision of the track, in which case the quickest results are often to be had simply by bouncing that effect from the rough mix or finding out the exact settings so that you can go about replicating it as closely as possible.

My final recommendation when dealing with client revisions is to keep backups of the project and mixdown files for each stage of revisions, because it's not uncommon for people to backtrack on a revision idea once they actually hear what it does to the mix in practice. "[I] save every two minutes," says Mark Lawson, "often doing a 'Save As,' so I end up with hundreds of different versions of each session."[96]And it should go without saying that keeping backups of all your work in general is essential. My maxim is this: files do not really exist unless they're saved in two different places at once, and they're only properly backed up when I can't stupidly delete them by mistake!

19.7 MASTERING

One of the questions small-studio mix engineers most often ask me is "do I need to get my mixes mastered?" I usually answer with a question of my own: "Why do *you* feel your mix needs mastering?"

19.7.1 Why Do You Need Mastering?

If the answer is that your mix isn't yet quite polished, warm, detailed, clear, exciting, or generally star-spangled enough to feel properly "finished" (even when loudness-matched against its commercial competitors), then my advice is to forget about mastering and go back to the mix instead. One of the biggest misconceptions among small-studio engineers is that the purpose of mastering is to make your mix sound finished. Yes, a mastering engineer may indeed be able to make your mix sound a fraction better, but if you'd be unwilling to release the mix with nothing more than loudness processing, then I would respectfully suggest that your mixing work simply isn't complete yet. Or let me put it another way: for years, small-studio engineers have regularly been sending me mixes they're dissatisfied with, and asking my advice about whether mastering's the remedy they need. And it *never* is—it's always the mix that's the real problem! In many of those cases I've even remixed their track for them (courtesy of *Sound on Sound* magazine's "Mix Rescue" column) to prove the point. So you'll have to forgive my taking a hard line on this: if you're hoping that mastering will finish your mix for you, it almost certainly won't. By the same token, although a bona fide mastering engineer can usually achieve more transparent loudness enhancement than a non-specialist mix engineer, again the difference is extremely unlikely to make or break the commercial potential of your work once you've spent a couple of hours experimenting with the techniques laid out in Section 19.2.2.

Where a mastering engineer really begins to add value, however, is in providing an additional set of (hopefully!) experienced ears with which you can cross-reference your own opinions about the balance, tonality, and dynamic range of the mix, bearing in mind the expectations of your target market. In this respect, mastering can be seen as a natural extension of your mix referencing process. The outcome of this professional collaboration may be that the mastering engineer is able to make all the necessary adjustments entirely with their own dedicated processing tools, but they may also perform the equally valuable service of alerting you to aspects of the sound you could better address by revisiting your mixdown project instead. And it's not a one-way street, either, because your willingness to supply mix revisions and "insider" feedback on draft masters will usually make the mastering engineer's job easier. If the master comes out better as a result, then the client will be happy and you'll both look good, so everyone's a winner!

> For years, small-studio engineers have regularly been sending me mixes they're dissatisfied with, and asking my advice about whether mastering's the remedy they need. And it *never* is—it's always the mix that's the real problem!

Another situation where I think mastering engineers really come into their own is when you step beyond the idea of processing a single mix. For example, processing several different mixes so that they sound like they belong side-by-side on the same album is a fiendishly difficult task, and requires a degree of specialized experience that only years of mastering records full time can bestow. A mastering engineer's knowledge of recording media, playback formats, and the latest mobile technology may also be well worth paying for, particularly in less orthodox situations, such as if you're mastering from an analog tape, restoring a damaged audio file, preparing your mix for pressing to vinyl, or wanting to optimize your sound for specific internet streaming services.

19.7.2 Choosing a Mastering Service

So if you've decided your mixes need mastering, how do you choose a good mastering service? Well, beyond doing some basic web research to find one that can work with the audio and media formats you require at a price you can afford, there are a couple of other baseline requirements. First, you need someone who can do at least as good a job as you can at bringing the mastered loudness to an appropriate level while minimizing negative sonic side effects. And, second, you need someone who will respect your mixing intentions with any further processing they apply. Both of these things are dead easy to check for yourself if the engineer in question will agree to test-master one track for you, something most mastering houses are happy to oblige with. All you have to do is line up the test master against your unmastered mix in a new DAW project, process your unmastered mix to match the subjective loudness of the test master using the best methods you have available, and then ask yourself whether the test master sounds worse. Although this might seem a pretty basic test, I've found it remarkably successful (sometimes depressingly so, in fact) at weeding out unsuitable candidates. Without the loudness bias interfering with the decision-making process, it's usually pretty obvious if the test master's loudness processing is causing too much distortion, or robbing low end, or ducking the drum levels excessively, and I'd expect any conscientious mastering engineer to take account of my preferences if I asked for a different balance of side effects than they initially chose.

In addition, assuming you've been careful about your mix referencing, I'd immediately eliminate any mastering engineer who radically changes any aspect of your mix during test-mastering without first contacting you to discuss why they feel the need to do so. This is what I mean about them respecting the work you've already done at mixdown: while you clearly want to be alerted if there's some glaring problem you've overlooked, it's difficult to develop a good working relationship with anyone who treats your mix as if it's broken, but at the same time implies you're too much of an idiot to help fix it!

Which brings me to the most important consideration, in my opinion: above all, you have to find a mastering engineer you can communicate with freely and easily. You're much more likely to get a great master if you both work as a team, and you can't have good team-work without excellent communication. To do their job, the mastering engineer needs to understand the aims of the project as a whole, factor in your sonic preferences, and engage appropriately with any revision requests—whether that's just interpreting how to implement them or carefully explaining some technical trade-off involved. But at the same time, you also want your colleague to feel comfortable discussing mix revisions and giving their own expert feedback on your work, because those things may massively help to improve the quality of your future productions. "If you want to get the best out of the mastering engineer," says Ian Shephard, "the key is communication."[97]

Given the opinions I've expressed so far, it should probably go without saying that I wouldn't recommend any mastering service that lacks that crucial element of human interaction. You'll find plenty of online mastering services who'll happily charge you to run your mix through their whizzy mastering chain while some remote and inaccessible engineer (or artificial-intelligence algorithm) handles the controls, but unless there's a serious conversation process involved you'll neither get a properly informed mastering job, nor learn much to improve your future mixing work. Let's not be pessimistic, though. When the artificial intelligence gets clever enough to query the validity of one of my references, ask me for a mix version with more snare, or suggest I reduce the subwoofer level for my next project, then maybe I'll change my tune. Although by that point the robots will probably have done me out of a job too . . .

> I wouldn't recommend any mastering service that lacks the crucial element of human interaction. Unless there's a serious conversation process involved you'll neither get a properly-informed mastering job, nor learn much to improve your future mixing work.

MIXING AN ALBUM

If you can mix a single piece of music successfully, then you should be able to mix a whole album without much more difficulty. It's human nature, though, to look for some "economies of scale" by copying settings from one song to the next. If the different tunes were recorded in a similar way, then you can often transplant plug-ins that deal with technical troubleshooting tasks such as reducing pick-noise, hiss, sibilance, or undesirable spill. You may even be able to get away with using similar EQ and compression settings for balance-processing, although you'll usually want to tailor those much more to the needs of each specific arrangement. It can also make a lot of sense to duplicate any general-purpose reverbs/delays and master-buss processing across all the songs on an album, as this helps give the different songs something of a "family sound." Another unifying strategy can be to identify some common arrangement elements that will function as a kind of "sonic anchor" for the production as a whole—things that appear in nearly every song, and which, if given a consistent sound signature, might conceptually glue all the songs together. The lead vocals are an obvious choice in a lot of cases, as they are usually very prominent in the mix, and consistency in this department is often desirable for commercial "brand recognition" purposes. However, any other important instrument might just as readily fulfill this role too: drums and bass for a rock band, perhaps; acoustic guitar or piano for a singer-songwriter; or the kick-drum samples for an EDM or hip-hop act.

But the copying approach can also be taken too far, in my opinion, and I think you'll waste your time if you try to match everything between different songs—it would take you ages and wouldn't make the mixes any better, because you'd be torn between making an instrument sound the same between songs and making it sound the way it needs to in that specific song. So by all means duplicate your plug-in settings for a couple of the most important elements in the arrangement, but then work from there, trusting yourself to "fill in the blanks" on each mix as you would normally do, and adjusting the duplicated plug-in settings as necessary to suit their new context. (Things such as compression and EQ settings tend to be very arrangement-dependent, for instance.) Frankly, though, I often prefer rebuilding all my processing from scratch anyway, rather than copying settings, because it encourages me to experiment with new approaches to similar problems, and frequently yields better solutions. Even if it doesn't, the exercise will still improve my mixing chops for future work.

Finally, it's important to realize that mixing an album is usually something of an iterative process, if only because some late-in-the-day mix decisions are better made by comparing all the different mixes. This is why I always set up a separate DAW project containing all my mix-in-progress bounce-downs, so I can switch between them to check for balance and overall-tonality inconsistencies that would be extremely tough to spot in any other way. Although this kind of work might be seen as trespassing on the realm of the mastering engineer, my thinking is that if my mix sonics require as little remedial mastering work as possible, the mastering engineer can spend more time concentrating on niceties.

19.7.3 Evaluating the Master

If you do get your work mastered professionally, then you will also need to evaluate the results. Despite all the technical expertise required for mastering, it's far from a purely technical task and inevitably involves the same kind of aesthetic guesswork as mixing, so it stands to reason that revision requests should usually be an integral part of the process. For single tracks, nothing beats the loudness-matched comparison process I mentioned a moment ago, especially if you make proper use of all the monitoring systems and techniques laid out in Part 1 of this book. For EPs and albums, however, once you've satisfied yourself of the engineer's baseline competence by scrutinizing the test master, I think it's important to "zoom out" perceptually, dissociate yourself from the technical nuts and bolts of the production process, and try to listen as a typical punter would.

For a start, think about how your target audience consumes music day-to-day, and try to judge the full-length master within those kinds of contexts, whether that's on a hi-fi system, through a club PA, or on cheapo headphones. Obviously, there's no end to the possible playback alternatives, so you have to make a few judgment calls there to make the best use of time. If you're mixing opera, then you probably don't need to lose much sleep over people listening through a single earbud, for instance.

I think it's also sensible to listen mostly top-to-tail, rather than skipping around between the tracks, not only because that gives a better perspective on the effectiveness of track-to-track transitions, but also because it encourages you to engage more with the music on a more emotional (rather than technical) level, as the general public will. Fundamentally, the best gauge of a successful full-length master is how it feels, and that judgment needs to happen on a subconscious level. The more you start playing small snippets out of context for intellectual reasons, the more you're likely to obsess about comparatively insignificant technical details at the expense of the big picture. If you're working with songs, another trick I find really useful for breaking out of the mix engineer's mindset is to actively concentrate on the lyrics, because that usually prevents me focusing unduly on sonic technicalities that non-specialist listeners simply won't notice in any conscious way.

CUT TO THE CHASE

- It's usually best to get any master-buss processing involved the moment all the tracks and effects in your first mix section are happening, so you can make any necessary balance compensations in response. Resist the temptation to use frequency-selective dynamics processing in this role. Parallel master-buss compression can help side-step undesirable processing artifacts or control the balance interactions between different musical parts. Master-buss EQ and stereo-width plug-ins are frequently useful for broad-brush changes at the mix-referencing stage.

- Both "bottom-up" and "top-down" mixing approaches have potential drawbacks, so most mixes will benefit from a compromise between the two.

- Building up the complete mix from your first completed mix section is primarily a question of multing, region-specific processing, and automation, which maintain appropriate balances in response to arrangement changes. Detailed automation rides are also essential for most productions, because they can troubleshoot otherwise insoluble balance problems, direct the listener's attention where you want it, ensure the maximum audibility of lead vocals/hooks, and enhance the apparent detail, energy, and emotion in the music itself.

- If you chicken out of referencing your mix properly, then you'll never learn how to make your mixes compete with commercial records. Be prepared to go through several stages of referencing and readjustments. Make sure to compare the mix's overall tonality and stereo spread; the balance, tone, and stereo image of the most important instruments; and the levels of any reverb and delay effects. Checking your work against a client's own rough mix is also sensible, especially if they've been living with it for a while.

- Applying loudness-maximization processing to your mixdown file can help you make meaningful comparisons against commercial masters, and loudness-matching algorithms also have an important role to play as loudness-normalized music distribution becomes more widespread. There are lots of methods of loudness maximization, and a combination of these techniques often gives the best results. When evaluating the suitability of the processing, remove the loudness boost from the equation and listen carefully for processing side effects using all your monitoring systems.

- As a last stage of referencing, build up a written snag list, because this helps focus your mind on what is necessary for completion. For best results, do this in the morning following a night's rest, listen to the track all the way through, avoid looking at visual displays while you're listening, and evaluate the validity of any snags on all your different monitoring hardware (including your favorite domestic playback systems) before actually implementing any changes.

- Even for the top professionals, mix revisions are a fact of life if you work on other people's music, so don't expect the client to approve your first-draft mix straightaway. If possible, try to communicate about the mix in terms of concrete audio examples from commercial records, and make sure to keep backups of every draft as you go in case anyone decides to backtrack on an idea.

- Don't rely on mastering to improve the sound quality—if you're dissatisfied with the sound, then your mixdown work isn't yet finished. The main reason to pay for mastering is to benefit from the specialist experience and listening skills of the engineer, particularly if you want to compile a coherent-sounding album from different mixes or optimize the sound quality for a variety of delivery formats. Good communication between you and the mastering engineer is essential, both to achieve the best master and to help you improve your future productions.

Assignment

- Apply master-buss processing to your mix if it's appropriate, and readjust the balance after that if necessary. Then export that section of the mixdown as a stereo audio file and import it to a new DAW project for referencing against relevant commercial tracks. If loudness differences are a problem, then consider using loudness-enhancement processing and/or loudness-normalization software at this stage for referencing purposes. If the overall tonality or stereo width of the mix is wrong, then consider correcting these things with further master-buss processing in your mixdown project, rather than messing with too many of your individual channel processors.
- When you've completed the referencing process and revised your first mix snippet, build up the rest of the arrangement sections in order of importance, engaging automation where necessary to achieve a full mix draft. Export an audio file of the full mix draft and reference against commercial productions again. To finish off, leave the mix overnight and build a snag list the following morning. Once those snags are dealt with, seek the client's feedback and implement any suggestions. When the mixdown's finally a wrap, generate any alternate mix versions and/or stems and make sure you've backed everything up.

Web Resources

On this book's companion website you'll find a selection of resources to support this chapter, including:

- special tutorial video content to accompany this chapter's assignment, in which I demonstrate different automation techniques, master-buss processes, and loudness enhancement tactics on a simple multitrack mix;
- a selection of audio examples to demonstrate automation, long-term dynamics adjustments, and loudness processing in practice;
- links to my favorite affordable loudness-processing and referencing software tools, as well as to some interesting audio shootouts between different mastering services;
- some mix case-studies from the *Sound on Sound* "Mix Rescue" column where I focus particularly on automation issues and the process of dealing with mix-revision requests;
- further reading about classic master-buss compressors, the loudness wars debate, and loudness-normalization developments.

 www.cambridge-mt.com/ms-ch19.htm

CHAPTER 20
Conclusion

In this book I've tried to cover the main areas of mix technique that small-studio users need to master to achieve commercial-level results. Essentially it all boils down to the following:

- setting up and learning to use your monitoring systems;
- organizing, comping, arranging, and correcting the raw recordings;
- balancing all the tracks and adjusting their timbres if necessary;
- adding sweetening effects, master-buss processing, and automation;
- referencing your work, and creating any necessary alternate mix versions.

I've deliberately taken a step-by-step approach in order to make sense of so many different subjects and also so that you can build up your own mix while working through the book. However, as I mentioned in Chapters 15 and 19, most real-world mix engineers don't actually follow a linear trajectory and instead react much more fluidly to the mix they're hearing from moment to moment. So while a structured approach can be reassuring at first, don't be afraid to deviate from that model as your confidence increases, because that's the key to developing your own unique mixing style. It's also worth pointing out that some of the processing and effects setups I've described may be a lot more useful than others for your own preferred styles of music, so once you begin to discover your own favorite treatments you should be able to stream-line your workflow significantly by incorporating them into some kind of mixing template project, so you don't have to construct them from scratch for each new mix.

I've covered a lot of ground in this book, and if you've reached this point you should already have improved your mixing by learning how to apply some of the top pro techniques within your own small studio. The only observation I still want to add is this: learning to mix takes time and effort. I've done what I can to speed up the process for you, but you'll still need to put in some hard graft if you're going to hold your own among the top names in the business. "I'm all for people getting into recording music, because it's a great thing to do," says Tony Platt, "but if that is your chosen career path then you should spend a bit of time and work out how to do it properly. When I say properly I'm not standing in judgment saying that's right or that's wrong . . . but it should be approached as a skill and you should take it as far as you possibly can. Don't approach it with the attitude of 'Oh well, that'll do,' because that's not good enough."[1] Tony Maserati stresses the point: "I studied very, very hard. I would be on the subway with my SSL manual, reading it and marking it. I tell my assistants now: it's great to record your friends' bands, but don't just do that. Take a tape or a file and the manuals and work through a room's gear—the reverbs, the plug-ins. Keep working it over and over. I'd sit there with the gear for hours. I think that helped me become a mixer, and to be able to come up with things that were creative and new."[2]

For some final words of encouragement, let me leave you with Al Schmitt: "Stick with what you're doing, hang in there, and keep doing it. . . . I can't tell you how many times guys get turned down and rejected and battered about, and then all of a sudden, wham!—they just pop through. I don't know a better way to enjoy your life than making a living doing something you really love to do."[3]

Music Studios and the Recording Process: An Overview

Throughout most of this book I make the assumption that you have a basic overview of what the music-production process entails. However, because I can't *rely* on that, I wanted to lay out exactly what background knowledge I'm expecting of the reader, and in the most condensed form I can. (Health Warning: anyone with a strong allergic reaction to sweeping generalizations should top up their medication now!)

I've put important pieces of technical jargon in bold-face to make them stand out. It's a fact of life that every engineer applies studio terminology slightly differently, so I hope that clarifying my own usage will help minimize confusion.

SOUND IN DIFFERENT FORMS

Sound is a series of pressure waves moving through the air at the speed of sound (roughly 343 meters/second). Any vibrating object can create these pressure waves, and when they reach your ear drums, those vibrate in sympathy, triggering the hearing sensation. Transducers such as pickups and microphones can convert air-pressure waves, or the vibrations that triggered them, into an electrical signal, representing their movements over time as voltage fluctuations—often displayed for inspection purposes as a waveform of voltage level (vertical axis) against time (horizontal axis). Other transducers can convert this electrical signal back into air-pressure waves for listening or **monitoring** purposes (via loudspeakers or headphones) or into some other form for storage/replay (e.g., the variable groove depths on a vinyl LP or the variable flux levels on a magnetic tape).

Once sounds have been converted to electrical signals, it becomes possible to process and combine them using all manner of electronic circuitry. In addition, the voltage variations of an electrical signal can also be represented as a stream of numbers in digital form, whereupon digital signal-processing (DSP) techniques can be applied to them. Transferring any signal between the **analog domain** (of electrical signals, vinyl grooves, magnetic flux, physical vibrations, and pressure waves) and the **digital domain** requires either an analog-to-digital converter (**ADC**) or a digital-to-analog converter (**DAC**). The fidelity of analog-to-digital conversion is primarily determined by two statistics: the frequency with which the analog signal's voltage level is measured (the **sample rate** or **sampling frequency**) and the resolution (or **bit depth**) of each measurement (or **sample**), expressed in terms of the length of the binary number required to store it.

SINEWAVES AND AUDIO FREQUENCIES

If an electrical signal's waveform looks chaotic, what you hear will usually feature noisy sounds, whereas repeating waveform patterns are heard as pitched events such as musical notes or overtones. However, before I say any more about complex sounds, it's useful first to understand the simplest repeating sound wave: a **sinewave tone**. Its pitch is determined by the number of times it repeats per second, referred to as its **frequency** and measured in Hertz (Hz). Roughly speaking, the human ear can detect sinewave tones across a 20Hz–20kHz frequency range (the **audible frequency spectrum**). Low-frequency tones are perceived as low-pitched, while high-frequency tones are perceived as high-pitched. Although a sinewave tone isn't exactly thrilling to listen to on its own, it turns out that all the more interesting musical sounds can actually be broken down into a collection of different sinewave tones. The mixture of different sinewave components within any given complex sound determines its timbre.

One way to examine these sinewave components is to use a **spectrum analyzer**, a real-time display of a sound's energy distribution across the audible frequency spectrum. On a spectrum analyzer, a simple sinewave tone shows up as a narrow peak, while real-world signals create a complex undulating plot. Narrow peaks in a complex spectrum-analyzer display indicate pitched components within the signal, while the distribution of energy across the frequency display determines the timbre of the sound—subjectively darker sounds are richer in low frequencies, whereas bright sounds tend to be strong on high frequencies. Although a single sinewave tone will be perceived as a pitched note, almost all real-world musical notes are actually made up of a **harmonic series** of related

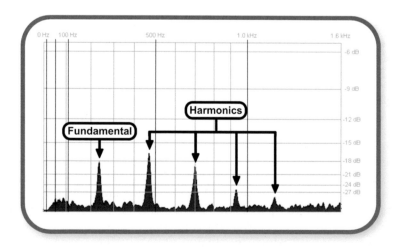

FIGURE A.1
On this spectrum analyzer display you can clearly see the fundamental and first four harmonics of a single pitched note projecting out of the full-frequency background noise.

sinewaves. The most low-frequency of these, the **fundamental**, determines the perceived pitch, while a series of **harmonics** at multiples of the fundamental's frequency determine the note's timbre according to their relative levels.

LOGARITHMIC SCALES FOR LEVEL AND PITCH

Human hearing perceives both level and pitch in a roughly logarithmic way— in other words, we compare levels and pitches in terms of ratios. For example, the perceived pitch interval between a note with its fundamental at 100Hz and a note with its fundamental at 200Hz is the same as that between notes with fundamentals at 200Hz and 400Hz. Similarly, when dealing with sound in electrical form, the perceived volume difference on playback between signals peaking at 100mV and 200mV is roughly similar to that between signals peaking at 200mV and 400mV. In recognition of this, both pitch and signal-level measurements are frequently made using a logarithmic scale. In the case of pitch, this is done in terms of traditional musical intervals: e.g., 200Hz is an octave above 100Hz, 200Hz is an octave below 400Hz, and so on. In the case of signal levels this is done using **decibels** (dB): e.g., 200mV is 6dB higher in level than 100mV, 200mV is 6dB lower in level than 400mV—or, to express it in another commonly used form, a +6dB level change ("+6dB gain") takes 100mV to 200mV, whereas a –6dB level change ("–6dB gain") takes 400mV to 200mV.

On their own, decibel values can only be used to indicate *changes* in signal level, which is why they are often used to label audio "gain controls" (such as faders) that are expressly designed for this purpose. However, it's important to remember that decibels (just like musical intervals) are always relative. In other words, it's meaningless to say that a signal level is "4.75dB," much as it's nonsensical to say that any isolated note is a major sixth, because the question is "4.75dB larger than what?" or "a major sixth above what?" Therefore, if you want to state absolute level values in terms of decibels, you need to express them relative to an agreed **reference level**, indicating this using a suffix. Common reference levels used for studio purposes include dBu and dBV (for electrical signals) and dBFS (for digital signals), but those are by no means the only ones out there.

FREQUENCY RESPONSE

Any studio device will alter the nature of sound passing through it in some way, however small, and the nature of any such effect on a signal's frequency balance is commonly expressed in terms of a **frequency response** graph, which shows the gain applied by the device across the frequency range. A device that left the frequency balance completely unchanged would show a straight horizontal frequency-response plot at the 0dB level. However, real-world equipment deviates somewhat from this ideal **flat response**—indeed, some devices deliberately warp their frequency-response curve for creative purposes.

THE MULTITRACK RECORDING PROCESS

Modern studio production revolves around the concept of **multitrack recording**, whereby you can capture different electrical signals on different recorder **tracks**, retaining the flexibility to process and blend them independently afterwards. Furthermore, multitrack recorders also allow you to **overdub** new signals to additional tracks while listening back to (**monitoring**) any tracks that have already been recorded, which enables complicated musical arrangements to be built up one instrument at a time if required.

An equally important cornerstone of the production process in many styles is the use of **synthesizers** (which generate audio signals electronically) and **samplers** (which can creatively manipulate selected sections of prerecorded audio). These can sometimes mimic the performances of live musicians, but more importantly provide the opportunity to design sounds that reach beyond the realms of the natural. Typically these devices are digitally controlled using **MIDI** (Musical Instrument Digital Interface) messages, which can be programmed/recorded in multitrack form, edited, and replayed using a **MIDI sequencer**.

Although the production workflow in different musical styles can contrast radically, many people in the studio industry find it useful to discuss the progress of any given project in terms of a series of notional "stages." Everyone has a slightly different view of what constitutes each stage, what they're called exactly, and where the boundaries are between them, but roughly speaking they work out as follows:

- **Preproduction** and **Programming**: The music is written and arranged. Fundamental synth/sampler parts may be programmed at this stage and musicians rehearsed in preparation for recording sessions.
- **Recording** (or **Tracking**): The instruments and vocals required for the arrangement are recorded, either all at once or piecemeal via overdubbing. Audio editing and corrective processing may also be applied during this stage to refine the recorded tracks into their final form, in particular when pasting together (**comping**) the best sections of several recorded **takes** to create a single **master take**. MIDI-driven synth and sampler parts are **bounced down** to the multitrack recorder by recording their audio outputs.
- **Mixing** (or **Mixdown**): All the recorded tracks are balanced and processed to create a commercial-sounding stereo mix.
- **Mastering**: The mixdown file is further processed to adapt it to different release formats.

The most common professional studio setup for the recording stage involves two separate rooms, acoustically isolated from each other: a **live room** where musicians perform with their instruments; and a **control room** containing the bulk of the recording equipment, where the recording engineer can make judgments about sound quality without the direct sound from the performers interfering with what he's hearing from his monitoring loudspeakers (or **monitors**). Where several performers are playing together in the live room, each with their own microphone, every mic will not only pick up the sound of the instrument/voice it's pointing at, but will also pick up some of the sound from the other instruments in the room—something variously referred to as **spill**, **leakage**, **bleed**, or **crosstalk**, depending on who you speak to! In some studio setups, additional sound-proofed rooms (**isolation booths**) are provided to get around this and improve the **separation** of the signals.

AUDIO SIGNALS AND MIXERS

A typical multitrack recording session can easily involve hundreds of different audio signals. Every audio source (microphones, pickups, synths, samplers) needs routing to its own track of the multitrack recorder, often through a **recording chain** of studio equipment designed to prepare it for capture. Each playback signal from the recorder will pass through its own **monitoring chain**, being blended with all the other tracks so that you can evaluate your work in progress via loudspeakers or headphones. Additional **cue/foldback mixes** may be required to provide personalized monitoring signals for each different performer during a recording session. Further mixes might also feed external (or **outboard**) effects processors, the outputs of which must be **returned** to the main mix so you can hear their results.

The way studios marshal all these signals is by using **mixers** (aka **mixing desks**, **boards**, or **consoles**). At its most basic, a mixer accepts a number of incoming signals, blends them together in some way, and outputs the resulting blended signal. Within the mixer's architecture, each input signal passes through its own independent signal-processing path (or **channel**), which is furnished with a set of controls (the **channel strip**) for adjusting the level and sound character of that signal in the mixed output. In the simplest of mixers, each channel strip may have nothing more than a fader to adjust its relative level for a single output mix, but most real-world designs have many other features besides this:

- If the **main/master mix** output is stereo (which it usually will be) then each mono channel will have a **pan control** (or **pan pot**) that adjusts the relative levels sent from that channel to the left and right sides of the main mix. If the mixer provides dedicated stereo channels, these may have a **balance control** instead, which sets the relative level of the stereo input signal's left and right signal streams.
- An independent **monitor mix** or **control-room mix** may be available for your studio loudspeakers. Although this will usually receive the master mix signal by default, you can typically also feed it with any subset of the input signals for closer scrutiny by activating per-channel **solo** buttons.

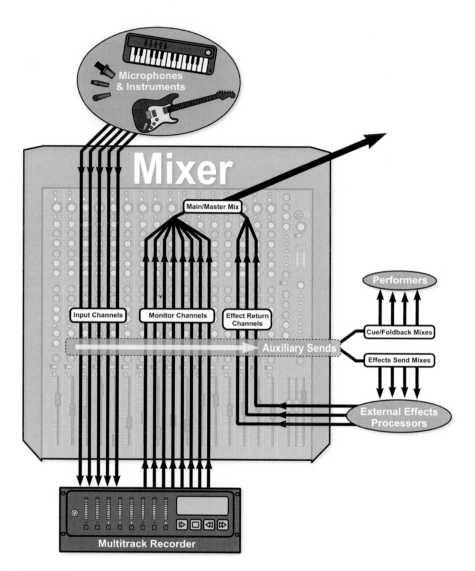

FIGURE A.2
Some of the basic functions of a mixer in the recording studio. Signals from microphones and instruments are relayed through the mixer's input channels to a multitrack recorder. Playback signals from the multitrack recorder are passed through the mixer's monitor channels on their way to the stereo main/ master mix. Auxiliary send facilities on each channel allow independent mixes to be sent to performers in the studio (cue/foldback mixes) and to external effects processors (effects send mixes), the latter returning their signals to the main/master mix.

- In addition to the faders that set each input signal's level in the main mix, there may be controls for creating further **auxiliary mixes** too—perhaps labeled as **cue sends** (for the purposes of foldback) and **effects sends** (for feeding external effects processors).

- There may be buttons on each channel strip that allow you to disconnect that channel from the main mix, routing it instead to a separate **group** or **subgroup** channel with its own independent output. This provides a convenient means of routing different input signals to different tracks of the multitrack recorder and of **submixing** several input signals together onto a single recorder track.
- Audio metering may be built in, visually displaying the signal levels for various channels as well as for the group, monitor, and master mix signals.

Mixer channels that are conveying signals to the multitrack recorder for capture are often referred to as **input channels**, whereas those that blend together the multitrack recorder's monitor outputs and send them to your loudspeakers/headphones are frequently called **monitor channels**. Some mixers just have a bunch of channels with identical functionality, and leave it up to you to decide which to use as input and monitor channels, while others have dedicated sections of input and monitor channels whose channel strip facilities are specifically tailored for their respective tasks. Another design, the **in-line mixer**, combines the controls of both an input channel and a monitor channel within the same channel strip. This is popular in large-scale studio setups because it creates a physically more

SIGNAL PROCESSING

Beyond simply blending and routing signals, multitrack production invariably involves processing them as well. In some cases this may comprise nothing more than "preamplifying" the signal to a suitable level for recording purposes, but there are several other processes that are frequently applied as well:

- Spectral Shaping: Audio **filters** and **equalizers** may be used to adjust the levels of different frequencies relative to each other;
- **Dynamics**: Tools such as **compressors**, **limiters**, **expanders**, and **gates** allow the engineer to control the level-contour of a signal over time in a semi-automatic manner;
- **Modulation** Effects: A family of processes that introduce cyclic variations into the signal. Includes effects such as **chorusing**, **flanging**, **phasing**, **vibrato**, and **tremolo**;
- Delay-based Effects: Another group of processes that involve overlaying one or more echoes onto the signal. Where these effects become complex, they can begin to artificially simulate the reverberation characteristics of natural acoustic spaces.

In some cases, such processing may be **inserted** into the signal path directly— rather than being fed from an independent effects send and then returned to the mix (a **send–return** configuration).

compact control layout, provides ergonomic benefits for advanced users, and allows the two channels to share some processing resources. (Plus there's the added benefit that it confuses the hell out of the uninitiated, which is always gratifying . . .)

Another specialized mixer, called a **monitor controller**, has evolved to cater for studios where several different playback devices and/or loudspeaker systems are available. It typically provides switches to select between the different audio sources and speaker rigs, as well as a master volume control for whichever speaker system is currently active.

REAL-WORLD STUDIO SETUPS: SOMETHING OLD, SOMETHING NEW

Although every recording studio needs to route, record, process, and mix audio signals, every engineer's rig ends up being slightly different, either by virtue of the equipment chosen, or because of the way the gear is hooked up. One defining feature of many systems is the extent to which digital technology is used. While there are still some people who uphold the analog-only studio tradition of the 1970s, the reliability, features, and pricing of DSP processing and data storage have increasingly drawn small studios toward hybrid systems. Standalone digital recorders and effects processors began this trend within otherwise analog systems, but the advent of comparatively affordable digital mixers and "studio in a box" digital **multitrackers** during the 1990s eventually allowed project studios to operate almost entirely in the digital domain, converting all analog signals to digital data at the earliest possible opportunity and then transferring that data between different digital studio processors losslessly. These days, however, the physical hardware units of early digital studios have largely been superseded by Digital Audio Workstation (**DAW**) software, which allows a single general-purpose computer to emulate all their routing, recording, processing, and mixing functions at once, connecting to the analog world where necessary via an **audio interface**: a collection of audio input/output (**I/O**) sockets, ADCs, and DACs.

A similar trajectory can be observed with synths and samplers. Although early devices were all-analog designs, microprocessors quickly made inroads during the 1980s as the MIDI standard took hold. The low data bandwidth of MIDI messages and the plummeting price of personal computing meant that computer-based MIDI sequencing was already the norm 20 years ago, but in more recent years the synths and samplers themselves have increasingly migrated into that world too, in the form of software **virtual instruments**. As a result, most modern DAW systems integrate MIDI sequencing and synthesis/sampling facilities alongside their audio recording and processing capabilities, making it possible for productions to be constructed almost entirely within a software environment. In practice, however, most small studios occupy a middle ground between the all-analog and all-digital extremes, combining old and new, analog and digital, hardware and software—depending on production priorities, space/budget restrictions, and personal preferences.

Web Resources

On this book's companion website you'll find a selection of resources to support this overview, including:

- an interactive version of this chapter peppered with dozens of links to more detailed further reading and some supporting audio/video materials;
- links to some well-maintained audio glossaries in case you need to look up any technical terms.

www.cambridge-mt.com/ms-basics.htm

Here's a list of all the engineers I've cited in this book, along with some of the most influential recordings they've been involved with.

ABBISS, JIM

Arctic Monkeys: *Whatever People Say I Am, That's What I'm Not*; Ladytron: *Witching Hour*; Sneaker Pimps: *Becoming X*; Editors: *The Back Room*; Kasabian: *Kasabian, Empire*; Adele: *19, 21*; KT Tunstall: *Tiger Suit*.

AINLAY, CHUCK

Trisha Yearwood: *Where Your Road Leads, Thinkin' about You, Everybody Knows, How Do I Live, Real Live Woman*; George Strait: *Somewhere Down in Texas, Pure Country, Blue Clear Sky, Carrying Your Love with Me*; Vince Gill: *High Lonesome Sound, The Key*; Dixie Chicks: *Wide Open Spaces*; Mark Knopfler: *Sailing to Philadelphia, The Ragpicker's Dream, Wag the Dog, Metroland*; Miranda Lambert: *Platinum*; Patty Loveless: *If My Heart Had Windows, Honky Tonk Angel*.

ALBINI, STEVE

The Pixies: *Surfer Rosa, Minotaur*; Nirvana: *In Utero*; Bush: *Razorblade Suitcase*; PJ Harvey: *Rid of Me*; Jimmy Page and Robert Plant: *Walking into Clarksdale*; Joan of Arc: *Life Like*; Manic Street Preachers: *Journal for Plague Lovers*; The Cribs: *Payola: 2002–2012, 24–7 Rock Star Shit, In the Belly of the Brazen Bull*; Robbie Fulks: *Gone Away Backward, Upland Stories*; The Wedding Present: *El Rey*; The Breeders: *Mountain Battles*; The Stooges: *The Weirdness*; Cheap Trick: *Rockford*.

ALEX DA KID

B.o.B: *B.o.B Presents The Adventures of Bobby Ray*; Cheryl: *A Million Lights*; Diddy: *Last Train to Paris*; Eminem: *Recovery, The Marshall Mathers LP 2*; Imagine Dragons: *Smoke + Mirrors, Demons, Evolve, Continued Silence*; Nicki Minaj: "Massive Attack," *The Pink Print*.

ALI, DEREK "MIXEDBYALI"

DJ Khaled: *Kiss the Ring*; Drake: *Take Care*; Dr. Dre: *Compton*; Snoop Dogg: *Bush*.

ARAICA, MARCELLA

Britney Spears: *Blackout, Circus, In the Zone, Femme Fatale*; Chris Brown: *X, Fortune*; Jennifer Lopez: *Love?*; Luke James: *Exit Wounds, Oh God*; Timbaland: *Shock Value*; Mariah Carey: *E = MC2*; Danity Kane: *Welcome to the Doll's House*; Keri Hilson: *In a Perfect World, No Boys Allowed*; Black Eyed Peas: *Monkey Business*; Usher: *Looking 4 Myself, Raymond v Raymond, Here I Stand.*

AVRON, NEAL

Disturbed: *Asylum, The Lost Children, Indestructible, The Sickness*; Fall Out Boy: *From under the Cork Trees, Infinity on High, Folie a Deux*; Weezer: *Make Believe*; Everclear: *So Much for the Afterglow*; Linkin Park: *A Thousand Suns, Minutes to Midnight*; The Wallflowers: *Bringing Down the Horse*; Lifehouse: *No Name Face*; Yellowcard: *Ocean Avenue*; Sara Bareilles: *Kaleidoscope Heart, What's Inside: Songs from Waitress.*

BALLARD, GLEN

David Benoit: *Believe*; Wilson Phillips: *Wilson Phillips*; Alanis Morissette: *Jagged Little Pill, Supposed Former Infatuation Junkie*; Michael Jackson: *Thriller, Bad, Dangerous*; POD: *Testify*; Anastacia: *Anastacia*; Goo Goo Dolls: *Let Love In*; Christina Aguilera: *Stripped*; No Doubt: *Return of Saturn*; Aerosmith: *Nine Lives*; Paula Abdul: *Forever Your Girl*; Ringo Starr: *Give More Love, Postcards from Paradise.*

BARRESI, JOE

Queens of the Stone Age: *Queens of the Stone Age, Lullabies to Paralyze*; Tool: *10,000 Days*; The Melvins: *The Bulls & The Bees/Electroretard, Stag, Honky*; Hole: *Celebrity Skin*; Limp Bizkit: *Chocolate Starfish & the Hotdog Flavored Water*; The Lost Prophets: *Start Something*; Skunk Anansie: *Stoosh*; Bad Religion: *The Empire Strikes First, New Maps of Hell, The Dissent of Man*; Buckcherry: *Rock N Roll, Confessions.*

BHASKER, JEFF

Beyonce: *4*; Bruno Mars: *24K Magic*; Dido: *Girl Who Got Away*; DJ Cover This: *All of the Lights*; Drake: *Thank Me Later*; Ed Sheeran: *X*; Eminen: *The Marshall Marthers LP 2*; Fun: *We are Young*; Jay-Z: *Run This Town, The Blueprint 3*; Kanye West: *My Beautiful Dark Twisted Fantasy*; Katy Perry: *Witness*; Mark Ronson: *Uptown Funk*; Rihanna: *Anti*; Taylor Swift: *Red.*

BIANCO, DAVE

Lianne La Havas: *Blood*; Miranda Lambert: *Platinum*; The Black Crowes: *Amorica, Shake Your Money Maker.*

BLAKE, TCHAD

Sheryl Crow: *The Globe Sessions, Be Myself*; Bonnie Raitt: *Souls Alike*; Phish: *Undermind*; Suzanne Vega: *Nine Objects of Desire, 99.9°F, Beauty and Crime*; Crowded House: *Farewell to the World*; Paul Simon: *Surprise*; Peter Gabriel: *Ovo, Long Walk Home, Up*; The Dandy Warhols: *Come Down, Odditorium*; Neil Finn: *One Nil, Try Whistling This*; Pearl Jam: *Binaural*; The Black Keys: *El Camino, Brothers*.

BOTTRELL, BILL

The Traveling Wilburys: *The Traveling Wilburys Volume 1*; Tom Petty: *Full Moon Fever*; Madonna: *Like a Prayer*; The Jacksons: *Victory*; Michael Jackson: *Bad, Dangerous*; Sheryl Crow: *Tuesday Night Music Club, Sheryl Crow, Icon*; Shelby Lynne: *I Am Shelby Lynne, Revelation Road*; Elton John: *Songs from the West Coast*; Pitbull: *Global Warming*.

BOTTRILL, DAVE

Birds of Tokyo: *Brace*; Blackbud: *From the Sky*; Circa Survive: *Appendage, Blue Sky Noise*; David Sylvian & Robert Fripp: *The First Day, Darshan*; dEUS: *The Ideal Crush, Keep You Close, Following Sea*; Flaw: *Through the Eyes, Endangered Species*; Lapko: *Freedom*; Stone Sour: *House of Gold & Bones-1, House of Gold & Bones-2*; The Getaway Plan: *Requiem*; Tool: *Salival, Lateralus*; The Smashing Pumpkins: *Oceania*.

BOWERS, DELBERT

Cheryl: *A Million Lights*; Christina Aguilera: *Lotus*; John Legend: *Love in the Future*; Giovanni James: *Whutcha Want*; Lady Gaga: *Artpop*; One Republic: *Native*; Pitbull: *Meltdown EP*; Shakira: *Shakira*; The Rolling Stones: *GRRR!*.

BOYER, ED

Committed: *Committed*; Home Free: *Full of Cheer, Crazy Life*; Pentatonix: *Pentatonix*.

BRADFIELD, ANDY

Robbie Robertson: *Contact from the Underworld of Redboy*; Rufus Wainwright: *Want One, Want Two, Release the Stars*; Josh Groban: *Awake*; Alanis Morissette: *Flavors of Entanglement*; Spice Girls: *Spice*; Faryl Smith: *Faryl*.

BRATHWAITE, LESLIE

Pharell Williams: "*Happy*"; Lil Uzi Vert: *Luv Is Rage 2*; Ludacris: *Ludaversal*; Future: *Honest*; Snoop Lion: *Reincarnated*; Keyshia Cole: *Point Of No Return*; Akon: *Freedom, I Wanna Love You, Konvicted, Lonely, Trouble*; Estelle: *True Romance*; Mogwai: *Central Belters, EP, Come to Die Young*; Nelly: *MO*; Minor Victories: *Minor Victories*.

BRAUER, MICHAEL

Coldplay: *Parachutes, X&Y, Viva La Vida, Mylo Xyloto*; Athlete: *Tourist*; The Fray: *The Fray*; Evans Blue: *The Pursuit Begins When This Portrayal of Life Ends*; Paolo Nutini: *Last Request*; John Mayer: *Continuum*; James Morrison: *Undiscovered*; The Kooks: *Inside In/Inside Out*; My Morning Jacket: *Evil Urges*; Travis: *The Boy with No Name*; The Doves: *Kingdom of Rust*; KT Tunstall: *Drastic Fantastic*; Idlewild: *Warnings/Promises*; Fountains of Wayne: *Traffic and Weather*; Aimee Mann: *Lost in Space*; Bob Dylan: *Lovesick*; Angélique Kidjo: *Eve*; Calle 13: *MultiViral*.

BUSH, BILLY

Fink: *Hard Believer, Perfect Darkness*; Foster the People: *Torches*; Garbage: *Why do you Love Me, Bleed Like Me, Cherry Lips, Not Your Kind of People, Strange Little Birds*; Jake Bugg: *Shangri La*; Neon Trees: *Picture Show*; The Naked and Famous: *In Rolling Waves, Passive Me Aggressive You*.

BUSH, STEVE

Stereophonics: *Word Gets Around, Performance & Cocktails, Just Enough Education to Perform*; Corinne Bailey Rae: *Corinne Bailey Rae*.

CAILLAT, KEN

Fleetwood Mac: *Rumours, Tusk, Mirage*; Colbie Caillat: *Coco, Breakthrough, All of You*; Christine McVie: *In the Meantime*; Lionel Ritchie: *Dancing on the Ceiling*; Michael Jackson: *Bad*.

CASTELLON, DEMACIO

Madonna: *Hard Candy, Celebration*; Nelly Furtado: *Loose, Mi Plan*; Rihanna: *Good Girl Gone Bad*; Monica: *After the Storm*; The Game: *The Documentary*; Fabri Febri: *Fenomeno*.

CHERNY, ED

The Rolling Stones: *Stripped, Bridges to Babylon, No Security*; Bonnie Raitt: *Nick of Time, Luck of the Draw, Longing in Their Hearts, Road Tested*; Bob Dylan: *Under the Red Sky, Unplugged*; Willie Nelson: *Summertime: Willie Nelson Sings Gershwin*.

CHICCARELLI, JOE

White Stripes: *Icky Thump*; Frank Zappa: *Sheik Yerbouti, Joe's Garage, Tinseltown Rebellion*; My Morning Jacket: *Evil Urges*; The Shins: *Wincing the Night Away*; Counting Crows: *The Desert Life*; Jamie Lawson: *Miracle of Love, Fall Into Me, Happy Accident*; Pink Martini: *A Retrospective*.

CHURCHYARD, STEVE

The Pretenders: *Learning to Crawl*; Counting Crows: *Recovering the Satellites*; Celine Dion: *Falling into You*; Ricky Martin: *Vuelve, Almas Del Silencio, Sound Loaded*; Shakira: *Laundry Service*; The Stranglers: *La Folie, Feline*; Big Country: *Wonderland*; Bryan Ferry: *Boys & Girls*; INXS: *Listen Like Thieves*; Kelly Clarkson: *Thankful, Meaning of Life*; Katy Perry: *Teenage Dream*; Jason Marz: *Love is a Four Letter Word*.

CLEARMOUNTAIN, BOB

The Pretenders: *Get Close*; Bryan Adams: *Into the Fire, Reckless, Cuts Like a Knife, So Far So Good, 18 till I Die, Room Service, 11*; Bruce Springsteen: *Born in the USA, Wrecking Ball*; The Rolling Stones: *Tattoo You*; Bon Jovi: *These Days, Bounce, Crush*; Roxy Music: *Avalon*; David Bowie: *Let's Dance*; INXS: *Kick, Full Moon, Dirty Hearts, Welcome to Wherever You Are*; The Corrs: *Talk on Corners, Forgiven Not Forgotten, Unplugged*; Robbie Williams: *Intensive Care*; Simple Minds: *Once Upon a Time, Black & White 050505, Graffiti Soul*; Sheryl Crow: *Wildflower*; Aimee Mann: *Whatever, Bachelor #2*; Rufus Wainwright: *Rufus Wainwright*.

CLINK, MIKE

Guns 'n' Roses: *Appetite for Destruction, GNR Lies, Use Your Illusion, The Spaghetti Incident*; Survivor: *Eye of the Tiger*; Megadeth: *Rust in Peace, Icon*; Mötley Crüe: *New Tattoo*; Whitesnake: *Whitesnake*; Shelter Dogs: *Take Me Home*.

COSTEY, RICH

Muse: *Absolution, Black Holes & Revelations, Drones*; Interpol: *Our Love to Admire*; Polyphonic Spree: *Together We're Heavy*; Franz Ferdinand: *You Could Have It So Much Better*; The Mars Volta: *Frances the Mute, Deloused in the Comatorium*; Audioslave: *Audioslave*; Weezer: *Make Believe*; Bloc Party: *Silent Alarm*; Doves: *Some Cities*; My Chemical Romance: *Three Cheers for Sweet Revenge*; Three Days Grace: *Three Days Grace*; Jimmy Eat World: *Futures*; POD: *Payable on Death*; Rage against the Machine: *Renegades*; Fiona Apple: *When the Pawn*; Jurassic 5: *Quality Control*; Nine Inch Nails: *With Teeth*; Blondie: *Pollinator*.

DAVIS, KEVIN "KD"

Coolio: "Gangsta's Paradise"; Destiny's Child: *The Writing's on the Wall, Survivor*; N'Sync: *No Strings Attached*; Pink: *Missundaztood*; Tupac Shakur: *Me against the World*; TLC: *3D*; Usher: *8701, Outkast: Speakerboxx/The Love Below*; John Legend: *Evolver*; Estelle: *All of Me*.

DE VRIES, MARIUS

Massive Attack: *Protection*; Bjork: *Debut, Post, Homogenic, Vespertine*; Madonna: *Bedtime Stories, Ray of Light*; U2: *Pop*; Mel C: *Northern Star*; David Gray: *White*

Ladder; Rufus Wainwright: *Want One, Want Two*; Josh Groban: *Awake*; David Bowie: *Nothing Has Changed*.

DECARLO, LEE

John Lennon: *Double Fantasy, Signature Box*; Black Sabbath: *Live Evil*; Quiet Riot: *Quiet Riot II*.

DJ PREMIER

Jeru the Damaja: *The Sun Rises in the East, Wrath of the Math*; Group Home: *Livin' Proof*; Nas: *Illmatic, It Was Written, I Am . . ., Stillmatic*; The Notorious BIG: *Ready to Die, Life after Death*; Jay-Z: *Reasonable Doubt, In My Lifetime, Hard Knock Life, The Life & Times of Shawn Carter*; Rakim: *The 18th Letter, The Master*; Black Eyed Peas: *Bridging the Gap*; D'Angelo: *Voodoo*; Snoop Dogg: *Paid tha Cost to Be da Bo$$*; Christina Aguilera: *Back to Basics*; Xzibit: *Man vs Machine*; Gang Starr: *No More Mr Nice Guy, Step in the Arena, Daily Operation, Hard to Earn, Moment of Truth, The Ownerz*; Dr. Dre: *Compton*.

DJ SWIVEL

Coldplay: *Kaleidoscope*; Fabolous: *Loso's Way, From Nothin' To Somethin'*; Jay Sean: *My Own Way, All or Nothing*; Jay-Z: *Magna Carta Holy Grail, The Blueprint 3, American Gangster*; The Chainsmokers: *Collage*.

DORFSMAN, NEIL

Dire Straits: *Brothers in Arms, Love over Gold*; Bruce Hornsby: *Scenes from the South Side*; Paul McCartney: *Flowers in the Dirt*; Bruce Springsteen: *Tracks, The River*; Sting: *Nothing Like the Sun, Brand New Day*; Bjork: *Medulla*.

DOUGLAS, JACK

Aerosmith: *Get Your Wings, Toys in the Attic, Rocks, Draw the Line, Rock in a Hard Place, Honkin' on Bob, Music from Another Dimension*; John Lennon: *Imagine, Double Fantasy, Icon*; Cheap Trick: *Cheap Trick, At Budokan, Standing on the Edge*.

DOUGLASS, JIMMY

Aaliyah: *One in a Million, Are You That Somebody*; Missy Elliott: *Supa Dupa Fly, Da Real World, Miss E . . . So Addictive* (including "Get Ur Freak On"); Aretha Franklin: *Young, Gifted & Black*; Jay-Z: *Hard Knock Life, 4:44*; Jodeci: *The Show, The Hotel, The After Party*; NERD: *In Search Of*; Timbaland & Magoo: *Timbaland and Magoo, Tim's Bio, Up Jumps the Boogie, Love 2 Love U, All Y'Al*; John Legend: *Wake Up!*; Pharrell Williams: *G I R L*; Justin Timberlake: *Justified, Futuresex/ Lovesounds, The 20/20 Experience*.

DR DRE

Dr Dre: *The Chronic, 2001, Compton*; Eminem: *The Slim Shady LP, The Marshall Mathers LP, The Eminem Show, Encore, Relapse, Recovery, Revival*; Kendrick Lamar: *The Recipe, Good Kid M.A.A.D. City, Compton*; Snoop Dogg: *Doggystyle, No Limit Top Dogg, The Last Meal, The Blue Carpet Treatment, Malice N Wonderland, More Malice, Doggy Stuff*; Busta Rhymes: *Genesis, The Big Bang*; 2Pac: *All Eyez On Me*; Jay-Z: *Kingdom Come*; Xzibit: *Restless/Man Versus Machine, Napalm*; The Game: *The Documentary, Jesus Piece, Purp and Patron*; Mary J Blige: *No More Drama, Love & Life*; Gwen Stafani: *Love Angel Music Baby*; 50 Cent: *Get Rich Or Die Tryin', The Massacre, Animal Ambition, Curtis, Before I Self Destruct, Compton King, Ghost Unit*.

DRESDOW, DYLAN "3D"

Black Eyed Peas: *Elephunk, Monkey Business, The END*; Flo Rida: *ROOTS*; The Game: *Doctor's Advocate*; Britney Spears: *Britney Jeans*; Black Eyed Peas: *The Beginning, Monkey Business/Elephunk*; Common: *Finding Forever*; Christina Aguilera: *Stripped*; Methodman & Redman: *Blackout, How High*; Prince: *HitNRun: Phase Two*.

DUDGEON, GUS

Elton John: *Elton John, Madman across the Water, Honky Chateau, Goodbye Yellow Brick Road, Caribou, Captain Fantastic and the Brown Dirt Cowboy*; David Bowie: *Space Oddity*; Lady Gaga: *Joanne*; Original Soundtrack: *Tommy*.

ELEVADO, RUSS

D'Angelo: *Voodoo*; Al Green: *Lay It Down*; Alicia Keys: *Songs in A Minor, Diary of Alicia Keys*; The Roots: *Phrenology, Game Theory*; Dru Hill: *Dru Hill*: Erykah Badu: *Mama's Gun*; Boney James: *Ride*; Blackalicious: *Blazing Arrow, The Craft*.

ELMHIRST, TOM

Adele: *19, 21, 25*, "Skyfall"; Amy Winehouse: *Frank, Back to Black*; Goldfrapp: *Black Cherry*; Lily Allen: *Alright, Still*; David Bowie: *The Blackstar, Legacy, No Plan*; Paolo Nutini: *These Streets*; Mark Ronson: *Version*; Sugababes: *Three*; Beck: *Morning Phase*; Lorde: *Melodrama*; Mark Ronson: *Uptown Special*; Emile Sande: *Our Version Of Events*; The Black Keys: *El Camino*; Ellie Goulding: *Halcyon Days*.

EMERICK, GEOFF

The Beatles: *Revolver, Sgt. Pepper's Lonely Hearts Club Band, Magical Mystery Tour, Abbey Road*; Paul McCartney: *Band on the Run, Flaming Pie*; Elvis Costello: *Imperial Bedroom, All This Useless Beauty*; Badfinger: *No Dice*; Robin Trower: *Bridge of Sighs*.

ENDERT, MARK

Chris Tomlin: *Burning Lights, And If Our God Is For Us . . .*; Christina Perri: *A Thousand Years*; Fiona Apple: *Tidal*; Gavin DeGraw: *Chariot*; Delta Goodrem: *Delta*; Maroon 5: *Songs about Jane, It Won't Be Soon before Long*; Miley Cyrus: *Breakout*; Madonna: *Ray of Light, Music*; Quietdrive: *When All That's Left Is You*; Secondhand Serenade: *Hear Me Now*; Rihanna: *Good Girl Gone Bad*; Sara Bareilles: *The Blessed Unrest*; Tracey Bonham: *Down Here*.

EVERETT, SHAWN

Alabama Shakes: *Sound & Color*; Julian Casablancas: *Tyranny*; The War on Drugs: *Strangest Thing, A Deeper Understanding, Holding On*; Weezer: *Death to False Metal, Hurley, Raditude*.

FIELDS, JOHN

Anastacia: *Resurrection*; Backstreet Boys: *Never Gone*; Busted: *Night Driver*; Daryl Hall: *Laughing Down Crying*; Dave Barnes: *Stories To Tell*; Miley Cyrus: *Breakout, The Time Of Our Lives*; Parachute: *The Way It Was, Losing Sleep*; Windsor Drive: *Wanderlust*.

FILIPETTI, FRANK

Danielle Brooks: *The Color Purple*; Foreigner: *40, Agent Provocateur, Inside Information*; Hole: *Celebrity Skin*; Korn: *Untouchables, Here to Stay*; Carly Simon: *Hello Big Man, Spoiled Girl, Coming around Again, Film Noir, Bedroom Tapes*; Barbra Streisand: *Higher Ground*; James Taylor: *Hourglass, That's Why I'm Here*; Kiss: *Lick It Up*; Bangles: *Everything*; Robert Lopez: *The Book of Mormon*; Survivor: *Too Hot to Sleep*.

FINN, JERRY

Alkaline Trio: *Crimson, Remains, Good Mourning*; Green Day: *Kerplunk, Dookie, Insomniac*; Smoking Popes: *Destination Failure*; MxPx: *The Ever Passing Moment, Ten Years and Running*; Blink-182: *Enema of the State, Take off Your Pants and Jacket, Blink-182*; Tiger Army: *Music from Regions Beyond*; +44: *When Your Heart Stops Beating*; AFI: *Sing the Sorrow, Decemberunderground, Crash Love*; Sparta: *Wiretap Scars*.

FITZMAURICE, STEVE

Depeche Mode: *Exciter, Freelove, I Feel Loved*; Frances: *Things I've Never Said*; Louis Berry: *Restless*; Sam Smith: *In The Lonely Hour, The Thrill Of It All*; Seal: *Kiss From A Rose*; The Frames: *The Cost, The Roads Outgrown*; U2: *All That You Can't Leave Behind*; Rag'n'Bone Man: *Human*; The Kooks: *Listen*.

FLOOD

Editors: *In This Light and on This Evening*; 30 Seconds to Mars: *This Is War*; Goldfrapp: *Seventh Tree*; PJ Harvey: *White Chalk, Is This Desire, To Bring You My Love*; The Killers: *Sam's Town*; U2: *The Joshua Tree, Achtung Baby, Pop, Zooropa, How to Dismantle an Atomic Bomb*; Smashing Pumpkins: *Mellon Collie and the Infinite Sadness, Machina/Machines of God, Adore*; Nine Inch Nails: *Downward Spiral, Pretty Hate Machine*; Depeche Mode: *Violator, Pop Will Eat Itself, Songs of Faith and Devotion*; Nick Cave and the Bad Seeds: *From Her to Eternity, The Firstborn Is Dead, Kicking against the Pricks, Your Funeral My Trial, Tender Prey, The Good Son*; Erasure: *Wonderland, The Circus*; U2: *Songs of Innocence*.

GABRIEL, PASCAL

Goldfrapp: *Head First*; Marina and the Diamonds: *The Family Jewels*; Ladyhawke: *Ladyhawke*; Miss Kittin: *BatBox*; Little Boots: *Nocturnes, Hands*; Dido: *No Angel*; Kylie Minogue: *Aphrodite, Fever*; S'Express: *Enjoy This Trip*; Will Young: *Echoes*.

GASS, JON

Paula Abdul: *Forever Your Girl*; Madonna: *Bedtime Stories*; Dru Hill: *Enter the Dru*; Usher: *My Way*; Whitney Houston: *Just Whitney*; Michael Jackson: *Invincible*; Mary J. Blige: *Mary*; Destiny's Child: *The Writing's on the Wall*.

GATICA, HUMBERTO

Michael Jackson: *Bad*; Celine Dion: *Celine Dion, D'eux, Falling into You, Let's Talk about Love, A New Day Has Come, One Heart*; Michael Bublé: *Call Me Irresponsible, Crazy Love*; Ricky Martin: *Vuelve*; Cher: *Believe*; Julio Iglesias: *Non Stop, Un Hombre Solo, Crazy, La Carretera, Tango*; Barbra Streisand: *The Mirror Has Two Faces, Higher Ground*; Selena Gomez: *For You*.

GHENEA, SERBAN

Adele: *25*; Katy Perry: *One of the Boys, Teenage Dream, Prism, Witness*; Ke$ha: *Animal*; Jason Derulo: *Jason Derulo*; Miley Cyrus: *The Time of Our Lives*; Britney Spears: *In the Zone, Circus*; Iyaz: *Replay*; Jay Sean: *All or Nothing*; Weezer: *Raditude*; The Fray: *The Fray*; Kelly Clarkson: *Stronger, All I Ever Wanted, Breakaway*; Pink: *I'm Not Dead, Funhouse*; Avril Lavigne: *The Best Damn Thing*; Black Eyed Peas: *Monkey Business*; Fergie: *The Dutchess*; Carrie Underwood: *Some Hearts*; Gwen Stefani: *This is What The Truth Feels Like*; Justin Timberlake: *Justified*; Mark Ronson: *Uptown Special, Version*; Sugababes: *Change*; Jill Scott: *Who Is Jill Scott? Beautifully Human, The Real Thing*; Dave Matthews Band: *Stand Up*; Robbie Williams: *Rudebox*; Usher: *My Way*; NERD: *In Search Of*; Taylor Swift: *Red, 1989, Reputation*; Bruno Mars: *24K Magic*.

GLOSSOP, MICK

Frank Zappa: *Shut up N Play Yer Guitar, Joe's Garage*; John Lee Hooker: *Don't Look Back*; Van Morrison: *Too Long in Exile, Hymns of the Silence, Enlightenment, Poetic Champions Compose, No Guru No Method No Teacher, Inarticulate Speech of the Heart, Into the Music, Wavelength*; The Waterboys: *This Is the Sea, The Whole of the Moon*; Tangerine Dream: *Rubycon, Ricochet*; The Wonder Stuff: *Never Loved Elvis, The Size of a Cow, Welcome to the Cheap Seats, Dizzy, Cursed with Insincerity*; Lloyd Cole: *Antidepressant, Music in a Foreign Language, Broken Record*.

GODRICH, NIGEL

Atoms for Peace: *Amok*; Radiohead: *The Bends, OK Computer, Kid A, Amnesiac, Hail to the Thief, In Rainbows, A Moon Shaped Pool*; Thom Yorke: *The Eraser*; Natalie Imbruglia: *Left of the Middle*; Beck: *Mutations, Sea Change, Guero, The Information*; The Divine Comedy: *Regeneration, Absent Friends*; Travis: *The Man Who, The Invisible Band, The Boy with No Name*; Paul McCartney: *Chaos and Creation in the Backyard*; Air: *Talkie Walkie, Pocket Symphony*; Pavement: *Terror Twilight*; REM: *Up*.

GOLDSTEIN, JASON

Kelly Rowland: *Ms Kelly*; Rihanna: *Music of the Sun*; Ludacris: *Theatre of the Mind*; The Roots: *Game Theory, Rising Down*; Tanya Morgan: *Rubber Souls*; Jay-Z: *The Blueprint*; R Kelly & Jay-Z: *Unfinished Business*; Nas: *Stillmatic*; Mary J. Blige: *Mary*; Beyoncé: *B'Day*.

GOSLING, JAKE

Christina Perry: *Head or Heart*; Ed Sheeran: *+, x, Song I Wrote With Amy*; Paloma Faith: *Fall to Grace*; Shania Twain: *Now*; Shawn Mendes: *Illuminate*; The Libertines: *Anthems for Doomed Youth*.

GUDWIN, JOSH

Coldplay/Rihanna: *Princess of China*; Celine Dion: *Loved Me Back to Life*; Dua Lipa: *Dua Lipa*; Fifth Harmony: *Fifth Harmony*; Jennifer Lopez: *Love?*; Justin Bieber: *Purpose, Believe*; Keyshia Cole: *Woman to Women, A Different Me*; Rihanna: *Loud*; The Knocks: *55, Love Me Like That*.

GUZAUSKI, MICK

Daft Punk: *Random Access Memories*; Eric Clapton: *Back Home*; Christina Aguilera: *Christina Aguilera*; Babyface: *The Day*; Backstreet Boys: *Millennium*; Boyz II Men: *II*; Toni Braxton: *Secrets* (including "Unbreak My Heart"); Herb

Alpert/Lani Hall: *Steppin' Out*; Brandy: *Never Say Never*; Mariah Carey: *Music Box* (including "Dream Lover" and "Hero"), *Butterfly, Charmbracelet*; Patti LaBelle: *Winner in You*; Pharrell Williams: *G I R L*; Jennifer Lopez: *On the 6, J.Lo, This Is Me . . . Then*; Barbra Streisand: *Higher Ground*; Leann Rimes: *You Light up My Life.*

HEAP, IMOGEN

Aphrodite: *Urban Junglist*; Hans Zimmer: *The Holiday* (soundtrack); Imogen Heap: *Ellipse, Headlock, Foiled, Sparks*; Jason Derulo: *Jason Derulo*; Jon Hopkins: *The Art of Chill 2, Contact Note*; Kelly Clarkson: *Wrapped in Red*; Taylor Swift: *1989*; William Orbit: *Odyssey.*

HODGE, STEVE

Janet Jackson: *Control, Rhythm Nation, Janet, The Velvet Rope, All for You, Damita Jo*; George Michael: *Faith*; Mariah Carey: *Rainbow, Share My World, Glitter, Butterfly, Charmbracelet*; Michael Jackson: *HIStory*; Toni Braxton: *More Than a Woman*; Stanley Clarke: *Time Exposed/Find Out!/Hideaway.*

HOFFER, TONY

Belle & Sebastian: *The Life Pursuit, Write About Love*; Foster The People: *Torches*; M83: *Junk, Hurry Up We're Dreaming*; ROMES: *ROMES, Believe EP*; The Fratellis: *In Your Sweet Time, Costello Music*; The Kooks: *Inside In/Inside Out, Junk of the Heart, Konk*; The Royal Concept: *Smile.*

HORN, BOB

Akon: *Trouble*; Ashanti: *The Declaration*; Brian Culbertson: *Another Long Night, Dreams*; Kelly Clarkson: *Meaning of Life*; Scarecrow & Tinmen: *Superhero, No Place Like Home*; Timbaland: *Timbaland Presents Shock Value*; The Cheetah Girls: *TCG*; Usher: *Versus, Raymond V Raymond*; Vivian Green: *The Green Room.*

HOSKULDS, S. "HUSKY"

John Legend: *Once Again*; Mike Patton: *The Solitude of Prime Numbers, Mondo Cane*; Norah Jones: *Come Away With Me, Don't Know Why*; Solomon Burke: *Don't Give Up on Me*; Turin Brakes: *Ether Song.*

HUART, WARREN

Ace Frehley: *Space Invader*; Daniel Powter: *Turn on the Lights, Best of Me*; James Blunt: *Some Kind of Trouble, I'll Be Your Man*; Mark Broussard: *Perfect to Me, A Life Worth Living*; The Fray: *The Fray, Scars & Stories, How to Save a Life*; Trevor Hall: *KALA, Unpack Your Memories, Chapter of the Forest.*

HYDE, MATT

ASG: *Blood Drive*; Big Jesus: *Oneiric*; Children of Bodom: *Relentless, Reckless Forever*; Deftones: *Gore, Koi No Yokan*; Hatebreed: *Perserverance*; Jonny Lang: *Turn Around*; Parkway Drive: *Atlas*; Monster Magnet: *Powertrip*; Pride Tiger: *The Lucky Ones*; Slayer: *God Hates Us All*; The 69 Eyes: *Back in Blood*.

JEAN, WYCLEF

Jay-Z: *4:44*; The Fugees: *Blunted on Reality, The Score*; Wyclef Jean: *The Carnival, The Ecleftic, Masquerade, The Preacher's Son*; Santana: *Supernatural*; Destiny's Child: *Destiny's Child*; Pras: *Ghetto Supastar, Win Lose or Draw*; Mya: *Fear of Flying*; Michael Jackson: *Blood on the Dance Floor*; Shakira: *Oral Fixation Vol. 2; She Wolf*.

JERKINS, RODNEY

Ashanti: *Declaration*; Beyonce: *B'day, Déjà vu, I Am . . . Sasha Fierce*; Britney Spears: *Femme Fatale, In the Zone, Oops . . . I did It Again*; Destiny's Child: *Say My Name*; Janet Jackson: *Make Me, Feedback, Discipline*; Justin Bieber: *Journals, Roller Coaster, Believe*; Lady Gaga: *Telephone, The Fame Monster, The Remix*; Mariah Carey: *The Art of Letting Go, The Emancipation of Mimi*; Mary J. Blige: *Enough Cryin', The One, The Breakthrough, I can Love You*; Michael Jackson: *Scream, Xscape, Immmortal, Invincible*; The Pussycat Dolls: *Doll Domination, When I Grow Up*; Whitney Houston: *The Essential, I Learned from the Best, If I Told You That, Heartbreak Hotel, My Love is Your Love*.

JOHNS, ANDY

Eric Johnson: *Up Close*; Led Zeppelin: *Led Zeppelin III, Led Zeppelin IV, Houses of the Holy, Physical Graffiti*; The Rolling Stones: *Sticky Fingers, Exile on Main Street, Goat's Head Soup*; Van Halen: *For Unlawful Carnal Knowledge*; Television: *Marquee Moon*; Jethro Tull: *Stand Up*.

JONES, GARETH

Erasure: *Wild!, Erasure, Cowboy, Other Peoples Songs, Light at the End of the World*; Depeche Mode: *Construction Time Again, Some Great Reward, Exciter*; Wire: *The Ideal Copy, A Bell Is a Cup . . . Until It Is Struck*.

JONES, QUINCY

Frank Sinatra: *L.A. Is My Lady, It Might as Well Be Swing*; Count Basie: *This Time by Basie*; Michael Jackson: *Off the Wall, Thriller, Bad*; Quincy Jones: *Walking in Space, Smackwater Jack, The Dude, Back on the Block, Q's Jook Joint, You've Got It Bad Girl*; Jennifer Lopez: *Love?*; George Benson: *Give Me the Night*; USA for Africa: "*We Are the World*."

JOSHUA, JAYCEN

Mariah Carey: *Memoirs of an Imperfect Angel, E = MC2*; Beyoncé: *I Am Sasha Fierce*; Jamie Foxx: *Intuition*; Sean Paul: *The Trinity*; Justin Bieber: *Purpose, Believe, My World 2.0*; Jay Sean: *Neon*; Jay-Z: *Magna Carta Holy Grail*; Katy Perry: *Teenage Dream*; R. Kelly: *Write Me Back*; Rihanna: *Rated R*; Mary J. Blige: *Growing Pains, Stronger with Each Tear*; Nikki Minaj: *Make Love*; Christina Aguilera: *Bionic*; New Kids on the Block: *The Block*; Soulja Boy: *Souljaboytellem. com*; Pussy Cat Dolls: *Doll Domination*; Usher: *Here I Stand*; Monica: *Still Standing*; Ashanti: *The Declaration*.

KADISH, KEVIN

Coldwater Jane: *Bring on the Love, Marionette*; Jason Mraz: *We Sing. We Dance. We Steal Things, Geek In the Pink, Wordplay*; Kris Allen: *Thank You Camellia*; Meghan Trainor: *Thank You, Title*; Miley Cyrus: *Can't be Tamed*; New Medicine: *Breaking The Model, Race You To The Bottom*.

KATZ, BOB

Paquito D'Rivera: *Portraits of Cuba*; The White Birch: *The Weight of Spring*; Olga Tañon: *Olga Viva, Viva Olga*; Sinead O'Connor: *Theology*.

KILHOFFER, ANTHONY

Adam Lambert: *For Your Entertainment*; Jennifer Lopez: *Do It Well, Como Ama una Mujer*; John Legend: *Evolver, Once Again, Get Lifted*; Kanye West: *Graduation, Late Registration, Fade, College Dropout, The Life of Pablo, Yeezus*; Kid Cudi: *Passion, Pain & Demon Slayin', Speedin' Bullet 2 Heaven, Make Her Say*; Marracash: *Status*; Rick Ross: *Teflon Don*; Rihanna: *Good Girl Gone Bad*; The Black Eyed Peas: *Elephunk*; The Pussycat Dolls: *PCD*.

KILLEN, KEVIN

Celtic Woman: *Lullaby*; David Bowie: *Blackstar, No Plan, Nothing Has Changed*; Peter Gabriel: *So*; U2: *War, The Unforgettable Fire, Rattle and Hum*; Elvis Costello: *Unfaithful Music, Spike, The Juliet Letters, Kojak Variety, North, Mighty Like a Rose, Cruel Smile*; Shakira: *Fijación Oral Vol. 1, Oral Fixation Vol. 2*; Tori Amos: *Under the Pink*; Bryan Ferry: *Boys and Girls*; Kate Bush: *The Sensual World, This Woman's Work*; Paula Cole: *This Fire, Courage*.

KING, JACQUIRE

Amber Rubarth: *A Common Case of Disappearing*; James Bay: *Chaos and the Calm*; Kings of Leon: *Aha Shake Heartbreak, Only by the Night, Come around Sundown*; Moon Taxi: *Daybreaker*; Norah Jones: *The Fall*; Robert Ellis: *The Lights from the Chemical Plant*; Tom Waits: *Blood Money, Mule Variations, Alice,*

Orphans: Brawlers, Bawlers and Bastards; Modest Mouse: *Good News for People Who Like Bad News.*

KIRWAN, ROB

Depeche Mode: *Delta Machine, Heaven, Soothe My Soul, Freelove, In Your Room*; Editors: *In This Light and On This Evening, You Don't Know Love*; Glasvegas: *Euphoria Take My Hand, Euphoric Heartbreak*; Hozier: *Hozier*; PJ Harvey: *The Hope Six Demolition Project*; Soulsavers: *It's Not How Far You Fall, It's the Way You Land*; Soulwax: *Any Minute Now*; Sneaker Pimps: *Bloodsport*; The Courteeners: *St. Jude*; The Horrors: *She Is the New Things, Strange House*; U2: *Zooropa, Pop.*

KRAMER, EDDIE

Jimi Hendrix: *Are You Experienced?, Axis: Bold as Love, Electric Ladyland, Band of Gypsys, Valleys of Neptune*; Led Zeppelin: *Led Zeppelin II, Led Zeppelin III, How the West Was Won, Houses of the Holy, Physical Graffiti, Coda*; Traffic: *Traffic*; The Nice: *Nice, Five Bridges*; Peter Frampton: *Frampton Comes Alive!*

LANGE, ROBERT JOHN "MUTT"

AC/DC: *Highway to Hell, Back in Black, For Those about to Rock We Salute You*; Def Leppard: *High 'n' Dry, Pyromania, Hysteria, Adrenalize*; Foreigner: *4*; Lady Gaga: *Born This Way*; The Cars: *Heartbeat City*; Bryan Adams: *Waking up the Neighbours, 18 till I Die*; Shania Twain: *The Woman in Me, Come on Over, Up*; The Corrs: *In Blue*; Nickelback: *Dark Horse*; Maroon 5: *Hands All Over*; Muse: *Drones.*

LANOIS, DANIEL

U2: *The Joshua Tree, Achtung Baby, All That You Can't Leave Behind, No Line on the Horizon*; Daniel Lanois: *Goodbye to Language, Flesh and Machine*; Peter Gabriel: *So, Us*; Bob Dylan: *Oh Mercy, Time Out of Mind.*

LAWSON, MARK

Arcade Fire: *The Suburbs, Funeral, Everything Now*; Bell Orchestre: *As Seen Through Windows*; Rare Essence: *Work the Walls, Body Snatchers*; Sarah Neufeld: *The Ridge, Never Were the Way She Was*; Will Butler: *Policy.*

LECKIE, JOHN

Pink Floyd: *Meddle, Discovery*; Radiohead: *The Bends*; Muse: *Showbiz, Origin of Symmetry*; The Stone Roses: *The Stone Roses*; The Verve: *A Storm in Heaven*; Kula Shaker: *K*; My Morning Jacket: *Z*; The Coral: *Butterfly House*; Cast: *All Change, Mother Nature Calls*; Doves: *Kingdom of Rust*; Paul McCartney: *Pure McCartney.*

LEHNING, KYLE

Randy Travis: *Storms of Life, Always & Forever, No Holdin' Back, This Is Me, Inspirational Journey, Rise and Shine, Around the Bend*; Dan Seals: *Stones, Harbinger, Rebel Heart*; Ronnie Milsap: *Heart and Soul*; Bryan White: *Bryan White, Between Now and Forever, Dowdy Ferry Road, Some Things Don't Come Easy*; Derailers: *Here Come the Derailers, Genuine*; Kenny Rogers: *Amazing Grace*.

LETANG, RENAUD

Feist: *Let It Die, The Reminder*; Alain Souchon: *C'est déjà ça*; Peaches: *Fatherf**ker*; Saul Williams: *Martyr Loser King, Volcanic Sunlight*.

LIPSON, STEVE

Annie Lennox: *Diva*; Frankie Goes to Hollywood: *Welcome of the Pleasuredome, Liverpool*; Paul McCartney: *Flowers in the Dirt*; Grace Jones: *Slave to the Rhythm*; Jordin Sparks: *Battlefield*; Boyzone: *Where We Belong, By Request*; Rachel Stevens: *Funky Dory*; Ronan Keating: *Fires*.

LORD-ALGE, CHRIS

James Brown: *Living in America*; Prince: *Batman* soundtrack; Joe Cocker: *Unchain My Heart*; Chaka Khan: *Destiny*; Green Day: *Nimrod, American Idiot, 21st Century Breakdown*; My Chemical Romance: *The Black Parade*; Stevie Nicks: *Trouble in Shangri-La*; POD: *Testify*; AFI: *Decemberunderground*; Darren Hayes: *Spin*; Creed: *Full Circle*; Sum 41: *Underclass Hero*; Switchfoot: *The Beautiful Letdown, Nothing Is Sound*; Slipknot: *Dead Memories, Sulfur*; Stone Temple Pilots: *Stone Temple Pilots*; Tina Turner: *Foreign Affair*; Breaking Benjamin: *Dark Before Dawn*; Carrie Underwood: *Storyteller*; Keith Urban: *Ripcord*; 5 Seconds Of Summer: *5 Seconds Of Summer*; Bruce Springsteen: *High Hopes*.

LORD-ALGE, TOM

Pink: *I'm Not Dead, Funhouse*; Avril Lavigne: *Let Go, The Best Damned Thing, Under My Skin*; Blink-182: *Blink-182, Take off Your Pants and Jacket*; Steve Winwood: *Back in the High Life, Roll with It*; Live: *Throwing Copper*; Crash Test Dummies: *God Shuffled His Feet*; Hanson: *Middle of Nowhere*; The Rolling Stones: *GRRR!, Bridges to Babylon*; Santana: *Supernatural*; McFly: *Radio:Active*; Marilyn Manson: *The Dope Show, Mechanical Animals*; Sum 41: *All Killer No Filler, All the Good S**t*; The Fratellis: *Here We Stand*; Korn: *Life Is Peachy*; Weezer: *Weezer, Maladroit*.

MALOUF, BRIAN

Everclear: *Sparkle and Fade*; David Gray: *White Ladder*; Lit: *A Place in the Sun*; Madonna: *I'm Breathless*; Pearl Jam: *Ten*; Slaughter: *Stick It to Ya*; Smokey

Robinson: *Smokey & Friends*; Joel Crouse: *Even the River Runs*; Sabrina Carpenter: *Can't Blame a Girl for Trying*; Big Time Rush: *24/Seven*.

MARASCIULLO, FABIAN

T-Pain: *Epiphany*; Birdman: *5 Star Stunna*; Toni Braxton: *More Than a Woman*; 50 Cent: *Before I Self Destruct*; TLC: *3D*; J.Lo: *Ain't It Funny*; Trey Songz: *Trey Day*; Monica: *After the Storm*; Lil Wayne: *Tha Carter, Tha Carter II, Tha Carter III*; Flo Rida: *Mail on Sunday*; Chris Brown: *Exclusive, Despicable Me 3: Original soundtrack*; R. Kelly: *12 Nights of Christmas, The Buffer*; DJ Khaled: *I Changed A Lot*.

MARROQUIN, MANNY

Kanye West: *College Dropout, Late Registration, Graduation, 808s and Heartbreaks*; Duffy: *Rockferry*; Rihanna: *Good Girl Gone Bad* (including "Umbrella"), *Talk That Talk, Unapologetic, Anti*; Alicia Keys: *Songs in A Minor, Unplugged, Diary of Alicia Keys, As I Am*; Whitney Houston: *My Love Is Your Love*; Usher: *Here I Stand*; Mary Mary: *Thankful, Get Up*; John Mayer: *Continuum*; Janet Jackson: *Damita Jo*; Common: *Be*; Natasha Bedingfield: *Unwritten*; Mariah Carey: *The Emancipation of Mimi*; Pitbull: *Global Warming*; John Legend: *Get Lifted*; Faith Evans: *The First Lady*; Sisqo: *Return of the Dragon*; Bruno Mars: *Unorthodox Jukebox*; Linkin Park: *Recharged, Living Things*; Justin Bieber: *Purpose*; Nicki Minaj: *The Pinkprint*.

MASERATI, TONY

Mary J. Blige: *What's the 411?, My Life, Share My World, Ballads, No More Drama*; Mariah Carey: *Daydream, Butterfly*; Destiny's Child: *Survivor*; Faith Evans: *Faith, Keep the Faith*; R. Kelly: *R, TP2.com*; John Legend: *Once Again*; Black Eyed Peas: *Elephunk, Monkey Business*; Jennifer Lopez: *J.Lo*; Alicia Keys: *Diary of Alicia Keys*; Lady Gaga: *Artpop*; Jason Mraz: *Yes!, Love Is a Four Letter Word*; Ed Sheeran: "*Perfect.*"

MASSENBURG, GEORGE

Little Feat: *Feats Don't Fail Me Now, Hoy-Hoy, Let It Roll, Shake Me Up*; Linda Ronstadt: *Mas Canciones, Frenesi, Feels Like Home, Dedicated to the One I Love, We Ran*; Lyle Lovett: *Joshua Judges Ruth*; Bonnie Raitt: *Nine Lives*; Toto: *The Seventh One*; Earth, Wind, & Fire: *Gratitude, That's the Way of the World, Spirit, I Amh All N All*; Journey: *Trial by Fire, When You Love a Woman*; Livingston Taylor: *Last Alaska Moon*.

MELDAL-JOHNSON, JUSTIN

Air: *10,000 Hz Legend, Twenty Years*; Beck: *Morning Phase, The Information, Guero, Midnight Vultures, Mutations*; Dido: *Safe Trip Home*; Garbage: *Absolute Garbage, Not Your Kind of People, Bleed Like Me*; Goldfrapp: *Seventh Tree*; Marianne Faithful: *Kissin' Time*; M83: *Junk, Hurry Up We're Dreaming, Midnight City*; Tori Amos: *From the Choigirl Hotel, Boys for Pele, Hey Jupiter*.

MEYERSON, ALAN

A. R. Rahman: *Couples Retreat*; Daft Punk: *Tron: Legacy*; Disney: *Disney's Winnie the Pool* (Original Score); Hanz Zimmer: *Dark Knight* (Original motion picture soundtrack), *Sherlock Holmes* (Original Score), *Angels & Demons* (Original motion picture soundtrack); James Newton Howard: *I am Legend* (Original motion picture soundtrack), *The Last Airbender*; Ramin Djawadi: *Iron Man* (original motion picture soundtrack), *Safe House*, *Clash of the titans* (2010 soundtrack).

MONTAGNESE, CARLO "ILLANGELO"

Alicia Keys: *Here*; Elijah Blake: *Shadows & Diamonds*; Illangelo: *History of Man*; Katy Perry: *Witness*; Majid Jordan: *Something About You, My Love*; Post Malone: *Stoney*; Romeo Santos: *Golden*; The Weeknd: *Beauty Behind the Madness*; Travis Scott: *Rodeo*.

MOSELEY, ADAM

Abandoned Pools: *Armed to the Teeth*, *The Reverb EP*, April Wine: *Nature of the Beast*; Eddie Hernandez: *The Yankles*; John Cale: *Circus Live*, *Shifty Adventures in Nookie Wood*; Lucybell: *Lumina*, *Salvame la Vida*, *Comeindo Fuego*; Wolfmother: *Cosmic Egg*.

MOULDER, ALAN

Elastica: *Elastica*; The Jesus and Mary Chain: *Honey's Dead*, *Automatic*; Marilyn Manson: *Portrait of an American Family*; My Bloody Valentine: *Glider*, *Loveless*, *Tremolo*; Nine Inch Nails: *The Downward Spiral*, *The Fragile*, *The Perfect Drug*; The Smashing Pumpkins: *Siamese Dream*, *Mellon Collie & the Infinite Sadness*, *Machina/The Machines of God*; U2: *Pop*; Foo Fighters: *Wasting Light*; Trent Reznor/Atticus Ross: *The Girl with the Dragon Tattoo* (original motion picture soundtrack); Led Zeppelin: *Celebration Day*.

MURPHY, SHAWN

Braveheart, *Dances with Wolves*, *E.T. the Extra-Terrestrial*, *Ghost*, *Jurassic Park*, *Men in Black*, *Pretty Woman*, *Saving Private Ryan*, *The Sixth Sense*, *Star Wars: Episode I–The Phantom Menace*, *Episode II–Attack of the Clones*, *Titanic*, *The Post*, *Ice Age: Collision Course*, *Star Wars: The Last Jedi*, *Star Wars: The Force Awakens*, *How To Train Your Dragon 2*, *The Hunger Games: Mockingjay, Part 1*, *After Earth*, *Ice Age: The Continental Drift*, *Snow White & the Huntsman*, *Rio*.

NEEDHAM, MARK

Blue October: *Home*; Chris Isaak: *Live at the Fillmore*, *Beyond the Sun*, *Always Got Tonight*; Imagine Dragons: *Continued Silence*, *Night Visions*; Elton John: *The Road to El Dorado* (Original Soundtrack); Fleetwood Mac: *Say You Will*; Neon Trees: *Habits*; Newsboys: *Love Riot*; P!nk: *The Truth About Love*; Shakira: *Sale el Sol*; The Killers: *Hot Fuss*, *Sawdust*, *Mr. Brightside*.

NICHOLS, ROGER

Steeley Dan: *Can't Buy a Thrill, Countdown to Ecstasy, Pretzel Logic, Gaucho, Aja, Two against Nature, Everything Must Go.*

NIEBANK, JUSTIN

Marty Stuart: *Country Music*; Keith Urban: *Be Here*; Patty Loveless: *On Your Way Home, Dreamin' My Dreams*; Vince Gill: *Next Big Thing, These Days*; Shakira: *Shakira*; Sheryl Crow: *Feels Like Home*; Taylor Swift: *Red.*

OGILVIE, DAVE

Fake Shark Real Zombie: *Meeting People is Terrible*; Jakalope: *Things That Go Jump in the Night, It Dreams, Born 4*; Killing Joke: *Democracy*; Left Spine Down: *Caution*; Marilyn Manson: *Beautiful People, Antichrist Superstar*; Queensryche: *Rage for Order*; Raggedy Angry: *How I Learned to Love Our Robot Overlords*; The Birthday Massacre: *Walking With Strangers, Under Your Spell, Imaginary Monsters*; Carly Rae Jepsen: "Call Me Maybe."

OLSEN, KEITH

Fleetwood Mac: *Fleetwood Mac*; Foreigner: *Double Vision*; Scorpions: *Crazy World*; Whitesnake: *Whitesnake, Slide It In, Slip of the Tongue.*

ORTON, ROBERT

Lady Gaga: *The Fame*; New Kids on the Block: *The Block*; Carly Rae Jepsen: *Kiss*; The Police: *Certifiable*; Tatu: *200 km/h in the Wrong Lane, Dangerous and Moving*; Pet Shop Boys: *Fundamental, Concrete*; Sean Paul: *Imperial Blaze, Full Frequency*; Pixie Lott: *Turn It Up*; Little Boots: *Hands*; Lana Del Rey: *Paradise, Ultraviolence*; Enrique Iglesias: *Sex And Love*; Nicole Scherzinger: *Killer Love.*

PADGHAM, HUGH

Genesis: *Abacab, Genesis, Invisible Touch*; Phil Collins: *Face Value, Hello, I Must Be Going!, No Jacket Required, . . . But Seriously, Both Sides, Take a Look at Me Now*; The Police: *Ghost in the Machine, Synchronicity*; Sting: *Nothing Like the Sun, The Soul Cages, Ten Summoner's Tales, Mercury Falling*; Peter Gabriel: *Peter Gabriel*; XTC: *Black Sea*; David Bowie: *Legacy, Nothing Has Changed.*

PALMER, TIM

Cutting Crew: *Broadcast*; Faith Hill: *Cry*; Ozzy Osbourne: *Prince of Darkness, Memoirs of Madman, Down to Earth*; Pearl Jam: *Ten*; Porcupine Tree: *In Absentia*; Robert Plant: *Now & Zen, Fate of Nations, Shaken 'N Stirred*; The Mighty Lemon Drops: *World Without End*; Tin Machine: *Tin Machine I, Tin Machine II, You Belong in Rock N' Roll*; U2: *All That You Can't Leave Behind.*

PANUNZIO, THOM

U2: *Rattle & Hum*; Deep Purple: *The Battle Rages On, A Fire in the Sky*; Black Sabbath: *Reunion*; Ozzy Osbourne: *Live at Budokan*; Willie Nile: *Golden Down*; Jeff Healey Band: *Cover to Cover, See the Light*; Motörhead: *Hammered*; Bruce Springsteen: *Born to Run*; New Found Glory: *Coming Home*; Nelly Furtado: *The Spirit Indestructible*.

PARR, STEVE

Rob Lane: *John Adams* (original TV soundtrack); LTJ Bukem: *Planet Earth*; Studio Voodoo: *Club Voodoo*.

PARSONS, ALAN

The Beatles: *Abbey Road*; Pink Floyd: *Dark Side of the Moon*; Al Stewart: *Year of the Cat*; Paul McCartney: *Wild Life, Red Rose Symphony, Pure McCartney*; The Hollies: *Hollies, "He Ain't Heavy, He's My Brother"*; Ambrosia: *Somewhere I've Never Travelled*; Alan Parsons Project: *Tales of Mystery and Imagination, Pyramid, Eve, The Turn of a Friendly Card, Eye in the Sky*.

PENSADO, DAVE "HARD DRIVE"

Destiny's Child: *Survivor*; Beyoncé: *I Am Sasha Fierce, "Check on It"*; Nelly Furtado: *Loose*; Mary J. Blige: *The Breakthrough, Stronger with Each Tear*; Black Eyed Peas: *Behind the Front, Bridging the Gap, Monkey Business*; Christina Aguilera: *Christina Aguilera, Stripped*; Justin Timberlake: *Justified*; Mya: *Moodring*; Sugababes: *Sweet 7*; P!nk: *Try This*; Pussycat Dolls: *PCD, Doll Domination*; Jill Scott: *Golden Moments*; Earth: *Now, Then & Forever, Holiday*.

PILEGAARD, MORTEN "PILO"

Brandon Beal: *Truth*; Lukas Graham: Blue Album (including "7 Years"); Morten Pilegaard: *My Little Pony* (Original Motion Picture Soundtrack), *Acoustic Hits*.

PLATT, TONY

Bob Marley: *Catch a Fire, Burnin'*; Toots & the Maytals: *Funky Kingston*; Aswad: *Aswad*; AC/DC: *Highway to Hell, Back in Black*; Foreigner: *4*; Boomtown Rats: *The Fine Art of Surfacing*; Anathema: *Eternity*; Buddy Guy: *Damn Right, I've Got the Blues, Sweet Tea*.

POWER, BOB

Erykah Badu: *Baduizm*; A Tribe Called Quest: *Midnight Marauders, The Low End Theory, People's Instinctive Travels, Beats, Rhymes and Life*; D'Angelo: *Brown Sugar*; India Arie: *Acoustic Soul*; Meshell Ndegeocello: *Plantation Lullabies, Peace Beyond Passion, Cookie: The Anthropological Mixtape, Comfort Woman*; De La Soul:

Buhloone Mind State, De La Soul Is Dead; Jungle Brothers: *Jbeez Wit Da Remedy, Red Hot & Blue*; Run DMC: *Down With The King*.

POWER, STEVE

Robbie Williams: *Life Thru a Lens, I've Been Expecting You, Sing When You're Winning, Swing When You're Winning, Escapology*; Babylon Zoo: *The Boy with the X-Ray Eyes* (including "Spaceman"); Babybird: *Ugly Beautiful* (including "You're Gorgeous"); Busted: *A Present for Everyone*; Delta Goodrem: *Mistaken Identity*; McFly: *Motion in the Ocean*.

PRICE, BILL

The Sex Pistols: *Never Mind the Bollocks*; The Clash: *The Clash, Give 'Em Enough Rope, London Calling, Sandinista!*; The Pretenders: *Pretenders, Pretenders II*; Elton John: *Too Low for Zero*; Pete Townshend: *Empty Glass*; The Jesus & Mary Chain: *Darklands*; Babyshambles: *Down in Albion*.

PUIG, JACK JOSEPH

Snow Patrol: *Eyes Open*; Goo Goo Dolls: *Let Love In*; Black Eyed Peas: *Monkey Business*; Fergie: *The Dutchess*; Mary J. Blige: *The Breakthrough*; Pussy Cat Dolls: *PCD*; Stereophonics: *You've Got to Go There to Come Back*; Sheryl Crow: *C'mon C'mon*; The Rolling Stones: *Forty Licks, A Bigger Bang, Biggest Mistake*; Green Day: *Warning*; No Doubt: *Return of Saturn*; Hole: *Celebrity Skin*; Weezer: *Pinkerton*; Jellyfish: *Spilt Milk, Bellybutton*; Eric Clapton: *Forever Man*.

RAMONE, PHIL

Paul Simon: *There Goes Rhymin' Simon, Still Crazy after All These Years*; Bob Dylan: *Blood on the Tracks*; Sinead O'Connor: *Am I Not Your Girl?*; Billy Joel: *52nd Street, Glass Houses, The Nylon Curtain, The Bridge*.

RONSON, MARK

Adele: *19*; Amy Winehouse: *Back to Black*; Bruno Mars: *Locked Out of Heaven*; Christina Aguilera: *Back to Basics*; Mark Ronson: *Here Comes the Fuzz, Version, Record Collection, Uptown Special* (including "Uptown Funk"); Lady Gaga: *Million Reasons, Perfect Illusion, Joanne*; Lily Allen: *Alright, Still*; Robbie Williams: *Rudebox*; Duran Duran: *All You Need is Now, Paper Gods*.

ROSSE, ERIC

Tori Amos: *Little Earthquakes, Under the Pink*; Lisa Marie Presley: *To Whom It May Concern*; Anna Nalick: *Wreck of the Day*; Sara Bareilles: *Little Voice*; Mary Lambert: *Heart on My Sleeve, Secrets*; Nerina Pallot: *Fires*.

SAMUELS, HARMONY

Ariane Grande: *The Way, Yours Truly, My Everything*; Chris Brown: *Fortune, F.A.M.E*; Fantasia: *Side Effects of You*; Fifth Harmony: *Reflection, Fifth Harmony, Better Together*; Jennifer Lopez: *A.K.A*; Kelly Rowland: *Talk a Good Game*; Keyshia Cole: *11:11 Reset, Women to Love, Enough of No Love*; Ne-Yo: *R.E.D*; Michelle Williams: *Journey To Freedom*.

SARAFIN, ERIC "MIXERMAN"

Pharcyde: *Bizarre Ride II The Pharcyde*; Acetone: *York Blvd*; Ben Harper: *Fight for Your Mind, Burn To Shine, The Will to Live*; Barenaked Ladies: *Everything To Everyone*; Hilary Duff: *Most Wanted, Hilary Duff*.

SAVIGAR, KEVIN

Bob Dylan: *Down in the Groove*; Jamie Walters: *Hold On*; Josh Gracin: *Josh Gracin*; Peter Frampton: *Gold*; Randy Newman: *Toy Story* (Original Soundtrack); Rod Stewart: *Time, Another Country, Vagabond Heart, Body Wishes, Tonight I'm Yours*; The Cheetah Girls: *The Cheetah Girls* (Original Soundtrack).

SCHEINER, ELLIOT

Steely Dan: *Aja, Gaucho, Two against Nature*; Donald Fagan: *Nightfly*; Billy Joel: *Songs in the Attic*; Fleetwood Mac: *The Dance*; Roy Orbison: *Black and White Night*; John Fogerty: *Premonition*; Van Morrison: *Moondance*; Bruce Hornsby & the Range: *The Way It Is*.

SCHILLING, ERIC

Gloria Estefan: *Into the Light*; Ricky Martin: *Sound Loaded*; Julio Iglesias: *Quelque Chose de France*; Jon Secada: *Secada*; Bacilos: *Caraluna, Sin Vergüenza*.

SCHLEICHER, CLARKE

Martina McBride: *The Time Has Come, Evolution, Emotion, Martina*; Big & Rich: *Comin' to Your City, Between Raising Hell and Amazing Grace*; Dixie Chicks: *Wide Open Spaces, Fly*; Pam Tillis: *Thunder and Roses*; Sara Evans: *Born to Fly, Restless*; Mark Chesnutt: *Savin' the Honky-Tonk*; Taylor Swift: *Taylor Swift*; Lady Antebellum: *Own the Night, Golden*; Blake Shelton: *Texoma Shore, If I'm Honest*.

SCHMITT, AL

George Benson: *Breezin', Guitar Man*; Steely Dan: *Aja, FM (No Static at All)*; Toto: *Toto IV*; Natalie Cole: *Unforgettable*; Diana Krall: *When I Look in Your Eyes, The Look of Love*; Eleonara Fagan: *To Billie With Love from Dee Dee Bridgewater*; Paul McCartney: *Kisses on the Bottom, Live Kisses*; Ray Charles: *Genius Loves*

Company; Jefferson Airplane: *After Bathing at Baxter's, Crown of Creation, Volunteers*; Luis Miguel: *Amarte Es un Placer*.

SEAY, ED

Dolly Parton: *White Limozeen*; Martina McBride: *The Time Has Come, The Way That I Am, Wild Angels, Evolution*; Ty Herndon: *What Mattered Most*; Pam Tillis: *Put Yourself in My Place, Homeward Looking Angel*; Lionel Cartwright: *Chasin' the Sun*; Whitney Wolanin: *Honesty, Run, Run Rudolph*; Hank Williams: *A Country Boy Can Survive*.

SELBY, ANDY

Adam Lambert: *Trespassing*; B. Reith: *The Forecast EP, Now is Not Forever*; Danny Gokey: *Rise, Hope in Front of Me*; Javier Colon: *Gravity, Come Through for You*; Josh Groban: *Stages, Awake, Noel*; Karmin: *Pulses*; Krystal Meyers: *Make Some Noise*; Mat Kearney: *Bullet, Just Kids, Young Love, Nothing Left to Lose*; Natalie Grant: *Hurricane, Be One*; Salvador: *Into Motion, Worship Love*.

SERLETIC, MATT

Matchbox Twenty: *Yourself or Someone Like You, Mad Season, Exile on Mainstream, More Than You Think You Are*; Carlos Santana: *Supernatural* (including "Smooth"); Taylor Hicks: *Taylor Hicks*; Courtney Love: *America's Sweetheart*; Rob Thomas: *Cradlesong, Something to Be*; Stacie Orrico: *Stacie Orrico*; Gloriana: *Gloriana, A Thousand Miles Left Behind, Three*; Collective Soul: *Hints, Allegations, and Things Left Unsaid, Collective Soul*.

SHEBIB, NOAH "40"

Action Bronson: *Mr. Wonderful*; Alicia Keys: *The Element of Freedom, VH1 Storytellers*; Drake: *Take Care, One Dance, Views, What a Time to be Alive, So Far Gone*; Jamie Foxx: *Best Night of My Life*; Lil Wayne: *I'm Single, I Am Not a Human Being, The Carter IV*.

SHEPHARD, IAN

Mastering engineer who has worked with Keane, Tricky, The Orb, Deep Purple, King Crimson, Ozric Tentacles, Andy Weatherall, The Las, New Order, Culture Club, Porcupine Tree, Leslie Garret, and The Royal Philharmonic Orchestra.

SHIPLEY, MIKE

Nickelback: *Dark Horse*; Faith Hill: *Breathe*; Def Leppard: *High 'n' Dry, Pyromania, Hysteria, Adrenalize*; The Cars: *Heartbeat City*; Shania Twain: *The Woman in Me, Come on Over, Up*; The Corrs: *In Blue*; Maroon 5: *Hands All Over*; Alison Krauss: *Paper Airplane*; 12 Stones: *Beneath the Scars*.

SHOEMAKER, TRINA

Amy Ray: *Lung of Love, Goodnight Tender*; Sheryl Crow: *Sheryl Crow, The Globe Sessions, C'mon C'mon*; Queens of the Stone Age: *R*; Steven Curtis Chapman: *All Things New*; Dixie Chicks: *Home*.

SIDES, ALLEN

Chris Botti: *Impressions*; Joni Mitchell: *Both Sides Now*; Pussy Cat Dolls: *Doll Domination*; Phil Collins: *Testify*; Alanis Morissette: *"Uninvited"*; Goo Goo Dolls: *Dizzy up the Girl; Dead Man Walking*.

SIGSWORTH, GUY

Frou Frou: *Details*; Seal: *Seal*; Bomb the Bass: *Unknown Territory*; Bjork: *Post, Homogenic, Vespertine*; Imogen Heap: *I Megaphone*; Madonna: *Music, American Life*; Britney Spears: *In the Zone, Circus, Femme Fetale*; Sugababes: *Three, Taller in More Ways*; Alanis Morissette: *Flavors of Entanglement, Guardian, Havoc and Bright Lights*; Alison Moyet: *The Minutes*; Josh Groban: *Awake*.

SIMPSON, RIK

Coldplay: *Viva la Vida, Prospekt's March, Clocks, The Scientist, A Rush of Blood to the Head, Lost!, Mylo Xyloto, A Head Full of Dreams*; Morning Runner: *Gone Up in Flames*; Natalie Imbruglia: *Come to Life*; Some Velvet Morning: *Silence Will Kill You*.

SMITH, DON

The Rolling Stones: *Voodoo Lounge*; Ry Cooder: *Chavez Ravine, My Name Is Buddy*; Stevie Nicks: *Rock a Little, Trouble in Shangri-La*; The Tragically Hip: *Up to Here, Road Apples*; Tom Petty: *Long after Dark, Southern Accents, Full Moon Fever, The Last DJ*; Roy Orbison: *Mystery Girl*; Eurythmics: *Be Yourself Tonight*.

SMITH, FRASER T.

Adele: *21*; Craig David: *Born to Do It, Slicker Than Your Average, Trust Me*; Rachel Stevens: *Come and Get It*; Tinchy Stryder: *Catch 22, Third Strike*; Taio Cruz: *Rokstarr*; Cheryl Cole: *Three Words*; Ellie Goulding: *Lights*; Kano: *Home Sweet Home, London Town*; Beyoncé: *B'Day*; James Morrison: *Songs for You Truths for Me, The Awakening*; N-Dubz: *Uncle B*; Jennifer Hudson: *Jennifer Hudson*; Pixie Lott: *Turn It Up*; Chipmunk: *I Am Chipmunk*; Sam Smith: *In the Lonely Hour*; Gorillaz: *Humanz*; Linkin Park: *One More Light*.

STAUB, RANDY

Alice in Chains: *Black Gives Way to Blue, The Devil Put Dinosaurs Here*; Big Wreck: *Diggin In, A Speedy Recovery, You Don't Even Know, Grace Street*;

Nickelback: *Feed the Machine, Dark Horse, All the Right Reasons, Photograph, Someday, The Long Road, Silver Side Up* (including "How you Remind Me"); 3 Doors Down: *Away From the Sun, Seventeen Days*; Evanescence: *Evanescence.*

STAVROU, MIKE

Siouxie and the Banshees: *Join Hands, The Scream*; T.Rex: *Dandy in the Underworld*; Paul McCartney: *Pure McCartney.*

STENT, MARK "SPIKE"

Spice Girls: *Spice* (including "Wannabe"), *Forever*; Bjork: *Post, Homogenic, Vespertine*; Keane: *Hopes and Fears, Perfect Symmetry*; Gwen Stefani: *Love Angel Music Baby, The Sweet Escape*; Massive Attack: *Protection, Mezzanine, Heligoland*; Beyoncé: *I Am Sasha Fierce*; Lady Gaga: *The Fame Monster*; Black Eyed Peas: *Monkey Business*; Muse: *The Resistance*; Sade: *Soldier of Love*; Ciara: *Fantasy Ride*; Britney Spears: *In the Zone*; Oasis: *Standing on the Shoulder of Giants*; Janet Jackson: *Damita Jo*; Maroon 5: *It Won't Be Long before Soon*; U2: *Pop*; Goldfrapp: *Supernature, Head First*; Usher: *Raymond v. Raymond*; Bruce Springsteen: *Wrecking Ball, We Take Care of Our Own*; Ed Sheeran: *X, Divide*; Taylor Swift: *Red*; Julia Michaels: *Nervous System* (including "Issues"); One Direction: *Midnight Memories*; Biffy Clyro: *Opposites*; Coldplay: *Ghost Stories.*

STONE, AL

Jamiroquai: *Return of the Space Cowboy, Travelling without Moving, Synkronized*; Daniel Bedingfield: *Gotta Get through This*; Stereo MCs: *Connected*; Bjork: *Debut, Post*; Turin Brakes: *The Optimist*; Lamb: *Fear of Fours*; Eagle Eye Cherry: *Sub Rosa*; Spice Girls: *Spice.*

SUECOF, JASON

All That Remains: *Overcome*; August Burns Red: *Triple Play, Leveler, Constellations*; Austrian Death Machine: *Triple Brutal, Double Brutal, Total Brutal*; Battlecross: *Rise to Power, Will of Will*; Bury Your Dead: *Beauty and the Breakdown*; Chelsea Grin: *Evolve*; Motionless in white: *Infamous*; Daath: *The Concealers*; The Black Dahlia Murder: *DeFlorate, Nocturnal, Everblack*; Whitechapel: *Mark of the Blade.*

SWANN, DARRYL

Macy Gray: *On How Life Is, The Id, The Trouble with Being Myself, The World Is Yours.*

SWEDIEN, BRUCE

Michael Jackson: *Off the Wall, Thriller, Bad, Dangerous, Immortal, Xscape*; Quincy Jones: *Back on the Block, Q's Juke Joint*; The Jacksons: *Victory*; George Benson: *Give Me the Night*; Jennifer Lopez: *This Is Me . . . Then, Rebirth, Brave.*

TAN, PHIL

Mariah Carey: *Daydream, The Emancipation of Mimi*; Rihanna: *A Girl Like Me, Good Girl Gone Bad, Anti, Unapologetic*; Leona Lewis: *Spirit, Glassheart*; Sean Kingston: *Sean Kingston*; Ludacris: *Release Therapy*; Fergie: *The Dutchess*; Nelly: *Suit*; Gwen Stefani: *Love Angel Music Baby, The Sweet Escape*; Snoop Dogg: *R&G: The Masterpiece*; Usher: *My Way, Confessions, 8701*; Monica: *The Boy Is Mine*; Ciara: *The Evolution, Fantasy Ride*; Ne-Yo: *Year of the Gentleman*; Jennifer Hudson: *Jennifer Hudson*; Justin Bieber: *Under the Mistletoe, Believe*; Karmim: *Pulses*.

THOMAS, CHRIS

Pink Floyd: *Dark Side of the Moon, Discovery*; Dave Gilmour: *On an Island*; Razorlight: *Razorlight*; U2: *How to Dismantle an Atomic Bomb*; Pulp: *Different Class, This Is Hardcore*; INXS: *Listen Like Thieves, Kick, X*; The Pretenders: *Pretenders, Pretenders II, Learning to Crawl*; The Sex Pistols: *Never Mind the Bollocks*; Roxy Music: *For Your Pleasure, Stranded, Siren*; The Strypes: *Snapshot*; Mystery Jets: *Serotonin*.

TOWNSHEND, CENZO

Snow Patrol: *Eyes Open, A Hundred Million Suns*; Florence and the Machine: *Lungs*; Kaiser Chiefs: *Stay Together, Start the Revolution Without Me, Off with Their Heads*; Editors: *The Back Room, An End Has a Start*; U2: *No Line on the Horizon*; Bloc Party: *A Weekend in the City*; Babyshambles: *Shotter's Nation*; Late of the Pier: *Fantasy Black Channel*; Everything Everything: *A Fever Dream, Get to Heaven*; Amy Macdonald: *Under Stars*; A-Ha: *Cast in Steel*; Passenger: *Young as the Morning, Old as the Sea, Whispers*; Friendly Fires: *Pala*.

TSAI, SERGE

Shakira: *Oral Fixation Vol. 2* (including "Hips Don't Lie"), *She Wolf*; Nelly: *Suit*; Justin Bieber: *My World 2.0*; Wyclef Jean; *The Preacher's Son*; Luscious Jackson: *Electric Honey*; Estelle: *All of Me*.

VALENTINE, ERIC

Good Charlotte: *The Chronicles of Life and Death, The Young & the Hopeless, Youth Authority*; Lost Prophets: *Burn, Burn, Start Something*; Nickel Creek: *Why Should the Fire Die?, A Dotted Line*; Queens of the Stone Age: *Songs for the Deaf*; Slash: *Apocalyptic Love, Sahara, Slash*; Smash Mouth: *Summer Girl, You are My Number One, All Star, Astro Lounge*; Taking Back Sunday: *Louder Now, Twenty-Twenty Surgery*; Third Eye Blind: *Third Eye Blind*.

VAN DER SAAG, JOCHEM

Josh Groban: *Awake*; Andrea Bocelli: *Amore, Passione, Cinema*; Katherine Jenkins: *Believe, Bring Me to Life*; Seal: *Soul, Standards*.

VIRTUE, BRIAN

Audioslace: *Doesn't Remind Me, Out of Exile*; Chevelle: *Sci-Fi Crimes, Stray Arrows: A Collection of Favorites*; Crazy Town: *Revolving Door, Tales from the Darkside*; Deftones: *Saturday Night Wrist, B-Sides and Rarities*; Hawthorne Heights: *Zero*; Jane's Addiction: *Stars, True Nature, Just Because*; Thirty Seconds to Mars: *This is War, A Beautiful Lie, 30 Seconds to Mars*.

VISCONTI, TONY

T-Rex: *Electric Warrior, The Slider*; David Bowie: *Diamond Dogs, Young Americans, Heroes, Low, Scary Monsters, Heathen, Reality, Blackstar, No Plan*; Iggy Pop: *The Idiot*; The Moody Blues: *The Other Side of Life, Sur La Mer*; Thin Lizzy: *Bad Reputation, Live and Dangerous, Black Rose*.

WALKER, MILES

Betty Who: *Take Me When You Go, The Valley*; Coldplay: *A Head Full of Dreams*; Fifth Harmony: *Reflection, 7/27*; Jay Sean: *Sex 101, Neon, Hit the Lights*; Kygo: *Firestone, Cloud Nine, Stole the Show*; Kylie Minogue: *Kylie Christmas*; Meek Mill: *Wins and Losses, Young Black America, Issues*; Rich Gang: *Sho Me Love, Lifestyle*; Selena Gomez: *For You, Stars Dance, Revival*; U2: *Songs of Experience*.

WALLACE, ANDY

Nirvana: *Nevermind, From the Muddy Banks of the Wishkah*; Jeff Buckley: *Grace*; Linkin Park: *Hybrid Theory, Meteora, The Hunting Party*; Slayer: *Reign in Blood, South of Heaven, Seasons in the Abyss*; Blink-182: *Blink-182*; New Model Army: *Thunder & Consolation*; Sepultura: *Arise, Chaos AD*; Bad Religion: *Stranger Than Fiction*; Sonic Youth: *Dirty*; Rage against the Machine: *Rage against the Machine, Evil Empire*; The Cult: *Electric*; Biffy Clyro: *Only Revolutions, Puzzle*; Skunk Anansie: *Post Orgasmic Chill*; Foo Fighters: *There Is Nothing Left to Lose*; System of a Down: *Toxicity, Steal This Album, Mesmerize, Hypnotize*; Slipknot: *Iowa*; Stereophonics: *Just Enough Education to Perform*; Puddle of Mudd: *Come Clean, Life on Display*; Korn: *Untouchables*; Kasabian: *Empire*; Kelly Clarkson: *My December*; Kaiser Chiefs: *Off with Their Heads*; Avenged Sevenfold: *City of Evil, Avenged Sevenfold, Nightmare*; Rancid: *And Out Come the Wolves*; Ghost: *Meliora*; Avenged Sevenfold: *Hail to the King*.

WALLACE, MATT

David Baerwald: *Bedtime Stories*; Faith No More: *Angel Dust, Epic, The Real Thing, Introduce Yourself*; Maroon 5: *Songs About Jane*; New Monkees: *What I Want Single*; O.A.R: *All Sides, Shattered (The Car Around), XX*; Sons of Freedom: *Sons of Freedom, Gump*; The Replacements: *All Shook Down, Don't Tell a Soul, All for Nothing*; Trains: *Drops of Jupiter, Meet Virginia*.

WAY, DAVE

Christina Aguilera: *Christina Aguilera* (including "Genie in a Bottle"); Spice Girls: *Spice*; TLC: *Crazysexycool*; Michelle Branch: *The Spirit Room*; Macy Gray: *On How Life Is* (including "I Try"), *The Id*, *The Trouble with Being Myself*, *Big*; Michael Jackson: *Dangerous*, *Blood on the Dance Floor*, *Scream*; Taylor Hicks: *Taylor Hicks*; India Arie: *Acoustic Soul*; Savage Garden: *Affirmation*; Shakira: *Fijación Oral Vol. 1*; Fiona Apple: *Extraordinary Machine*; Toni Braxton: *Toni Braxton*; Sheryl Crow: *Wildflower*; Pink: *Missundaztood*; Destiny's Child: *Survivor*; Jennifer Paige: *Jennifer Paige* (including "Crush"); Boyz II Men: *Coolhighharmony*; Weird Al Yankovic: *Mandatory Fun*.

WHITE, JACK

Jack White: *Lazaretto*, *Blunderbuss*, *Boarding House Reach*, *Black Bat Licorice*, *Conquest*; The Dead Weather: *Dodge and Burn*, *Sea of Cowards*, *Horehound*; The Raconteurs: *Broken Boy Soldiers*, *Consolers of the Lonely*; The White Stripes: *Under Great White Northern Lights*, *Icky Thump*, *My Doorbell*, *Blue Orchid*, *Hotel Yorba*, *The Big Three Killed My Baby*.

WHITE, STUART

Alicia Keys: *The Element of Freedom*, *As I Am*; Beyonce: *Beyoncé*; Destiny's Child: *Love Songs*; Ed Sheeran: "Perfect"; Jay-Z: *4:44*, *Magna Carta Holy Grail*; Guns 'N' Roses: *Chinese Democracy*; K'NAAN: *Wavin' Flag*, *Troubadour*; Nicki Minaj: *Pink Friday: Roman Reloaded*.

WILTSHIRE, JAMES (OF THE FREEMASONS)

Beyonce: *Déjà vu*, *Beautiful Liar*; James Wiltshire: *Pump It Up*, *Zoo Summer Anthems*, *Café Mambo 2006*; Kylie Minogue: *The Best of Kylie Minogue*, *The Abbey Road Sessions*, *Aphrodite Les Folies*, *Wow*, *Boombox*; The Shapeshifters: *In the House*.

WORLEY, PAUL

Martina McBride: *The Time Has Come*, *The Way That I Am*, *Evolution*, *Emotion*, *Martina*; Pam Tillis: *Put Yourself in My Place*, *Homeward Looking Angel*, *Thunder and Roses*; The Dixie Chicks: *Wide Open Spaces*, *Fly*; Big & Rich: *Horse of a Different Color*, *Comin' to Your City*; Lady Antebellurn: *Need You Know*.

WRIGHT, TOBY

Slayer: *Divine Intervention*, *Soundtrack to the Apocalypse*; Alice in Chains: *Jar of Flies*, *Alice in Chains*, *Unplugged*; Korn: *Follow the Leader*; Metallica: *and Justice for All*; Mötley Crüe: *Girls, Girls, Girls*; Soulfly: *Primitive*; The Letter Black: *Rebuild*.

YOUNG GURU

Beyonce: *B'day*; Jay-Z: *4:44, Blueprint 3, American Gangster, Kingdom Come*; Linkin Park/Jay-Z: *Collision Course*; Snoop Dogg: *More Mallice*.

YOUTH

The Verve: *Urban Hymns*; Dido: *No Angel*; Dub Trees: *Celtic Vedic*; Embrace: *Good Will Out*; Heather Nova: *Oyster, Siren*; Crowded House: *Together Alone*; Primal Scream: *Riot City Blues*; The Fireman: *Strawberries Oceans Ships Forest, Rushes, Electric Arguments*; Killing Joke: *Revelations, Night Time, Pandemonium, Killing Joke*; Pink Floyd: *The Endless River*.

ZDAR, PHILIPPE

Cassius: *Go Up, Ibifornia, 15 Again, Au Reve, Fame*; Malante: *Whow!*; Phoenix: *Alphabetical/United, If I Ever Feel Better, Run Run Run*; Beastie Boys: *Hot Sauce Committee Part Two*.

ZOOK, JOE

Modest Mouse: *We Were Dead Before the Ship Even Sank, Stranger to Ourselves*; Sheryl Crow: *C'mon C'mon*; Courtney Love: *America's Sweetheart*; One Republic: *Dreaming Out Loud, Native*; Rancid: *Life Won't Wait*; Brooke Fraser: *Albertine*.

CHAPTER 1

1 Massey, H. (2000). *Behind the glass: Top record producers tell how they craft the hits* (Vol. I). Miller Freeman Books.
2 Robinson, A. (1997). Interview with Nigel Godrich. *Mix*, August.
3 Droney, M. (2003). *Mix masters: Platinum engineers reveal their secrets for success.* Berklee Press.
4 Tingen, P. (2010). Inside track. *Sound on Sound*, February.
5 Massey, H. (2009). *Behind the glass: Top record producers tell how they craft the hits* (Vol. II). Backbeat Books.
6 Langford, S. (2009). Inside track. *Sound on Sound*, October.
7 Massey, H. (2000). *Behind the glass: Top record producers tell how they craft the hits* (Vol. I). Miller Freeman Books.
8 Massey, H. (2009). *Behind the glass: Top record producers tell how they craft the hits* (Vol. II). Backbeat Books.
9 Massey, H. (2009). *Behind the glass: Top record producers tell how they craft the hits* (Vol. II). Backbeat Books.
10 Massey, H. (2000). *Behind the glass: Top record producers tell how they craft the hits* (Vol. I). Miller Freeman Books.
11 Massey, H. (2000). *Behind the glass: Top record producers tell how they craft the hits* (Vol. I). Miller Freeman Books.

CHAPTER 2

1 Senior, M. (2009). Interview with Bruce Swedien. *Sound on Sound*, November.
2 Tingen, P. (2011). Inside track. *Sound on Sound*, September.
3 Parsons, A. *Art & science of sound recording: Understanding & applying equalisation.* Video available from www.artandscienceofsound.com/
4 Katz, B. (2002). *Mastering audio: The art and the science.* Focal Press.
5 Parsons, A. *Art & science of sound recording: Understanding & applying equalisation.* Video available from www.artandscienceofsound.com/
6 Tingen, P. (2015). Inside track. *Sound on Sound*, June.
7 Tingen, P. (2007). Inside track. *Sound on Sound*, May.
8 Massey, H. (2000). *Behind the glass: Top record producers tell how they craft the hits* (Vol. I). Miller Freeman Books.
9 Owsinski, B. (2006). *The mixing engineer's handbook* (2nd ed.). Thomson Course Technology PTR.
10 *Pensado's Place* web TV show (available at www.pensadosplace.tv/), Episode 117.
11 Tingen, P. (2016). Inside track. *Sound on Sound*, January.
12 Tingen, P. (2015). Inside track. *Sound on Sound*, August.
13 Lockwood, D. (1999). Interview with Bob Clearmountain. *Sound on Sound*, June.
14 Tingen, P. (2008). Inside track. *Sound on Sound*, November.

CHAPTER 3

1 *Pensado's Place* web TV show (available at www.pensadosplace.tv/), Episode 270.
2 Tingen, P. (2016). Inside track. *Sound on Sound*, February.

3 Owsinski, B. (2006). *The mixing engineer's handbook* (2nd ed.). Thomson Course Technology PTR.
4 Owsinski, B. (2006). *The mixing engineer's handbook* (2nd ed.). Thomson Course Technology PTR.
5 Owsinski, B. (2006). *The mixing engineer's handbook* (2nd ed.). Thomson Course Technology PTR.
6 Owsinski, B. (2006). *The mixing engineer's handbook* (2nd ed.). Thomson Course Technology PTR.
7 Tingen, P. (2008). Inside track. *Sound on Sound*, September.
8 Tingen, P. (2007). Inside track. *Sound on Sound*, December.
9 Tingen, P. (2007). Inside track. *Sound on Sound*, January.
10 Massey, H. (2000). *Behind the glass: Top record producers tell how they craft the hits* (Vol. I). Miller Freeman Books.
11 Tingen, P. (2007). Inside track. *Sound on Sound*, February.
12 Tingen, P. (2016). Inside track. *Sound on Sound*, March.
13 Tingen, P. (2007). Inside track. *Sound on Sound*, June.
14 Massey, H. (2000). *Behind the glass: Top record producers tell how they craft the hits* (Vol. I). Miller Freeman Books.

CHAPTER 4

1 Elevado, R. Available from www.gearslutz.com/board/q-russell-elevado/115987-monitoring-dragon.html
2 Tingen, P. (2010). Inside track. *Sound on Sound*, March.
3 Lockwood, D. (1999). Interview with Bob Clearmountain. *Sound on Sound*, June.
4 Elevado, R. Available from www.gearslutz.com/board/q-russell-elevado/117088-my-mixing.html/
5 *Pensado's Place* web TV show (available at www.pensadosplace.tv/), Episode 163.
6 Buskin, R. (2004). Classic tracks. *Sound on Sound*, August.
7 Owsinski, B. (2006). *The mixing engineer's handbook* (2nd ed.). Thomson Course Technology PTR.
8 Lockwood, D. (1999). Interview with Bob Clearmountain. *Sound on Sound*, June.
9 Massey, H. (2000). *Behind the glass: Top record producers tell how they craft the hits* (Vol. I). Miller Freeman Books.
10 Elevado, R. Available from www.gearslutz.com/board/q-russell-elevado/117226-mixing-process-time-breaks.html
11 Massey, H. (2000). *Behind the glass: Top record producers tell how they craft the hits* (Vol. I). Miller Freeman Books.
12 Tingen, P. (2008). Inside track. *Sound on Sound*, June.
13 Owsinski, B. (2006). *The mixing engineer's handbook* (2nd ed.). Thomson Course Technology PTR.
14 Massey, H. (2009). *Behind the glass: Top record producers tell how they craft the hits* (Vol. II). Backbeat Books.
15 Lockwood, D. (1999). Interview with Bob Clearmountain. *Sound on Sound*, June.
16 Massey, H. (2000). *Behind the glass: Top record producers tell how they craft the hits* (Vol. I). Miller Freeman Books.
17 Tingen, P. (2014). Inside track. *Sound on Sound*, July.
18 Lockwood, D. (1999). Interview with Bob Clearmountain. *Sound on Sound*, June.
19 Tingen, P. (2007). Inside track. *Sound on Sound*, May.
20 Buskin, R. (2000). Classic tracks. *Sound on Sound*, May.
21 Massey, H. (2000). *Behind the glass: Top record producers tell how they craft the hits* (Vol. I). Miller Freeman Books.
22 Massey, H. (2000). *Behind the glass: Top record producers tell how they craft the hits* (Vol. I). Miller Freeman Books.
23 Owsinski, B. (2006). *The mixing engineer's handbook* (2nd ed.). Thomson Course Technology PTR.
24 *Pensado's Place* web TV show (available at www.pensadosplace.tv/), Episode 95.
25 *Pensado's Place* web TV show (available at www.pensadosplace.tv/), Episode 108.
26 Massey, H. (2000). *Behind the glass: Top record producers tell how they craft the hits* (Vol. I). Miller Freeman Books.
27 Tingen, P. (2008). Inside track. *Sound on Sound*, October.

28   *Pensado's Place* web TV show (available at www.pensadosplace.tv/), Episode 108.
29   Massey, H. (2000). *Behind the glass: Top record producers tell how they craft the hits* (Vol. I). Miller Freeman Books.
30   Stavrou, M. (2003). *Mixing with your mind: Closely guarded secrets of sound balance engineering revealed*. Flux Research Pty.
31   Owsinski, B. (2006). *The mixing engineer's handbook* (2nd ed.). Thomson Course Technology PTR.
32   Tingen, P. (2010). Inside track. *Sound on Sound*, February.
33   Massey, H. (2000). *Behind the glass: Top record producers tell how they craft the hits* (Vol. I). Miller Freeman Books.
34   Owsinski, B. (2006). *The mixing engineer's handbook* (2nd ed.). Thomson Course Technology PTR.
35   Owsinski, B. (2006). *The mixing engineer's handbook* (2nd ed.). Thomson Course Technology PTR.
36   Tingen, P. (2011). Inside track. *Sound on Sound*, September.
37   Droney, M. (2003). *Mix masters: Platinum engineers reveal their secrets for success*. Berklee Press.
38   Tingen, P. (2015). Inside track. *Sound on Sound*, October.
39   Owsinski, B. (2006). *The mixing engineer's handbook* (2nd ed.). Thomson Course Technology PTR.
40   Tingen, P. (2009). Inside track. *Sound on Sound*, December.

PART 2

1   Senior, M. (2003). Interview with Glen Ballard. *Sound on Sound*, March.
2   *Pensado's Place* web TV show (available at www.pensadosplace.tv/), Episode 127.
3   Tingen, P. (2010). Inside track. *Sound on Sound*, March.
4   Tingen, P. (2008). Inside track. *Sound on Sound*, November.
5   Tingen, P. (2007). Inside track. *Sound on Sound*, May.
6   Tingen, P. (2007). Inside track. *Sound on Sound*, December.
7   Daley, D. (2004). Interview with Tom Maserati. *Sound on Sound*, February.
8   Tingen, P. (2010). Inside track. *Sound on Sound*, February.

CHAPTER 5

1   Tingen, P. (2009). Inside track. *Sound on Sound*, November.
2   Tingen, P. (2015). Inside track. *Sound on Sound*, March.
3   Tingen, P. (2009). Inside track. *Sound on Sound*, September
4   Tingen, P. (2016). Inside track. *Sound on Sound*, February
5   Tingen, P. (2000). Interview with Tom Lord-Alge. *Sound on Sound*, April.
6   *Pensado's Place* web TV show (available at www.pensadosplace.tv/), Episode 127.
7   Tingen, P. (2014). Inside track. *Sound on Sound*, May
8   Tingen, P. (2008). Inside track. *Sound on Sound*, November.
9   Tingen, P. (2007). Inside track. *Sound on Sound*, May.
10   Tingen, P. (2000). Interview with Tom Lord-Alge. *Sound on Sound*, April.
11   Tingen, P. (2009). Inside track. *Sound on Sound*, March.
12   Tingen, P. (2009). Inside track. *Sound on Sound*, April.
13   Tingen, P. (2008). Inside track. *Sound on Sound*, July.
14   Tingen, P. (2014). Inside track. *Sound on Sound*, May.
15   Tingen, P. (1999). Interview with Spike Stent. *Sound on Sound*, January.
16   Droney, M. (1999). Interview with Mike Shipley. *Mix*, June.

CHAPTER 6

1   Daley, D. (2006). Interview with Serban Ghenea. *EQ*, May.
2   Tingen, P. (2007). Inside track. *Sound on Sound*, November.
3   *Pensado's Place* web TV show (available at www.pensadosplace.tv/), Episode 121.
4   Stavrou, M. (2003). *Mixing with your mind: Closely guarded secrets of sound balance engineering revealed*. Flux Research Pty.

5 Massey, H. (2009). *Behind the glass: Top record producers tell how they craft the hits* (Vol. II). Backbeat Books.
6 Tingen, P. (2016). Inside track. *Sound on Sound*, January.
7 Tingen, P. (2012). Inside track. *Sound on Sound*, March.
8 Massey, H. (2000). *Behind the glass: Top record producers tell how they craft the hits* (Vol. I). Miller Freeman Books.
9 Senior, M. (2000). Interview with Steve Bush. *Sound on Sound*, March.
10 *Pensado's Place* web TV show (available at www.pensadosplace.tv/), Episode 96.
11 Tingen, P. (2015). Inside track. *Sound on Sound*, August.
12 Daley, D. (2003). Vocal fixes. *Sound on Sound*, October.
13 Tingen, P. (2016). Inside track. *Sound on Sound*, January.
14 Tingen, P. (1998). Interview with Marius de Vries. *Sound on Sound*, September.

CHAPTER 7

1 Flint, T. (2001). Interview with Tony Platt. *Sound on Sound*, April.
2 Massey, H. (2009). *Behind the glass: Top record producers tell how they craft the hits* (Vol. II). Backbeat Books.
3 Inglis, S. (2001). Interview with Pascal Gabriel. *Sound on Sound*, April.
4 Inglis, S. (2001). Interview with Pascal Gabriel. *Sound on Sound*, April.
5 Massey, H. (2000). *Behind the glass: Top record producers tell how they craft the hits* (Vol. I). Miller Freeman Books.
6 Daley, D. (2006). Interview with Serban Ghenea. *EQ*, May.
7 Tingen, P. (2016). Inside track. *Sound on Sound*, January.
8 Doyle, T. (2007). Interview with Mark Ronson. *Sound on Sound*, May.
9 Tingen, P. (2011). Inside track. *Sound on Sound*, July.
10 Tingen, P. (2010). Inside track. *Sound on Sound*, April.
11 Owsinski, B. (2006). *The mixing engineer's handbook* (2nd ed.). Thomson Course Technology PTR.
12 *Pensado's Place* web TV show (available at www.pensadosplace.tv/), Episode 106.
13 Tingen, P. (2012). Inside track. *Sound on Sound*, October.
14 Tingen, P. (2014). Inside track. *Sound on Sound*, September.
15 Daley, D. (2004). Interview with Wyclef Jean. *Sound on Sound*, July.
16 Tingen, P. (2007). Inside track. *Sound on Sound*, November.
17 Bell, M. (1998). Interview with Steve Power. *Sound on Sound*, November.
18 Buskin, R. (2007). Classic tracks. *Sound on Sound*, August.
19 Massey, H. (2009). *Behind the glass: Top record producers tell how they craft the hits* (Vol. II). Backbeat Books.
20 Tingen, P. (2012). Inside track. *Sound on Sound*, October.
21 Flint, T. (2001). Interview with Marius de Vries. *Sound on Sound*, November.
22 *Pensado's Place* web TV show (available at www.pensadosplace.tv/), Episode 131.
23 *Pensado's Place* web TV show (available at www.pensadosplace.tv/), Episode 114.
24 Droney, M. (2003). *Mix masters: Platinum engineers reveal their secrets for success*. Berklee Press.

PART 3

1 Massey, H. (2000). *Behind the glass: Top record producers tell how they craft the hits* (Vol. I). Miller Freeman Books.

CHAPTER 8

1 *Pensado's Place* web TV show (available at www.pensadosplace.tv/), Episode 108.
2 Tingen, P. (2007). Inside track. *Sound on Sound*, July.
3 Tingen, P. (2010). Inside track. *Sound on Sound*, August.
4 Tingen, P. (2008). Inside track. *Sound on Sound*, March.

5 *Pensado's Place* web TV show (available at www.pensadosplace.tv/), Episode 95.

6 Lockwood, D. (1999). Interview with Bob Clearmountain. *Sound on Sound*, June.

7 Tingen, P. (2007). Inside track. *Sound on Sound*, May.

8 Tingen, P. (2007). Inside track. *Sound on Sound*, March.

9 Tingen, P. (2010). Inside track. *Sound on Sound*, March.

10 Massey, H. (2000). *Behind the glass: Top record producers tell how they craft the hits* (Vol. I). Miller Freeman Books.

11 Massey, H. (2000). *Behind the glass: Top record producers tell how they craft the hits* (Vol. I). Miller Freeman Books.

12 Massey, H. (2000). *Behind the glass: Top record producers tell how they craft the hits* (Vol. I). Miller Freeman Books.

13 Tingen, P. (2000). Interview with Tom Lord-Alge. *Sound on Sound*, April.

14 Massey, H. (2000). *Behind the glass: Top record producers tell how they craft the hits* (Vol. I). Miller Freeman Books.

15 Massey, H. (2000). *Behind the glass: Top record producers tell how they craft the hits* (Vol. I). Miller Freeman Books.

16 Owsinski, B. (2006). *The mixing engineer's handbook* (2nd ed.). Thomson Course Technology PTR.

17 Tingen, P. (2011). Inside track. *Sound on Sound*, August.

18 *Pensado's Place* web TV show (available at www.pensadosplace.tv/), Episode 220.

19 Massey, H. (2009). *Behind the glass: Top record producers tell how they craft the hits* (Vol. II). Backbeat Books.

20 Massey, H. (2000). *Behind the glass: Top record producers tell how they craft the hits* (Vol. I). Miller Freeman Books.

21 Tingen, P. (2010). Inside track. *Sound on Sound*, April.

22 Massey, H. (2000). *Behind the glass: Top record producers tell how they craft the hits* (Vol. I). Miller Freeman Books.

23 Nichols, R. (2006). Across the board. *Sound on Sound*, August.

24 Tingen, P. (1997). Interview with Flood. *Sound on Sound*, July.

25 Mixerman. (2010). *Zen and the Art of Mixing*. Hal Leonard Books.

26 Tingen, P. (2011). Inside Track. *Sound on Sound*, October.

27 *Pensado's Place* web TV show (available at www.pensadosplace.tv/), Episode 155.

28 *Pensado's Place* web TV show (available at www.pensadosplace.tv/), Episode 179.

29 Massey, H. (2009). *Behind the glass: Top record producers tell how they craft the hits* (Vol. II). Backbeat Books.

30 Massey, H. (2009). *Behind the glass: Top record producers tell how they craft the hits* (Vol. II). Backbeat Books.

31 Stavrou, M. (2003). *Mixing with your mind: Closely guarded secrets of sound balance engineering revealed*. Flux Research Pty.

32 Nichols, R. (2006). Across the board. *Sound on Sound*, August.

33 Izhaki, R. (2008). *Mixing audio: Concepts, practices and tools*. Focal Press.

34 *Pensado's Place* web TV show (available at www.pensadosplace.tv/), Episode 115.

35 *Behind The Speakers* web video (available at behindthespeakers.com/mixing-vocals-5-tricks/).

36 *Pensado's Place* web TV show (available at www.pensadosplace.tv/), Episode 246.

37 Tingen, P. (2007). Inside track. *Sound on Sound*, July.

38 Massey, H. (2009). *Behind the glass: Top record producers tell how they craft the hits* (Vol. II). Backbeat Books.

39 Buskin, R. (2007). Classic tracks. *Sound on Sound*, August.

40 Droney, M. (2003). *Mix masters: Platinum engineers reveal their secrets for success*. Berklee Press.

41 Massey, H. (2009). *Behind the glass: Top record producers tell how they craft the hits* (Vol. II). Backbeat Books.

42 Massey, H. (2000). *Behind the glass: Top record producers tell how they craft the hits* (Vol. I). Miller Freeman Books.

43 Tingen, P. (2010). Inside track. *Sound on Sound*, February.

44 Owsinski, B. (2006). *The mixing engineer's handbook* (2nd ed.). Thomson Course Technology PTR.

45 Daley, D. (2006). Interview with Steve Hodge. *Sound on Sound*, November.

46 Tingen, P. (2012). Inside track. *Sound on Sound*, January.
47 Massey, H. (2009). *Behind the glass: Top record producers tell how they craft the hits* (Vol. II). Backbeat Books.

CHAPTER 9

1 Tingen, P. (2012). Inside track. *Sound on Sound*, January.
2 *Pensado's Place* web TV show (available at www.pensadosplace.tv/), Episodes 98 and 146.
3 Tingen, P. (2016). Inside track. *Sound on Sound*, June.
4 Massey, H. (2000). *Behind the glass: Top record producers tell how they craft the hits* (Vol. I). Miller Freeman Books.
5 *Pensado's Place* web TV show (available at www.pensadosplace.tv/), Episode 179.
6 Owsinski, B. (2006). The mixing engineer's handbook (2nd ed.). Thomson Course Technology PTR.
7 *Pensado's Place* web TV show (available at www.pensadosplace.tv/), Episode 136.
8 Tingen, P. (2000). Interview with Tom Lord-Alge. *Sound on Sound*, April.
9 Tingen, P. (2016). Inside Track. *Sound on Sound*, March.
10 Tingen, P. (2005). Interview with Joe Barresi. *Sound on Sound*, July.
11 Massey, H. (2009). *Behind the glass: Top record producers tell how they craft the hits* (Vol. II). Backbeat Books.
12 Tingen, P. (2008). Inside track. *Sound on Sound*, November.
13 Droney, M. (2003). *Mix masters: Platinum engineers reveal their secrets for success*. Berklee Press.
14 Tingen, P. (2011). Inside Track. *Sound on Sound*, October.
15 Tingen, P. (2010). Inside track. *Sound on Sound*, February.
16 *Pensado's Place* web TV show (available at www.pensadosplace.tv/), Episode 178.

CHAPTER 10

1 Tingen, P. (2007). Inside track. *Sound on Sound*, July.
2 Bruce, B. (1999). Interview with Al Stone. *Sound on Sound*, December.
3 Tingen, P. (2007). Inside track. *Sound on Sound*, April.
4 Tingen, P. (2007). Inside track. *Sound on Sound*, November.
5 Tingen, P. (2016). Inside Track. *Sound on Sound*, January.
6 Tingen, P. (2015). Inside Track. *Sound on Sound*, December.

CHAPTER 11

1 *Pensado's Place* web TV show (available at www.pensadosplace.tv/), Episode 151.
2 Hatschek, K. (2005). *The golden moment: Recording secrets from the pros*. Backbeat Books.
3 *Pensado's Place* web TV show (available at www.pensadosplace.tv/), Episode 141.
4 Massey, H. (2000). *Behind the glass: Top record producers tell how they craft the hits* (Vol. I). Miller Freeman Books.
5 Massey, H. (2009). *Behind the glass: Top record producers tell how they craft the hits* (Vol. II). Backbeat Books.
6 Massey, H. (2000). *Behind the glass: Top record producers tell how they craft the hits* (Vol. I). Miller Freeman Books.
7 Tingen, P. (2007). Inside track. *Sound on Sound*, May.
8 Tingen, P. (2008). Inside track. *Sound on Sound*, March.
9 Tingen, P. (2015). Inside track. *Sound on Sound*, July.
10 *Pensado's Place* web TV show (available at www.pensadosplace.tv/), Episode 178.
11 Tingen, P. (2008). Inside track. *Sound on Sound*, February.
12 *Pensado's Place* web TV show (available at www.pensadosplace.tv/), Episode 129.
13 *Pensado's Place* web TV show (available at www.pensadosplace.tv/), Episode 151.
14 *Pensado's Place* web TV show (available at www.pensadosplace.tv/), Episode 141.
15 Tingen, P. (2007). Inside track. *Sound on Sound*, August.

16 Tingen, P. (2007). Inside track. *Sound on Sound*, August.
17 Massey, H. (2000). *Behind the glass: Top record producers tell how they craft the hits* (Vol. *I*). Miller Freeman Books.
18 Massey, H. (2009). *Behind the glass: Top record producers tell how they craft the hits* (Vol. *II*). Backbeat Books.
19 Droney, M. (2003). *Mix masters: Platinum engineers reveal their secrets for success*. Berklee Press.
20 Tingen, P. (2007). Inside track. *Sound on Sound*, January.
21 *Pensado's Place* web TV show (available at www.pensadosplace.tv/), Episode 276.
22 Droney, M. (2003). *Mix masters: Platinum engineers reveal their secrets for success*. Berklee Press.
23 Massey, H. (2000). *Behind the glass: Top record producers tell how they craft the hits* (Vol. *I*). Miller Freeman Books.
24 Tingen, P. (2008). Inside track. *Sound on Sound*, April.
25 Droney, M. (2003). *Mix masters: Platinum engineers reveal their secrets for success*. Berklee Press.
26 *Pensado's Place* web TV show (available at www.pensadosplace.tv/), Episode 127.

CHAPTER 12

1 *Pensado's Place* web TV show (available at www.pensadosplace.tv/), Episode 183.
2 Tingen, P. (2016). Inside track. *Sound on Sound*, June.
3 Tingen, P. (2012). Inside track. *Sound on Sound*, September.
4 Tingen, P. (2015). Inside track. *Sound on Sound*, February.
5 *Pensado's Place* web TV show (available at www.pensadosplace.tv/), Episode 129.
6 Tingen, P. (2003). Interview with Rich Costey. *Sound on Sound*, December.
7 Daley, D. (2006). Interview with Serban Ghenea. *EQ*, May.
8 Massey, H. (2000). *Behind the glass: Top record producers tell how they craft the hits* (Vol. *I*). Miller Freeman Books.
9 Buskin, R. (1997). Interview with Jon Gass. *Sound on Sound*, August.
10 Tingen, P. (2016). Inside track. *Sound on Sound*, May.
11 Tingen, P. (2008). Inside track. *Sound on Sound*, December.
12 Tingen, P. (2014). Inside track. *Sound on Sound*, March.
13 Tingen, P. (2010). Inside track. *Sound on Sound*, March.
14 Tingen, P. (2011). Inside track. *Sound on Sound*, October.
15 *Pensado's Place* web TV show (available at www.pensadosplace.tv/), Episode 129.
16 Barbiero, M. (2005). Interview with Andy Wallace. *Mix*, October.
17 Tingen, P. (2014). Inside track. *Sound on Sound*, July.
18 Tingen, P. (2014). Inside track. *Sound on Sound*, September.
19 *Pensado's Place* web TV show (available at www.pensadosplace.tv/), Episode 281.
20 Tingen, P. (2007). Interview with DJ Premier. *Sound on Sound*, July.

CHAPTER 13

1 Droney, M. (2003). *Mix masters: Platinum engineers reveal their secrets for success*. Berklee Press.
2 Tingen, P. (2009). Inside track. *Sound on Sound*, July.
3 Tingen, P. (2007). Inside track. *Sound on Sound*, December.
4 Buskin, R. (1997). Interview with Jon Gass. *Sound on Sound*, August.
5 Tingen, P. (2007). Inside track. *Sound on Sound*, December.
6 Tingen, P. (2009). Inside track. *Sound on Sound*, February.
7 Droney, M. (2003). *Mix masters: Platinum engineers reveal their secrets for success*. Berklee Press.
8 Tingen, P. (2011). Inside track. *Sound on Sound*, July.
9 Tingen, P. (2016). Inside track. *Sound on Sound*, May.
10 Tingen, P. (2016). Inside track. *Sound on Sound*, June.
11 Tingen, P. (2013). Inside track. *Sound on Sound*, October.
12 Massey, H. (2000). *Behind the glass: Top record producers tell how they craft the hits* (Vol. *I*). Miller Freeman Books.

13 Tingen, P. (2016). Inside track. *Sound on Sound*, June.
14 Sherbourne, S. (2009). Interview with Imogen Heap. *Sound on Sound*, December.
15 Tingen, P. (2016). Inside track. *Sound on Sound*, January.
16 Tingen, P. (2011). Inside track. *Sound on Sound*, September.
17 *Pensado's Place* web TV show (available at www.pensadosplace.tv/), Episode 283.
18 Tingen, P. (2016). Inside track. *Sound on Sound*, March.
19 Tingen, P. (2015). Inside track. *Sound on Sound*, October.
20 Tingen, P. (2014). Inside track. *Sound on Sound*, May.
21 Tingen, P. (2014). Inside track. *Sound on Sound*, September.

CHAPTER 14

1 Buskin, R. (2004). Classic tracks. *Sound on Sound*, September.
2 Tingen, P. (2007). Inside track. *Sound on Sound*, April.
3 Tingen, P. (2016). Inside track. *Sound on Sound*, March.
4 Tingen, P. (2007). Inside track. *Sound on Sound*, April.

CHAPTER 15

1 Tingen, P. (2007). Interview with DJ Premier. *Sound on Sound*, July.
2 Tingen, P. (2008). Inside track. *Sound on Sound*, August.
3 Doyle, T. (2009). Interview with Youth. *Sound on Sound*, March.
4 Tingen, P. (2009). Inside track. *Sound on Sound*, December.
5 Tingen, P. (2015). Inside track. *Sound on Sound*, December.
6 Tingen, P. (2012). Inside track. *Sound on Sound*, January.
7 Massey, H. (2000). *Behind the glass: Top record producers tell how they craft the hits* (Vol. I). Miller Freeman Books.
8 Massey, H. (2000). *Behind the glass: Top record producers tell how they craft the hits* (Vol. I). Miller Freeman Books.
9 *Pensado's Place* web TV show (available at www.pensadosplace.tv/), Episode 265.

PART 4

1 Owsinski, B. (2006). *The mixing engineer's handbook* (2nd ed.). Thomson Course Technology PTR.
2 Elevado, R. Available from www.gearslutz.com/board/q-russell-elevado/117088-my-mixing.html/

CHAPTER 16

1 Massey, H. (2000). *Behind the glass: Top record producers tell how they craft the hits* (Vol. I). Miller Freeman Books.
2 Massey, H. (2000). *Behind the glass: Top record producers tell how they craft the hits* (Vol. I). Miller Freeman Books.
3 Massey, H. (2000). *Behind the glass: Top record producers tell how they craft the hits* (Vol. I). Miller Freeman Books.
4 Owsinski, B. (2006). *The mixing engineer's handbook* (2nd ed.). Thomson Course Technology PTR.
5 Hoskulds, S. (2016). Mixing Ornette Coleman's Last Album. *Sound on Sound*, March.
6 Droney, M. (2003). *Mix masters: Platinum engineers reveal their secrets for success*. Berklee Press.
7 Massey, H. (2000). *Behind the glass: Top record producers tell how they craft the hits* (Vol. I). Miller Freeman Books.
8 Droney, M. (2003). *Mix masters: Platinum engineers reveal their secrets for success*. Berklee Press.
9 Sherbourne, S. (2009). Interview with Imogen Heap. *Sound on Sound*, December.
10 Massey, H. (2000). *Behind the glass: Top record producers tell how they craft the hits* (Vol. I). Miller Freeman Books.

11 Massey, H. (2000). *Behind the glass: Top record producers tell how they craft the hits* (Vol. I). Miller Freeman Books.
12 Senior, M. (2003). Interview with Glen Ballard. *Sound on Sound*, March.
13 *Pensado's Place* web TV show (available at www.pensadosplace.tv/), Episode 163.
14 Bruce, B. (1999). Interview with Al Stone. *Sound on Sound*, December.
15 Inglis, S. (2001). Interview with Guy Sigsworth. *Sound on Sound*, March.
16 *Pensado's Place* web TV show (available at www.pensadosplace.tv/), Episode 241.
17 Massey, H. (2000). *Behind the glass: Top record producers tell how they craft the hits* (Vol. I). Miller Freeman Books.
18 Daley, D. (2005). Interview with Manny Marroquin. *Sound on Sound*, May.
19 Daley, D. (2005). Interview with Manny Marroquin. *Sound on Sound*, May.
20 Droney, M. (2003). *Mix masters: Platinum engineers reveal their secrets for success*. Berklee Press.
21 Tingen, P. (2003). Interview with Elliot Scheiner. *Sound on Sound*, August.
22 Tingen, P. (2014). Inside track. *Sound on Sound*, October.
23 Senior, M. (2003). Interview with S. Husky Hoskulds. *Sound on Sound*, July.
24 *Pensado's Place* web TV show (available at www.pensadosplace.tv/), Episode 129.
25 Massey, H. (2000). *Behind the glass: Top record producers tell how they craft the hits* (Vol. I). Miller Freeman Books.
26 Flint, T. (2001). Interview with Tony Platt. *Sound on Sound*, April.
27 Tingen, P. (2005). Interview with Steve Albini. *Sound on Sound*, September.
28 *Pensado's Place* web TV show (available at www.pensadosplace.tv/), Episode 179.
29 *Pensado's Place* web TV show (available at www.pensadosplace.tv/), Episode 140.

CHAPTER 17

1 *Pensado's Place* web TV show (available at.pensadosplace.tv/), Episode 250.
2 *Pensado's Place* web TV show (available at www.pensadosplace.tv/), Episode 183.
3 Owsinski, B. (2006). *The mixing engineer's handbook* (2nd ed.). Thomson Course Technology PTR.
4 Tingen, P. (1999). Interview with Spike Stent. *Sound on Sound*, January.
5 *Pensado's Place* web TV show (available at www.pensadosplace.tv/), Episode 166.
6 Tingen, P. (2008). Inside track. *Sound on Sound*, January.

CHAPTER 18

1 Tingen, P. (2016). Inside track. *Sound on Sound*, June.
2 Tingen, P. (2014). Inside track. *Sound on Sound*, July.
3 *Pensado's Place* web TV show (available at www.pensadosplace.tv/), Episode 122.
4 Tingen, P. (2016). Inside track. *Sound on Sound*, June.
5 Barbiero, M. (2005). Interview with Andy Wallace. *Mix*, October.
6 *Pensado's Place* web TV show (available at www.pensadosplace.tv/), Episode 147.

CHAPTER 19

1 Barbiero, M. (2005). Interview with Andy Wallace. *Mix*, October.
2 Owsinski, B. (2006). *The mixing engineer's handbook* (2nd ed.). Thomson Course Technology PTR.
3 Tingen, P. (2014). Inside track. *Sound on Sound*, July.
4 Tingen, P. (2008). Inside track. *Sound on Sound*, November.
5 Tingen, P. (1999). Interview with Spike Stent. *Sound on Sound*, January.
6 Tingen, P. (2008). Inside track. *Sound on Sound*, December.
7 *Pensado's Place* web TV show (available at www.pensadosplace.tv/), Episode 127.
8 Tingen, P. (2015). Inside track. *Sound on Sound*, October.
9 *Pensado's Place* web TV show (available at www.pensadosplace.tv/), Episode 178.
10 Tingen, P. (2012). Inside track. *Sound on Sound*, January.

11   Tingen, P. (2016). Inside track. *Sound on Sound*, May.
12   *Pensado's Place* web TV show (available at www.pensadosplace.tv/), Episode 166.
13   Tingen, P. (2014). Inside track. *Sound on Sound*, May.
14   Tingen, P. (2011). Inside track. *Sound on Sound*, September.
15   Tingen, P. (2014). Inside track. *Sound on Sound*, July.
16   *Pensado's Place* web TV show (available at www.pensadosplace.tv/), Episode 250.
17   *Pensado's Place* web TV show (available at www.pensadosplace.tv/), Episode 127.
18   *Pensado's Place* web TV show (available at www.pensadosplace.tv/), Episode 166.
19   Tingen, P. (2008). Inside track. *Sound on Sound*, July.
20   Tingen, P. (2015). Inside track. *Sound on Sound*, July.
21   Tingen, P. (2016). Inside track. *Sound on Sound*, July.
22   Tingen, P. (2014). Inside track. *Sound on Sound*, May.
23   Tingen, P. (2011). Inside track. *Sound on Sound*, September.
24   Tingen, P. (2016). Inside track. *Sound on Sound*, June.
25   Tingen, P. (2016). Inside track. *Sound on Sound*, June.
26   Tingen, P. (2009). Inside track. *Sound on Sound*, July.
27   Tingen, P. (2015). Inside track. *Sound on Sound*, March.
28   Tingen, P. (2016). Inside track. *Sound on Sound*, June.
29   Massey, H. (2000). *Behind the glass: Top record producers tell how they craft the hits* (Vol. I). Miller Freeman Books.
30   Owsinski, B. (2006). *The mixing engineer's handbook* (2nd ed.). Thomson Course Technology PTR.
31   Lockwood, D. (1999). Interview with Bob Clearmountain. *Sound on Sound*, June.
32   *Pensado's Place* web TV show (available at www.pensadosplace.tv/), Episode 136.
33   Tingen, P. (2007). Inside track. *Sound on Sound*, May.
34   Tingen, P. (2007). Inside track. *Sound on Sound*, September.
35   Droney, M. (2003). *Mix masters: Platinum engineers reveal their secrets for success*. Berklee Press.
36   Tingen, P. (2011). Inside track. *Sound on Sound*, September.
37   *Pensado's Place* web TV show (available at www.pensadosplace.tv/), Episode 136.
38   *Pensado's Place* web TV show (available at www.pensadosplace.tv/), Episode 147.
39   *Pensado's Place* web TV show (available at www.pensadosplace.tv/), Episode 105.
40   Tingen, P. (2014). Inside track. *Sound on Sound*, July.
41   Tingen, P. (2011). Inside track. *Sound on Sound*, July.
42   Anderson, J. (2007). Interview with Mike Shipley. *EQ*, November.
43   Daley, D. (2006). Interview with Steve Hodge. *Sound on Sound*, November.
44   *Pensado's Place* web TV show (available at www.pensadosplace.tv/), Episode 127.
45   Tingen, P. (2014). Inside track. *Sound on Sound*, July.
46   Tingen, P. (2007). Inside track. *Sound on Sound*, May.
47   Tingen, P. (2014). Inside track. *Sound on Sound*, July.
48   Tingen, P. (2006). Interview with Jim Abbiss. *Sound on Sound*, September.
49   Barbiero, M. (2005). Interview with Andy Wallace. *Mix*, October.
50   *Pensado's Place* web TV show (available at www.pensadosplace.tv/), Episode 135.
51   Tingen, P. (2015). Inside track. *Sound on Sound*, October.
52   Massey, H. (2009). *Behind the glass: Top record producers tell how they craft the hits* (Vol. II). Backbeat Books.
53   Tingen, P. (2000). Interview with Tom Lord-Alge. *Sound on Sound*, April.
54   Droney, M. (2003). *Mix masters: Platinum engineers reveal their secrets for success*. Berklee Press.
55   Massey, H. (2009). *Behind the glass: Top record producers tell how they craft the hits* (Vol. II). Backbeat Books.
56   Tingen, P. (2016). Inside track. *Sound on Sound*, February.
57   Buskin, R. (1997). Interview with Jon Gass. *Sound on Sound*, August.
58   Tingen, P. (2011). Inside track. *Sound on Sound*, August.
59   Massey, H. (2009). *Behind the glass: Top record producers tell how they craft the hits* (Vol. II). Backbeat Books.
60   *Pensado's Place* web TV show (available at www.pensadosplace.tv/), Episode 230.

61 Tingen, P. (2007). Inside track. *Sound on Sound*, May.
62 Massey, H. (2000). *Behind the glass: Top record producers tell how they craft the hits* (Vol. *I*). Miller Freeman Books.
63 Lockwood, D. (1999). Interview with Bob Clearmountain. *Sound on Sound*, June.
64 Massey, H. (2000). *Behind the glass: Top record producers tell how they craft the hits* (Vol. *I*). Miller Freeman Books.
65 Tingen, P. (2014). Inside track. *Sound on Sound*, September.
66 Droney, M. (1999). Interview with Mike Shipley. *Mix*, June.
67 Tingen, P. (2016). Inside track. *Sound on Sound*, June.
68 Tingen, P. (2011). Inside track. *Sound on Sound*, September.
69 Tingen, P. (2015). Inside track. *Sound on Sound*, October.
70 Tingen, P. (2011). Inside track. *Sound on Sound*, October.
71 Tingen, P. (2015). Inside track. *Sound on Sound*, June.
72 Tingen, P. (2014). Inside track. *Sound on Sound*, September.
73 Tingen, P. (2015). Inside track. *Sound on Sound*, June.
74 Tingen, P. (2011). Inside track. *Sound on Sound*, September.
75 Tingen, P. (2015). Inside track. *Sound on Sound*, October.
76 Tingen, P. (2011). Inside track. *Sound on Sound*, July.
77 Massey, H. (2009). *Behind the glass: Top record producers tell how they craft the hits* (Vol. *II*). Backbeat Books.
78 Lockwood, D. (1999). Interview with Bob Clearmountain. *Sound on Sound*, June.
79 Tingen, P. (2012). Inside track. *Sound on Sound*, January.
80 Robinson, A. (1997). Interview with Nigel Godrich. *Mix*, August.
81 Owsinski, B. (2006). *The mixing engineer's handbook* (2nd ed.). Thomson Course Technology PTR.
82 Buskin, R. (2007). Interview with Gareth Jones. *Sound on Sound*, February.
83 Tingen, P. (2007). Inside track. *Sound on Sound*, February.
84 Tingen, P. (2014). Inside track. *Sound on Sound*, July.
85 Tingen, P. (2015). Inside track. *Sound on Sound*, December.
86 Senior, M. (2009). Interview with Bruce Swedien. *Sound on Sound*, November.
87 Tingen, P. (2014). Inside track. *Sound on Sound*, December.
88 Tingen, P. (2007). Inside track. *Sound on Sound*, February.
89 Massey, H. (2009). *Behind the glass: Top record producers tell how they craft the hits* (Vol. *II*). Backbeat Books.
90 Massey, H. (2009). *Behind the glass: Top record producers tell how they craft the hits* (Vol. *II*). Backbeat Books.
91 Tingen, P. (2014). Inside track. *Sound on Sound*, May.
92 *Pensado's Place* web TV show (available at www.pensadosplace.tv/), Episode 200.
93 *Pensado's Place* web TV show (available at www.pensadosplace.tv/), Episode 250.
94 *Pensado's Place* web TV show (available at www.pensadosplace.tv/), Episode 166.
95 Massey, H. (2009). *Behind the glass: Top record producers tell how they craft the hits* (Vol. *II*). Backbeat Books.
96 Tingen, P. (2014). Inside track. *Sound on Sound*, March.
97 Senior, M. (2015). Mix Rescue. *Sound on Sound*, August.

CHAPTER 20

1 Flint, T. (2001). Interview with Tony Platt. *Sound on Sound*, April.
2 Droney, M. (2003). *Mix masters: Platinum engineers reveal their secrets for success*. Berklee Press.
3 Massey, H. (2000). *Behind the glass: Top record producers tell how they craft the hits* (Vol. *I*). Miller Freeman Books.

Picture Credits

CHAPTER 1

Figure 1.2: Images courtesy of ADAM Audio®, KRK Systems®, Behringer®, and Yamaha®.

Figure 1.3: Waterfall plots courtesy of Philip Newell, Keith Holland, and Julius Newell, "The Yamaha NS10M: Twenty Years a Reference Monitor. Why?" Research paper for the Institute of Acoustics. (You can find the whole article at www.cambridge-mt.com/ms-ch1.htm)

Figure 1.4: Ibid.

Figure 1.5: Ibid.

Figure 1.6: Image courtesy of Simon-Claudius Wystrach (www.simonclaudius.com).

CHAPTER 2

Figure 2.6: Image courtesy of Avantone Pro®.

Figure 2.8: Images courtesy of Beyerdynamic® and Sennheiser®.

CHAPTER 4

Figure 4.1: Image courtesy of www.shutterstock.com.

Index

Page numbers in *italics* refer to figures.

A

Abba 128
Abbiss, Jim 346
absorptive acoustic treatment 19
absorptive material, thickness of 20
A/B switching plug-ins 75–76
ABX testing 262–263
a capellas 83
AC/DC 77
acoustically neutral fabric 28
acoustic foam 31, 33
acoustic guitar recording, compressing *175*
acoustic treatment: absorption, in moderation 20; diffusers, on reflective surfaces 22; foam, in moderation 19–21; reflections, dealing with 18–22, *20*; spending on 33
ADAM: A7X *6*; S2A *8*
adding atmosphere 306
Adele 37
Ainlay, Chuck 3, 57, 67, 82, 203, 205
air gap 26, 30
Albini, Steve 294
album, mixing 362
Alex Da Kid 109
Ali, Derek "MixedByAli" 39, 267, 352
"All Four Seasons" (Sting) 25
all-pass filters 154
alternate mix versions 81–82, 85
"American Life" 36
amplification, speakers with built-in *4*
analogue/digital mixing, hybrid 184
analogue equalizer designs 212
analogue recording, compared to digital 190
Anastacia 126, *126*
Antares Auto-Tune 133
antinode 23
Aphex Aural Exciter process 227
Apple Logic, built-in compressor *168*
Araica, Marcella 204, 357

arrangement: breathing life into 123–128; importance of 119; lack of light and shade in 124; static *125*
artificial reverberation *see* reverb(s)
ATC SCM20A, waterfall plot *8*
"A Team, The" (Sheeran) 267
atmosphere, adding 306
attack and release times 177–179
"attack" bump 283, *283*
Attack Time control: on a compressor 178, 181; on expanders and gates 188
audio crackles 111
audio editing techniques, for timing adjustment 105–109
audio files: editing out silences in 93; highlighting sections for special emphasis 94; unprocessed raw for each track 89
audio-frequency modulation effects 226
Audionamix ADX Trax Pro *265*, 266
Augilera, Christina 74
Aural Exciter 227
Auratone 5C Super Sound Cube 9, 36–45; attributes of 37–39; frequency response of *38*; headphones and 46, 48; substitutes for 43–45, 49, 51; usefulness in poor acoustics 39; waterfall plot *11*
automated EQ 216–218
automated pitch correctors *112*
automatic attack and release times 180
automatic pith corrector *114*
automation: creating buildup 343; dealing with isolated blemishes 345; fixing bad compression 347; for long-term mix dynamics 340–343; for perfecting mix balance 345–347

auto-panning 316
Avantone Mix Cube *44*
averaging the room 55–56
Avron, Neal 133, 301

B

background noise: adding 282; low-pass filtering and 264; receding as monitoring levels increase 61; reducing unwanted 187, 246; unintended consequences of removing 122–123
background parts 136, 350
Back in Black (AC/DC) 77
backing vocal parts, multing 95, *95*
backward temporal masking 106
bad compression, fixing with automation 347
balance: arriving at initial 161; building one track at a time 261; building raw 133–162; comparing 336; compression and 171–172, 184–185; of ensembles 155–157; as first priority 261–262; involving trade-off 144; judging exact levels 39–40; role of 131–132; *see also* balancing
balance judgments 39–42
balance refinements, subtle 346
balancing: blend reverb 279–282; fluent 261; fundamental questions 184–185; for imaginary multitrack drum recording 157–162; multimiked ensembles 152–155; multimiked instruments 151–152; procedures 162; with shelving EQ 201–202; size reverb 284–286; tasks 139–151; tempo-driven 194–195; *see also* balance
Ballard, Glen 88, 138, 283
Bandwidth control 140, 202–203

Barresi, Joe 116, 169, 181
barrier matting 29–30
basement rooms, concrete-walled 26
bass-balance versions 82
bass guitar 6, 55, 57, 60, 83, 90, 91, 125, 223, 243, 306, 324, 347
bass instruments: balancing 39; equalizing 203–204; reference tracks for 74; as subject of postmix prevarication 81–82
bass line, as a second melody 128
bass notes: first harmonic of 6; rebalancing individual 345; tail ends of 104
bass reflex speaker designs 5
bass response, listening to 55
bass-synth presets 58
bass traps 26–30, 28, 31
"Battlefield" (Sparks) 126
Behringer: B2031A 6; TRUTH B2031 8
Beyerdynamic DT880 Pro headphone 48
Bhasker, Jeff 123, 126–127
Bianco, Dave 212
Bieber, Justin 331
"Big Girls Don't Cry" (Fergie) 192
Black Eyed Peas 238
"Black Magic" (Little Mix) 126
Blake, Tchad 199
blank slate, starting with 89–90
blend delay 298
blending, by means other than reverb 282
blend reverb 275–282; applying across all tracks 278; balancing 279–282; creating 294; enhancing 271
blocking speaker ports 54–55, 54, 61
Blue Cat Flanger 318
bookshelves, using as diffusers 22
bottom-up approach 329–330, 334–335
Bottrell, Bill 65
Bottrill, Dave 145
boundary effects 21–22
Bowers, Delbert 170, 245, 249, 305, 337, 354
"Bowtie" (Outkast) 76
Boyer, Ed 46, 109, 116, 122, 194, 251
Boz Digital Transgressor 220
Bradfield, Andy 355
Brainworx BX cleansweep 139
Brand New Day (Sting) 77

Brathwaite, Leslie 90, 95, 251, 331, 337, 358
Brauer, Michael 51, 88, 90, 182, 324
breaks: during correction process 116, 117; during mixing 66
"Breathe In, Breathe Out" (West) 76, 309, 310
breathing compression 240–241
breath noises 122
brightness, as distance cue 280
Brothers In Arms (Dire Straits) 36
buildups, mixing 343
Burke, Solomon 76
Bush, Billy 65, 285
Bush, Steve 114
buss compression 80

C
CableGuys VolumeShaper 195
cable summing stereo headphone output to mono 44–45, 45
Caillat, Ken 124, 152
calibrated reference monitoring levels 69
camouflaging edit points 105–107
carpets, as limp-mass bass traps 29
Cascada 74
case studies, multitrack drums 157–162
Castellon, Demacio 93, 332
CD racks, as diffusers 22
CDs, burning your mix to 355–356
ceiling-mounted acoustic-foam panels 20
Celemony Melodyne 111, 114, 115, 115, 219, 231, 265
central sounds, in mono 39
chambers, reverb 290
chart music, commanding attention continuously 127–128
chart-ready lead vocals, level consistency demanded 183
chattering, gate 189
cheap ported speakers, coping with 54–55
Cherny, Ed 208
Chiccarelli, Joe 55, 56, 66, 136, 172
chordal synth pads 233–234
choruses: double 126, 129; drop 126, 129; reverb/delay levels between verses and 341–342; stereo width between verses and 342; subjective importance of each 133–134
chorusing 233, 317

Christian Knufinke SIR2 289
Churchyard, Steve 111
clashing timbres 124
clean slate, starting with 89–90
Clearmountain, Bob 51, 65, 66, 67, 68, 135, 143, 171, 241, 294, 340, 351, 355
click track, drummer recording along to 101
Clink, Mike 66, 122, 135, 137
clipping 336, 336
close-miking 275
coaxial monitors 17
Cockos: ReaFir 253; Reaper 30, 105
Collins, Phil 256
color-coding 91–92
coloring tools, highlighting special moments 94
colors, speaking faster than words 91–92
Comber, Neil 88
comb filtering: acoustic treatment and 19, 22; Auratones and 39; effects of 15, 139, 183; EQ-based widening and 312; in mono 42; phase and 13–14; resistance to acoustics and 39; of stereo files in mono 42; as tonal tool 233; turning effects to your advantage 153; between wet and dry sounds 287; between woofer and tweeter 15
commercial appeal, enhancing 138
commercial CDs, mix referencing against 70, 330–331
commercial releases, seeking out unmastered mixes of 74
communicating using musical examples 70
comping 119–123, 128
compression: approaches to use of 74; balance and 184–185; coloristic uses of 177; of individual tracks 173; of lead vocals 183; master-buss processing and 321–326; multilayered approach 177; as not the answer 172–173; parallel 181–184; presets 171–172; ratio 174–176; reasons for using 169–170; refining settings for 173–181; tracks needing 169–170; with two controls 165–173

compression artifacts, produced by ports 9
compressors: designed to offer high-ration compression 175; gain-reduction actions compared to duckers *257*; group-channel 173; inserting expander before or after 188; main control types *167*; models of 170–171; parameters for 174–181; reducing dynamic range 165–166; in series 177; side-chain equalization of 240; sonic differences between models 322
comp sheets 121–122, *121*
computer DAW mixing system 80
cone excursions, increasing for lower frequencies 58
ConeFlapper audio examples 58
confirmation bias 262–263
constructive criticism, seeking out 69–70
continuous audio, siting an edit point in 105–106
control-room modes 22–30
corner frequency 139, 200, 201
corrective audio editing, speeding up 116–117
corrective processing, how much to use 99
Costey, Rich 84, 135, 203, 224, 251, 331, 347, 352
CPU: deactivating plug-ins for muted tracks 76; maxing out *137*; pitch shifting and 314; power required by reverb 273
criticism, interpreting 70
crossfades: concealing transitions with 105; immediately before percussive attack 105–106; placing gaps in audio 105; types of 108
crossover 15, 16
crossover electronics 39
crossover frequencies 15, 39, 245, 247
cut-and-crossfade audio editing 117
cut-off frequency 139–140
cuts, effects of EQ 207–211

D

D16 Redopter *335*
data-compressed formats 78

Davis, Kevin 79
DAW systems: coloring tools 94; ending up with a fresh project file 90; with pitch correctors 113–116; time-stretching built in 107–108
DC (0Hz), effect on a mix file's waveform 59–61, *60*
DDMF NYCompressor 242
DeCarlo, Lee 269, 323
decibels per octave 140
dedicated monitor controller 65
de-essing: with automated threshold 354; with frequency-selective dynamics 250; multiple approaches to 267; reverb returns 281
delay effects: for blend 298; comparing 339; ducked 299; essential controls 297–301; Haas 312–314; resonant "ringing" of 298–299; for rhythmic energy 301; for size 298; for spread 302; for sustain 298–299; tempo-matched 300–301; for tone 298–299; using in stereo 301–302
Delay Time Control 233
detail, adding 127–128
detailed fader rides 343–354
detuned multioscillator synth patches 232
de Vries, Marius 116, 127
Dido 120, 121, 126, *126*, *136*
Die Krupps 290
diffusion 290–282
digital algorithmic reverb processors 274
digital aliasing distortion *225*
digital convolution reverb processors 274
digital MIDI instruments, pitch 109
DI recording, extra advantages of 152
Direct/Effect Mix control 273
directional microphones, boosting recorded bass levels 245
Dire Straits 36
distance: EQ for 216; for listening 347
distorted slapback trick 299
distortion: for loudness enhancement 335–336; as a mix tool 223–227; produced

by ports 9; as a send effect 226–227; unwanted 323
distortion and resonance artifacts, evaluating impact of 50
distortion devices, adding new harmonics 224
Dorfsman, Neil 152
double-tracked parts 104, 117
double tracks, generating fake 305, *305*, 319
Douglas, Jack 60, 135, 152, 228, 287, 290
Douglass, Jimmy 74, 79, 84, 135, 146, 190, 294
down-the-line transmission compression 322
drafts, around the microphone 58
"Drag Me Down" (One Direction) 126
Drake 110
Dr Dre 74, 267
Dresdow, Dylan 238, 307, 337
Drones (Muse) 347
drop chorus 126, 129
drummer, curse of the talentless 103
drum(s): adding samples for 230, *230*; compression 179; emphasizing transients 190; gating to reshape envelope 190; high-pass filtering and 140; multitrack as a case study 157–162; starting editing process from 100–101; time-aligning 159; triggering 229–231
drums, multitrack 157–162
DSP calculation method 143
ducking 257–258, 321
Dudgeon, Gus 145, 211, 348
dynamic equalizers 247–252
dynamic low-pass filter 246
dynamic mixer automation 364
dynamic noise reduction 246
dynamic range: compressors reducing 165–166; ears reducing at high volumes 67; increasing 187, 280; as too wide 147
"dynamics processes" family 187
dynamic stereo widening effects 317–318

E

ears: adaptive capabilities of 78; fighting your own 64–70; processing only three things at once 124; trusting over eyes 116

echo-based effects 297–298
editing, power of 264
edit points, positioning 105–107, *105*, 117
effects: across a range of styles 74, 76; as makeup 269
Eiosis E² De-esser *250*
electrical supply, induced noise from 205
electronic tuner 109
Elevado, Russ 64, 65, 66, 270
Elmhirst, Tom 37, 83, 205, 206, 251, 332, 337, 341, 352, 354
Emerick, Geoff 41, 282
Eminem 76
emotional impact, maximizing 351, 354
emulated mixing-console saturation *225*
Endert, Mark 341
"endstop" examples, tracks useful as 74, 76
ensembles: balance and tone 155–157; multimiked 152–157, 214–215; time-aligning 159
EQ curves: showing peaking-filter cuts *203*; showing shelving-filter cuts *201*
equal-gain crossfade 108
equalization (EQ): advanced tools for 215–220; automated and matching 216–218, *217*; basic tools and techniques for 200–211; of bass instruments 203–204; chaining plug-ins 212; compensating for frequency masking 198; correcting effects of room modes 27, 33; cuts rather than boosts 207–209, *209*; for distance 216; fine-tuning decisions 201–202; good habits 207–211; importance for balancing 199; independent, for periodic and nonperiodic components 219–220, *220*; for length 214; limits of 220–221; master-buss processing and 326; of multimiked recordings 211–215; overuse of 197; pitch-tracking and MIDI-triggered 218–219; presets of no use at mixdown 198;

purpose of 221; rebalancing premixed audio 264–265; room modes and 27
equal-loudness contours *68*
equal-power crossfade 108
equipment, choosing 3–5
essential groundwork 89–96
Eventide Fission 220
Everett, Shawn 204, 336
expansion: for balancing purposes 188; before compression 195–196; increasing dynamic range 187
experimentation, all processing as 267
external side-chain input 255

F

Fabfitter Timeless *300*
fader levels 145
faders: compared to compressors 165; control scale *144*; feeling unstable 146; hardware fader controllers 352–353; judging precise levels 145; quality time with 162; setting the agenda 261
fake double-tracks, generating 319
fake gated-reverb effects 291
features, in chart music 128
feedback howlarounds, on live recordings 345
Feedback Level control 297–298
Fergie 192
Fields, John 89
"Fight Song" (Platten) 126
Filipetti, Frank 267
filter slopes 140
filter types, offered by an equalizer 200
final version, nailing down 354–359
Finn, Jerry 82
first mode, of guitar string resonance 22–23
Fitzmaurice, Steve 124, 230, 251, 337, 352, 354
flanging 233, 317
Fleetwood Mac 124
floating point processing 143
Flood 69, 141
fluent balancing 261
flutter echoes 30
Flux, Stereo Tool *149*
foam, covering entire room in 21
foam platforms, between speaker and mounting surface 12

foam spacer blocks, behind main foam panels 21
Foley ambiences 307
foot tapping, vibrations of 58
formants 113
formats, data-compressed 78
Freeware: compressors *171*; distortions *225*; level meters 57, *57*; limiters *176*; modulation plug-in *318*; for reverb 273; spectrum analyzers 55, 56, *310*; stereo utility plug-in 308–309, *308*; stereo vectorscope plug-ins *149*
frequency-domain balance problems 197
frequency extremes 67, *68*
frequency masking 198–200
frequency response: correction plug-ins for headphones 47; creating a peak or trough in 202; effects of porting on 5–6, *5*; sculpting 212
frequency-selective dynamics processing 237, *239*
frequency-selective stereo width control 309
frequency-selective transient enhancement 238
frequency sharp-shooting 205
front-back depth 271
full-band "bottom-up" squeeze 334
full-band dynamics processors 237–242, 265
full-band limiting 335
full-band "top-down" squeeze 334, *334*
Fun 126
furniture, as reflectors 21

G

Gabriel, Pascal 120, 121
Gabrielle 74
gain management, with plug-ins 170
gain pumping: creating kick-triggered 258; excessive 323
gain reduction: initial layer of 183; metering 171, 175–176; producing distortion 180
gap-filling fixes, built-in processing for 107
Gass, Jon 82, 228, 241, 348
gate, chattering 189
gated reverb 291
Gatica, Humberto 123, 137, 339

gating 187, 190, 246
gating action, refining 238–240
gear, not an excuse 266
gearslutz.com 73, *73*
gems: calling on buried 349; unveiling hidden 93–94
generic layout, for mix projects 91
Ghenea, Serbab 104, 122, 225
Glossop, Mick 26–27, 249
Godrich, Nigel 3, 355
Goldstein, Jason 192, 256
Good Girl Gone Bad (Rihanna) *136*
Gosling, Jake 267
Graham, Lukas 306
graphical symbols 92
graphic EQ 206
Groban, Josh 48
groove and timing 100–104
grotbox speakers 50–51, 52
group channel, EQing 213
grouping tasks 151
Gudwin, Josh 228, 245, 285, 331
Guzauski, Mick 84, 198, 279–280, 341
GVST: GChours *318*; GClip *225*

H

Haas delays 312–314, *314*
"Happy" (Williams) 95
haptic devices 59
hard-knee compressor designs 178
hardware fader controllers 352–353, *353*
hardware monitor controlling 65
harmonic density, increasing 223–224, 226
harsh-sounding vocal notes 249
Headlines (Drake) 110
headphones: care when listening to 67; emulating loudspeaker monitoring using 49–50; frequency-response correction plug-ins for 47; importance of 52; not affected by monitoring environments 46; as supplementary monitoring system 46–50; top-of-the-range *48*; working primarily on 48
Head Related Transfer Function (HRTF) 50
Heap, Imogen 251, 281
hearing: fatigue 66; importance of 1; risk of damage 66
"Heathens" (Twenty One Pilots) 126

"Here with Me" (Dido) 120
hierarchical mixing approach 134–135
hi-fi equipment, purpose of 4
hi-fi speakers 4
high-density mineral-fiber panels 28
high end, adding in the digital domain 326
high-frequency enhancement technique 227
high-frequency reflection, reintroducing 29
high-pass filtering: for each individual mic 153; inserting 158–159; reassessing during the mix 215; removing unwanted low frequencies 59–60
high-quality plug-ins, as more CPU-hungry 137
high-resolution peak/average metering 57
high-resolution spectrum analyzer 205, *205*
high shelving filter *201*
hi-hat spill 160, 187, 189–190, 240, 345
Hodge, Steve 156, 216, 345
HOFA 4U Meter Fader & MS Pan *57*
Hoffer, Tony 123
hold time, on expanders 189, 190
"Hold Your Hands Up" (Cascada) 74
hollow frame, of dedicated speaker stands 12
Honor Roll of Dynamic Recordings 73, 77, 79
Horn, Bob 183, 204, 331
Horrortone 43
Hoskulds, S. Husky 280, 290
Hourglass (Taylor) 267
"Housewife" (Dr Dre) 74
Huart, Warren 349
human auditory system 64
"Hunter" (Dido) 126
hybrid analogue/digital mixing 184
Hyde, Matt 337
hypercompressed CDs 73
hysteresis 189, 196

I

icons 91–92, *92*
IK Multimedia: Amplitube *110*; T-RackS *210*
image spread, adjusting 305
Imbruglia, Natalie 74

"I'm Outta Love" (Anastacia) 126
"Independence Day" (Gabrielle) 74
independent EQ, for periodic and nonperiodic components 219–220, *220*
inertia, presenting by mounting hardware 11
"Infidelity" (Skunk Anansie) 7, 74
infinity:1 ration 187
initial balance, making or breaking a record 153
in phase sine waves 13–14
input gain 167, *167*
Input Gain controls 167, 176
instincts: playing tricks on you 72; as unreliable 161–162
instrumental mix 83
instrumental solo, mix version without 83
instruments: changing relative importance of 137–138; importance of as genre dependent 137; with low-end transients 323; with low-frequency components 6; mixing in order of importance 135–138, 162, 221; multimiked 151–155, 211–214; removing completely 265–266; spreading the image of multimiked 152
interference, constructive low-frequency 21
interleaved file, converting split-stereo into 148
"In the Air Tonight" (Collins) 256
"in the box": DAW mixing systems 91; working entirely in 80
intro, importance of 135
inverted waveform polarity, in stereo 42
Izhaki, Roey 145
Izotope: Insight *56*; Neutron 218

J

Jackson, Michael 36, *37*
Jaz-Z 74
Jean, Wyclef 124
Jerkins, Rodney 128
Johns, Andy 76, 123, 135, 340
Jones, Gareth 356
Jones, Norah 74
Jones, Quincy 37, 357
Joshua, Jaycen 88, 90, 135, 216, 328, 332, 345

K

Kadish, Kevin 224
Katz, Bob 37, 73, 79
Keane 76
keyboard shortcut support, in a
    third-party application 108
keyed multiband dynamics 258–259
keyed parallel gate channel, polarity
    inverted 259
key tracks, in a production 138
kick drums: adding low-end welly
    to 256; components delayed
    58; dealing with low end of
    6; fitting together with bass
    203–204; layering sample
    against 229
Kilhoffer, Anthony 74
Killen, Kevin 67, 267
King, Jacquire 297, 326
Kirwan, Rob 89, 337
Klangfreund LUFS Meter 333
Knee control 178
Kramer, Eddie 214
Krauss, Alison 123
KRK Rokit 8 G3 6

L

Lacinato ABX/Shootouter 263, 263
Lange, Robert John "Mutt"
    347, 351, 354
Lanois, Daniel 357
Lawson, Mark 228, 359
layered sounds, polarity and phase
    relationships between 157
layers, tightening timing of each 103
LCR panning 141
lead vocal levels, as hard to get right 81
lead vocals: building up via comping
    119; compressing 183;
    dynamic EQ seeing most
    use on 250; level of 145;
    multing 94; preventing long
    echo and reverb effects from
    swamping 256; pushing up
    first couple of syllables of 349;
    slapback delay and 299; stereo
    widening 314; timing of 104,
    106–107
Leckie, John 79, 132, 199, 351
LED bar graphs 171
Lehning, Kyle 357
Leslie-style rotary speaker
    emulation 317
Letang, Renaud 214
level balancing *see* balancing

level compensation, for a
    compressor 168
level detection, peak versus
    average 181
level envelope, of each drum hit 190
levels, setting 143–146
level window, setting a fader in 145
LFSineTones audio file 9, 12, 24, *25*,
    26, 54
Life for Rent (Dido) 136
lightweight walls, rooms with 26
limiters 176, *176*
limp-mass bass traps 29–30, 33
linear phase equalizers 210, *210*
Lipson, Steve 109, 119
listener's attention, directing
    348–350
listening: importance of 1, 2; shifting
    perspective 55, 206–207;
    subjective versus objective
    63, 85
listening position, aiming speakers
    directly at 12–13
Little Mix 126
live performances: comping several
    takes 128; mixing with
    machine elements 104; tuning
    and timing discrepancies 99
lo-fi playback, reference point for 50
LongSound Microverb VST *287*
long-term mix dynamics 133,
    340–343
Lookahead control 195
Lord-Alge, Chris 40, 69, 88, 91, 135,
    203, 341, 346, 350
Lord-Alge, Tom 90, 91, 135, 177, 347
loud listening, inherent excitement
    of 67
loudness bias 182, 185, 208, 262,
    263, 291, 313–314, 322, 328,
    331, 333, 337, 361
loudness-matching 69
loudness matching *333*
loudness maximization 332–334
loudness-normalized playback 338
loudness processing 330–338;
    common options 332–333; for
    comparison purposes 78–79;
    "loudness wars" debate 77
loudspeaker monitoring, emulating
    with headphones 49
"Love Lockdown" (West) 123
low end, restricting in your mix 61
low-end buildup, avoiding excessive
    61–62, 256

low-end damage limitation 53–62
low-end enhancements 228–232
low-end resonance 25, 39
low frequencies, removing unwanted
    139–140, 162
low-frequency background noise 38
low-frequency monitoring 53
low frequency output: of
    headphones 46; of small
    ported speakers 5
low-frequency pitfalls, working
    around 53
low-pass filtering 140
low shelving filter *201*, 241
lyrics: creating "clean" version of 82;
    maximizing audibility of 350

M

Mackie HR824 10, *10*
Madonna 36, 76, 78
magic moments, keeping an ear out
    for 120–121
mainstream listening devices 36
Makeup Gain 166–169
Malouf, Brian 128
Marroquin, Manny 57, 88, 240, 241,
    287, 342
Masasciullo, Fabian 264
Maserati, Tony 88, 216, 244,
    245, 368
masked edits *106*
masking: counteracting effects of
    293; frequencies 198–200;
    as a perceptual
    phenomenon 346
Masking Meter 218
Massenburg, George 55, 70, 267
Massey L2007 limiter 337
master-buss processing 84, 321–330
mastering 359–363
mastering service, choosing 360–361
Mastering the Mix Reference 75, *75*
matched-waveform edits
    106–107, *107*
matching EQ 216–218, *217*
McVie, Christine 124
Melda: MCompare 75, *75*;
    MFreeformPhase *154*;
    MStereoScope *149*
Meldal-Johnson, Justin 358
melodic instrument solos, time
    correction of 106–107
metering 56–57
metering time response 56
Meyerson, Alan 143

Middle and Sides (M&S): encoding 307–308; processing 308–309; recording 307; stereo encoding/decoding 319
Middle signal 307
MIDI instruments 89–90
MIDI subsynth 231
MIDI synth paths: replacing from source files 111; stress on computer from 89; as substitutes for EQ 235
MIDI-triggered EQ 218–219
midrange: critical sounds in 15; as crucial 37; focusing attention on 37–38, 38; in relation to Auratone 43–44
mineral-fiber bass traps 26–28, 28, 31
minimum-phase DQ designs 210
mistakes, learning from 339
mix balance, perfecting 345–347
mix-buss compression see buss compression
mixdown file, retaining unprocessed version 337
mixdown stage, as a separate task 89
mix dynamics 133–135, 134
mix engineers: cutting and adding parts 134–135; mastering their own productions 76–77
mix(es): adding instruments in order of importance 133–138, 162, 221; building in stages 133; building one instrument at a time 133–134; building up sections in order of importance 162; creating alternate versions of 81–82, 85; playing to other people 69–70; restarting 264; switching between instantaneously 330–331
mixing an album 362
mix preparation: comping and arrangement 119–129; groundwork for 89–96; importance of 87–88; timing and tuning adjustments 99–117
mix recall 81
mix referencing: art of 70–79; choosing material for 71–72; getting best out of 77–79; process of 85, 338–340, 354–359

mix revisions 364
mix stems 83–85
mix tonality 74, 76, 78, 326
modulated pitch shifting 318; see also vibrato
modulated stereo enhancements 316–318
Modulation Depth control 226
modulation effects: distancing instruments 282; high-speed 226; for stereo width 318; for tonal change 233
"Money Money Money" (Abba) 128
monitor controller 64–65
monitoring, good enough 30–32
monitoring levels 66–69
monitoring practices, good 144
monitoring systems, switching frequently 64–65
monitoring tool, best suited for low-end 60–61
monitors see speakers
monitor-selection buttons 64–65
mono: feeding speakers into 44–45; listening back in 39–42; relevance of 42–43; unpleasant surprises switching to 41–42
mono balance shift, drawing attention to center 41
MonoBalanceShift audio file 40–41, 52
mono compatibility: balance judgments and 39–42, checking individual 288; expensive reverbs outclassing cheap 277
mono delays, uses of 301–302
mono feed 44, 45
mono prefader auxiliary send 44
mono recordings, panning 141–143, 152
mono signals, expanding stereo width 313–314
Montagnese, Carlo "Illangelo" 194, 266, 357
Moseley, Adam 317, 342
Moulder, Alan 69, 228
mounting hardware for speakers 10–12
mounting spikes, inverted 12
MS adjustment plug-in 318
multiband buss compression 323–324, 325–326, 335
multiband compression 325–326

multiband crossovers 327, 335
multiband dynamics 243–247
multiband loudness enhancement 335
multiband processing, do-it-yourself 244
multimiked ensembles 152–157, 214–215, 308
multimiked instruments 151–157, 162, 211, 213–214
multimiked recordings, equalizing 211–215
multing 94–96, 95; compared to automation 340; slicing to balance at independent levels 194; vocal sections 183
multi-purpose monitors 31
multispeaker surround setup 4
multitap delay 302
multitrack drums, case study 151–157
multitrack files, listening through to individual 96
mumbled words: compression rebalancing 165–166, 166; reducing difference between clear and 169
Mundt, Thomas, LoudMax 176
Murphy, Shawn 211
Muse 347
music, tapping into emotional flow 134
musical arrangement: breathing life into 123–128; differentiating sections of 125; importance of 119
musical contrast, opportunities for 284
musical intelligence, for pitch-detection and correction 113–114
musical sections, mixing in order of importance 162
musical style, volume levels for 67
musical styles, needing help of compression 170
musical timeline, adjusting 127
"Music Sounds Better with You" (Stardust) 242
mute button 129, 134–135, 135

N
narrow-bandwidth peaking cuts 203–206
navigating the mix project 90–92, 96
navigation, enhancing 90–92

nearfield monitoring system(s): affordable two-way ported 6; compared to headphones 48–49; description of 3–4; full-range 61; low-end resonance problems 53; quality of 31–32; reasonable small-studio 45; strengths of 36; with two drivers 15; *see also* speakers
Needham, Mark 251
Never Mind the Bollocks (Sex Pistols) 256
"New Shoes" (Nutini) 74
"New York" compression 181
Nichols, Roger 140, 145, 216
Niebank, Justin 137, 144, 359
No Angel (Dido) *126*
noisy sections, as good spots for edits 105
notches, on project-studio drum recordings 205
notch filter 215
notching, applications of 205–206
note durations, subjective 214
note pitch-center offsets 115
Not That Kind (Anastacia) *126*
NuBi3 Spinner LE *316*
Nutini, Paolo 74

O
objective decisions, delivered by mix referencing 71
Ogilvie, Dave 224
Olsen, Keith 21
one-cone design, of the Auratone 38
One Direction 126
180 degrees out of phase 13
one-note bass 7
One Republic 66
opposition panning 141–142
optimum listening area, increasing size of 12–15
order, for mixing instruments and sections 133–138
Orton, Robert 92
"Other Side of the Coin" (Burke) 76
Outkast 76
out of phase 14–15
out-of-phase waveform 13–14
out-of-polarity speakers 17
out-of-polarity stereo components 41–42
out-of-time note, stumbling feelings indicating 102

out-of-tune note, encapsulating emotion 99
Output Gain control 168
overall ensemble mic(s), using 155–156
overdubs, ordering over the Internet 127
overhead mics, balancing with 153

P
Padgham, Hugh 25, 125, 256, 267
pads, requiring automation adjustments *341*, 343
Palmer, Tim 172, 340, 342
panning: decisions for multimiked instruments 154–155; mono recordings 141–143, 152; panning laws 142; technical issues 162; toward extremes 141
panning mono recordings 141–143
Panunzio, Thom 66
parallel compression 181–184
parallel master-buss configuration 324–325
parallel processing: for distortion 226–227; equalizing return channels 238; with multiband dynamics units 245, 247; reducing undesirable side effects of fast compression 185; setup for expansion and gating 185–186
parallel room surfaces 22
parametric EQ 206
Parr, Steve 347
Parsons, Alan 70, 135, 278, 282
partial phase cancellation 13–14
passive radiators 10
paving slab, underneath each speaker 11
Peak/Average control, for a compressor 181
peaking filters: adding in 202–203; causing more destructive phase shifts 209
peak reduction 166, *167*
Pensado, Dave 57, 74, 135, 145, 182, 212, 238, 266, 279, 347
percussion sounds, removing discordant pitched resonances in 215
percussive bass sounds, losing bass content 181

periodic and nonperiodic components, independent EQ for 219–220, *220*
phantom images: difficulty in real-world mixing 35; illusion of 35; making instruments sound less solid 311; none in single-speaker mono 39
phase and comb filtering 13–14
phase and polarity issues during balancing: addressing mismatches *150*; exploiting cancellations 151, 156; issues in mono 148–150
Phase control, for tremolo effects 195
phase-manipulation tools, complex 154, *154*
phase response 208–209
phase rotators 154, 158
phasing effects 233, 317
physical resonance, in a speaker stand 11
Pilegaard, Morten "Pilo" 224, 306
"Pillowtalk" (Zayn) 306
ping-pong delay 302
pink noise 147
PinkNoise file 13
pitch correction: adjusting the correction algorithm 115; applying only when needed 115; audio fidelity of 99; automated and prescanning 113–116; detection, taking guesswork out of 114–115; offsets, applying to offending notes 113; pre-scanning software *115*
pitch-correction processors 113, *115*
pitching and groove, tightness of both 99
pitch-shifted delay patch 314–315, *315*
pitch shifting 235, 314
pitch-shifting, for density 231
pitch/time-correction processors, audio fidelity of 99
pitch-tracking EQ 218–219
plates and springs 274
Platt, Tony 119, 290, 367
Platten, Rachel 126, *126*
playback timeline *see* timeline, dividing
Plug & Mix LS-Rotator *316*

plug-in delay-compensation routines 184
plyboard, gluing foam to 20
PMC, transmission-line system 10
PMC LB1BP 10, *10*
polarity 14, 42, *152*, 157
polarity button *152*
polarity/phase relationship 153
polarity reversal 14, 150
polyrhythmic delay times, choosing 301
ported speakers 5–9, *7*, 31, 54–55
porting, side effects of 6–10
porting frequency 54
ports, blocking 54–55, *54*, 61
positioning speakers 12–17, 21–22, 32–33, 44–45
postdynamics EQ 221
post-fader sends 273, 297
Power, Bob 145
Power, Steve 124
Precedence Effect 159
predelay, adding 278, 314–315
predynamics EQ 221
preemptive low-end processing 59–61
premasking *105*, 106, 117
Premier, DJ 231, 262
premixed audio, rebalancing 264–266, *265*
presets: compression 173; EQ 198; reverb 276–277, 282
Price, Bill 256
Primacoustic Recoil Stabilizer 12
processed effect ("wet" signal) 181, 273
programmable tremolo units *195*
project reconnaissance 93–96
proximity effect 245
public appearance (PA) mix 82–83
Puig, Jack Joseph 37, 109, 124, 135, 192
pumping 241–242, 321
Pussy Cat Dolls 76, *78*

Q

Q control 140, 202
quality-control procedures 88
Q value 202–203

R

radio station, compression from 322
Ramone, Phil 46, 276, 277, 352
Range control, on expander 189–190

rappers: comping performances 122; mix automation for 354
Ratio control: adjusting 178–179; on a compressor 175–176; on an expander 187
real reverb 290
real-world mixing, not a linear process 262
reamping 227
rebalancing premixed audio 264–266, *265*
"Red Dress" (Sugababes) 136
reference instrument, for purposes of timing correction 100, 104
reference recordings 4, 70–79
reference tracks: for different mix decisions 74; editing out highlights 78; getting best out of 77–79; for learning attributes of new rooms 79; relating to musical styles 73–74; selecting 71–77; sources of suggested 73–74
referencing: checklist 338–340; exporting mix for *329*; goal of 340; loudness processing and 330–338; stages of 363–364; as ultimate bang-per-buck studio tool 71
reflectors 19
reflex loaded speaker designs 5
Regeneration control 297–298
region-specific processing 345
"Rehab" (Winehouse) 206
relative timing, importance of 101–102, *101*
Release Time control: on a compressor 178, 179, 181; on an expander 188, 190
Repeats control 297–298
repetition 127, 128
resonance, as tonal tool 233
Resonance controls 140, 202
resonances 11–12; room 22–30, 55
resonance side effects for monitoring 7–9, *8*
resonant filter ringing 210
resonant frequency: of the first room mode 22–23; of ported monitors 61
resonant qualities, of the throat, nose, and mouth 248–249
reverb/delay levels, between verses and choruses 341
reverb modulation 288

reverb return 233, 280–281, *281*, 308
reverb(s): approaches to use of 74; with an "attack" bump 283, *283*; for blend 275–282; combining different reverb patches 293; comparing effects 339; controls on 273–275, *273*; designs, background on 274; enhancements of 271–273; gated 291; juggling different reverb enhancements 292–294; multifaceted nature 272; plug-ins 272–273, *272*; reverb level versus reverb length 288; as send effects 233; for size 282–286; for spread 292; for sustain 290–292; for tone 286–290; unmasking with 293
revision requests 356–359
rides *see* detailed fader rides
Righteous Brothers 265
Rihanna *136*
ringing, caused by porting 7–8
ring modulation, adding growl to lead vocals 226
Roland DS50A, waterfall plot *8*
Ronson, Mark 122
room acoustics, importance of 16–17
room modes: affecting lower half of audio spectrum 24–25; equalization to correct effects of 27, 33; generating nodes and antinodes 24; non-parallel walls to avoid 25; reducing impact of 25–26; relating to room dimensions 26
room resonances 22–30, *23*, 55
room tone 282, 307
Rosse, Eric 56
rotary speaker emulation 316–317, *316*
rough mix, referencing 331–332
rough mixes, used on albums 132
rubberized barrier matting 29
Rumours (Fleetwood Mac) 124

S

Sample Magic Magic AB 75
Samuels, Harmony 127
Sarafin, Eric "Mixerman" 141
Savigar, Kevin 55, 90, 348
scaffolding, removing 261–264
Scheiner, Elliot 280, 290

Schilling, Eric 22, 26
Schleicher, Clarke 152
Schmitt, Al 30, 46, 290, 368
Schwa: Dyno *191*; Schope *205*
sealed-box speaker design 40
Seay, Ed 281, 356
second resonant mode, in a guitar
    string 23
Selby, Andy 48, 116
send-return effect configuration 181,
    273, 297
Sennheiser HD650 *48*
Serletic, Matt 348
set-and-forget approach, of pitch
    correction 113–114
"7 Years" (Graham) 306
Sex Pistols 256
Shebib, Noah "40" 110, 198, 204
Sheeran, Ed 267
"She Has No Time" (Keane) 76
shelving filters: attributes of
    200–201; balancing with
    201–202; basics of 200–201;
    several active at once 202
Shephard, Ian 361
Shipley, Mike 95, 123, 245, 343,
    345, 352, 354
shock tactics 64–65
Shoemaker, Trina 19
sibilants 240–241; *see also*
    de-essing
side-chain EQ, faking 241
side-chain equalization 238–239,
    240, 253
side chains 238–239, 255
side effects: of buss compression
    323–324; of compressor
    attack and release times
    180–181; of down-the-line
    transmission compression
    322; of loudness processing
    332–333; of master-buss
    compression 325; of pitch
    correction 100; of resonance
    in monitoring systems 6–9, *8*;
    of speaker porting 5–9
Sides, Allen 43, 55, 66, 81, 156
sides, placing speakers on 16
Sides signal 307–308, 309, 310, 327
Sigsworth, Guy 281, 285
Simpson, Rik 60, 177, 251, 256
sine-wave signal 13–14
single-driver construction 39, *39*
single-speaker mono 39–41
size delay 297

size reverb: adding predelay to 283;
    adding sustain 283–284;
    balancing 284–286; creating
    294; making audible as an
    effect 284–286
Skunk Anansie 7, 74
slapback delay 298, 299
slap-bass part, compressing 174
Slope control, of a compressor 175
small improvements, adding up 266
Smith, Don 298
Smith, Fraser T. 89
Smith. Sam 126
Smyth Research 50
snagging, effective 355–356
snag list 355, 356, 364
snare ambience, supplementing
    229–230
snare-drum compression, different
    settings for 179
snare-drum samples, with low-
    frequency rumble 58
snare sound, boosting in overhead
    mics 256
soft furnishing, damping
    reflections 21
soft-knee compression 178
Solid State Logic X-Phase *154*
solo button, success of mix EQ and
    199, *199*
solo-out mixes 82–83
Sonalksis FreeG *57*
song sections: displaying as bars or
    markers 92; naming *93*
Sonic Anomaly Unlimited *176*
sonic importance, evaluating
    133–134
Sonsible Entropy:EQ+ 220, *220*
Sony boombox, mixing using 51
"Sorry" (Madonna) 76, *78*
soundonsound.com 73, *73*
sound quality, overall judgement of
    72–77
SoundRadix SurferEQ plug-in
    218–219, *219*
"South Out To My Ex"
    (Little Mix) 126
spaced-pair microphone
    techniques *41*
spaced-pair stereo recordings 41–42
Sparks, Jordin 126
speaker cone excursions 57–59, 62
speaker positioning 12–17, 44–45
speakers: blocking ports 54, *54*, 61;
    with built-in amplification

4; EQing to compensate for
    low-end tip-up 22; feeding
    in mono 44–45; finding to
    fulfill Auratone role 43–45;
    mounting 33; with only one
    driver *39*; out of polarity
    with each other 17; personal
    preference regarding 4; ported
    5–9, *7*, 31, 54–55; with ports
    at the rear 22; positioning
    12–17, 21–22, 32–33;
    switching 64–65; on their
    sides 16; unported 9; *see
    also* nearfield monitoring
    system(s)
speaker spacing 33
speaker stands: dedicated 10–12;
    filled with sand 31
Spears, Britney 74
special effect reverbs 285
special effects, impact diminishing
    with familiarity 285
specialized filter shapes 215–216
"special sauce" 327–328
spectral decay plot 7–8, *8*, *10*
spectral dynamics 252–253
spectrum analysis 56–57, *56*
split stereo files 148
SPL Transient Designer Plus *194*
spot mics 155–156
spread delay 302
spread reverb 272, 292, 295
spring reverb 287
"Square Dance" (Eminem) 76
SSL X-Verb *281*
stable balance, as purpose of
    compression 185
Stardust 242
static arrangements *125*
static stereo enhancements
    310–315
stationary note, rebalancing 205
Staub, Randy 156, 169, 331, 355
Stavrou, Mike 79, 109, 145
"Stay With Me" (Smith) 126
Steinberg Cubase 148; built-in 31-
    band graphic EQ plug-in *206*;
    Gate *240*
stems, mixing to 83, 85
Stent, Mark "Spike" 18, 81, 88,
    95, 153, 182, 297, 299, 326,
    332, 358
stereo adjustment and metering
    utilities 148–150, *149*
stereo balance offsets, correcting 149

stereo enhancements: adding widescreen background texture 307; at mixdown 305; plug-ins 307; of reverb effects 292
stereo enhancer plug-ins 307
stereo files, digital clipping on 46
stereo image: adjusting width 307–309; comparing 339; crumbling outside sweet spot 42; destabilizing center of 16; important musical elements in the middle of 141–143; of stereo files 280; stereo panning 139; widening mono tracks 310–315; widening with delay effects 302; widening with reverb effects 292
stereo imaging 42–43, *42*
stereo manipulation 326–327
stereo metering tools 309–310
stereo mic pairs, within an ensemble setup 152
stereo mixer channels 148
stereo monitoring 15–17, 33, 39–40
stereo vectorscope 309–310
Sting 25, 77
Stone, Al 191, 285
"Stop & Stare" (One Republic) 66
studio models of headphones, versus hi-fi 47
studio monitoring headphones, top-of-the-range 47
subbass synthesizer part 61, 231–232
subharmonic synthesizer 228
subjective activity, listening as 63
subjective length, of a note 214
subjective loudness *336*
subjective mix decisions, as not teachable 63
subjective versus objective results 63–85
subjective volume differences, between tracks 78–79
submixing 83, 85, 90
subsonic energy, buildup of inaudible 57
subsynth, beefing up existing fundamental frequency 232
subwoofers 19
Suecof, Jason 347
Sugababes 136
"Sunrise" (Jones) 74
supplementary monitoring systems 35–36

surround monitoring 4
surround sound, reworking a track for 85
sustain delay 299–300
sustained monophonic lines, editing 111
sustain reverb 272, 290–292, 295
SVS technology 50
Swann, Darryl 146
Swedien, Bruce 37, 357
sweeting effects 269
sweet spot 12–15, 42
Swivel, DJ 141, *182*, 229, 231
symbols, speaking faster than words 91–92, *92*
Syncho Arts Revoice 110, *114*
synth pads 233–234, 308, *341*
synth waveform, selecting *232*

T
"Taking over the World" (Pussy Cat Dolls) 76, *78*
TAL Reverb 4 *287*
Tan, Phil 60, 356
Tannoy Reveal, waterfall plot *8*
tape hiss 282, 307, 319
tape saturation *225*
Taylor, James 267
tempo-driven balancing 194–195
tempo-driven dynamics processing 196
tempo-related predelay time 283
tempo-synced delay 299–300
tempo-synchronized gates 194–195
third mode, of a guitar string 23
Thomas, Chris 256
three in a row test 127
Threshold control: on a compressor 166, *167*, 171; on an expander 187
threshold-dependent transient processing 192–193
threshold-independent transient processing 191–192
Thriller (Jackson) 36, *37*
Timberlake, Justin 294
time, as money 138
time-aligning drumkits and ensembles 159
time-delay, between channels in stereo 41–42
time-domain problems, creating fader instability 197
time-domain processing, about prediction 344–345

timeline, dividing 92
timeline markers, setting 96
time smearing 210
timing: adjustment 105–109; editing techniques for 105–109; edit on simplest level 105; groove and 100–104; as a relative perception 101–102; role of time stretching 107–108; specialist software for manipulating 114; tightening 103–104
timing and tuning adjustments 99–117
tonal delay 298–299
tonal EQ 212
tonal imbalances, ear compensating for 64
tonal reverb, with phase relationships 295
tonal shaping, creative 131
tonal tools, comb filtering and resonance as 233
Tonebooster: Module *300*; TB_BusCompressor *334*; TB-Flx *215*
top-down approach 329–330, 334
top-to-tail takes, recording numerous full 120
"Torn" (Imbruglia) 74
total phase cancellation 13–14
Townshend, Cenzo 65, 88, 135, 229
"Toxic" (Spears) 74
track count, limiting 91
track icons *92*
track layout, standardizing 90, 96
tracks: blending too well 280; color-coding 91–92; labeling 91–92; listening through 93–94; maximum number of 90–91; needing compression 169–170; organizing 90–92, *91*; panning off-center 143; recorded as separate overdubs 275; recording single one phase at a time 120; reducing number including low-end information 61, 62; sacrificing less important 207; targeting for tuning correction 111–112
traffic rumble, vibrations of 58
transient processors: different brands of 196; multiband 244; threshold-dependent 191–192; threshold-independent 192–193

transient-rich instruments 182
transients: enhancing 190–192; loss of attack on prominent 323; producing distracting flams or stereo ricochets 281
transmission-line monitor design 10
transparent gain reduction 180
tremolo effect 194–195, *195*
triangle wave, using 232
triggering, samples 229–230
troubleshooting, intelligent 344–345
Tsai, Serge 60
tuning: adjustment 109–116, *110*; correction 113–116; mix implications 109–110, 117; specialist software for manipulating 114
turbulence noise, produced by ports 9
Twenty One Pilots 126

U

2.1 system, creating 19
U-he Uhbik-T *195*
"Umbrella" (Rihanna) 136
unblending, tricks 280
unity gain position, leaving fader at 144
Universal Audio: Cambridge Equalizer *139*; LA3A *167*
unmasking, with reverb 293
unmastered mixes of commercial releases 79
unmatched delay times 301, 303
unported speakers 9
unprocessed sound ("dry" or "direct" signal) 273
unstable fader, spotting 146
upward expander (or decompressor) 191

V

Valentine, Eric 198, 204
valve distortion *225*
van der Saag, Jochem 92

variable-knee clipping *225*
vectorscope 149
vertical monitor alignment 14–15
vibrato 107, 113, 226, 318, 319, 375
vinyl noise 244, 282, 307, 319
Viper ITB Vee Spring Verb *287*
Virtue, Brian 136
Visconti, Tony 135, 171, 267, 350
vocal-out mix 82–83
vocal(s): automating EQ 354; intelligibility and fader automation 351; option of completely replacing 82; "p," "b," and "w" sounds 57–58; reference tracks for exposed 74; rides 350–351, 354; vocal comping 122
vocal-up mix 81
volume levels, listening at a wide variety of 67–68
Voxengo: LF Max Punch *228*; MSED 308, *308*; OldSkoolVerb *287*; PHA-979 *154*; SPAN *56*, *310*; TransGainer *194*; Tube Amp *225*
VU-style moving-coil meters 171

W

Walker, Miles 300, 331, 332, 358
Wallace, Andy 67, 72, 229, 305, 317, 322, 323, 332, 342, 346, 356
Wallace, Matt 115
wall reflection points 20
waterfall plot 7–8; ADAM S2A *8*; ATC SCM20A *8*; Auratone 5C Super Sound Cube *11*; Behringer TRUTH B2031 *8*; Mackie HR824 *10*; PMC LB1BP *10*; Roland DS50A *8*; Tannoy Reveal *8*; Westlake BBSM5 *8*; Yamaha NS10 *11*

waveform cycles, crossfading over 106–107
waveguide 14–15
Waves: L2 limiter 337; LoAir *228*
Way, Dave 4, 214
"We Are Young" (Fun) 126
West, Kanye 76, 123, 309, *310*
Westlake BBSM5, waterfall plot *8*
Wet/Dry Mix controls 182, 190, 273
"What Do You Mean?" (Bieber) 331
White, Jack 290
White, Stuart 55
"White Flag" (Dido) 136
widening delay returns 315
Wildfire (Platten) 126
Williams, Pharell 95, 337
Wilson, Brian 43
Wiltshire, James 19
Winehouse, Amy 206
"Wings" (Little Mix) 126
wooden frame, around mineral fiber panels 28
woofer cone excursions, becoming visible 57
Worley, Paul 161
worst-case listening scenarios 52
Wright, Toby 182

Y

Yamaha: HS8 *6*; NS10 monitor 9, *11*, 40
Yohng, George, Wi Limiter *176*
Young Guru 85, 204, 224, 229, 265, 290
Youth 264
"You've Lost the Lovin' Feelin'" (Righteous Brothers) 265

Z

Zayn 306
Zdar, Philippe 136, 348
Zook, Joe 66
Zynaptiq Unmix: Drums *265*, 266